Quantum Computer Science

An Introduction

In the 1990s it was realized that quantum physics has some spectacular applications in computer science. This book is a concise introduction to quantum computation, developing the basic elements of this new branch of computational theory without assuming any background in physics. It begins with a novel introduction to the quantum theory from a computer-science perspective. It illustrates the quantum-computational approach with several elementary examples of quantum speed-up, before moving to the major applications: Shor's factoring algorithm, Grover's search algorithm, and quantum error correction.

The book is intended primarily for computer scientists who know nothing about quantum theory but would like to learn the elements of quantum computation either out of curiosity about this new paradigm, or as a basis for further work in the subject. It will also be of interest to physicists who want to learn the theory of quantum computation, and to physicists and philosophers of science interested in quantum foundational issues. It evolved during six years of teaching the subject to undergraduates and graduate students in computer science, mathematics, engineering, and physics, at Cornell University.

N. David Mermin is Horace White Professor of Physics Emeritus at Cornell University. He has received the Lilienfeld Prize of the American Physical Society and the Klopsteg award of the American Association of Physics Teachers. He is a member of the U.S. National Academy of Sciences and the American Academy of Arts and Sciences. Professor Mermin has written on quantum foundational issues for several decades, and is known for the clarity and wit of his scientific writings. Among his other books are *Solid State Physics* (with N. W. Ashcroft, Thomson Learning 1976), *Boojums all the Way Through* (Cambridge University Press 1990), and *It's about Time: Understanding Einstein's Relativity* (Princeton University Press 2005).

"This is one of the finest books in the rapidly growing field of quantum information. Almost every page contains a unique insight or a novel interpretation. David Mermin has once again demonstrated his legendary pedagogical skills to produce a classic."

Lov Grover, Bell Labs

"Mermin's book will be a standard for instruction and reference for years to come. He has carefully selected, from the mountain of knowledge accumulated in the last 20 years of research in quantum information theory, a manageable, coherent subset that constitutes a complete undergraduate course. While selective, it is in no sense "watered down"; Mermin moves unflinchingly through difficult arguments in the Shor algorithm, and in quantum error correction theory, providing invaluable diagrams, clear arguments, and, when necessary, extensive appendices to get the students successfully through to the end. The book is suffused with Mermin's unique knowledge of the history of modern physics, and has some of the most captivating writing to be found in a college textbook."

David DiVincenzo, IBM T. J. Watson Research Center

"Mermin's book is a gentle introduction to quantum computation especially aimed at an audience of computer scientists and mathematicians. It covers the basics of the field, explaining the material clearly and containing lots of examples. Mermin has always been an entertaining and comprehensible writer, and continues to be in this book. I expect it to become the definitive introduction to this material for non-physicists."

Peter Shor, Massachusetts Institute of Technology

"Textbook writers usually strive for a streamlined exposition, smoothing out the infelicities of thought and notation that plague any field's early development. Fortunately, David Mermin is too passionate and acute an observer of the cultural side of science to fall into this blandness. Instead of omitting infelicities, he explains and condemns them, at the same time using his experience of having taught the course many times to nip nascent misunderstandings in the bud. He celebrates the field's mongrel origin in a shotgun wedding between classical computer scientists, who thought they knew the laws of information, and quantum physicists, who thought information was not their job. Differences remain: we hear, for example, why physicists love the Dirac notation and mathematicians hate it. Worked-out examples and exercises familiarize students with the necessary algebraic manipulations, while Mermin's lucid prose and gentle humor cajole them toward a sound intuition for what it all means, not an easy task for a subject superficially so counterintuitive."

Charles Bennett, IBM T. J. Watson Research Center

Quantum Computer Science
An Introduction

N. David Mermin

Cornell University

CAMBRIDGE
UNIVERSITY PRESS

In memory of my brother, Joel Mermin

You would have enjoyed it.

CAMBRIDGE UNIVERSITY PRESS
Cambridge, New York, Melbourne, Madrid, Cape Town, Singapore, São Paulo

Cambridge University Press
The Edinburgh Building, Cambridge CB2 8RU, UK

Published in the United States of America by Cambridge University Press, New York

www.cambridge.org
Information on this title: www.cambridge.org/9780521876582

First published 2007

Printed in the United Kingdom at the University Press, Cambridge

A catalog record for this publication is available from the British Library

ISBN 978-0-521-87658-2 hardback

Contents

Preface

It was almost three quarters of a century after the discovery of quantum mechanics, and half a century after the birth of information theory and the arrival of large-scale digital computation, that people finally realized that quantum physics profoundly alters the character of information processing and digital computation. For physicists this development offers an exquisitely different way of using and thinking about the quantum theory. For computer scientists it presents a surprising demonstration that the abstract structure of computation cannot be divorced from the physics governing the instrument that performs the computation. Quantum mechanics provides new computational paradigms that had not been imagined prior to the 1980s and whose power was not fully appreciated until the mid 1990s.

In writing this introduction to quantum computer science I have kept in mind readers from several disciplines. Primarily I am addressing computer scientists, electrical engineers, or mathematicians who may know little or nothing about quantum physics (or any other kind of physics) but who wish to acquire enough facility in the subject to be able to follow the new developments in quantum computation, judge for themselves how revolutionary they may be, and perhaps choose to participate in the further development of quantum computer science. Not the least of the surprising things about quantum computation is that remarkably little background in quantum mechanics has to be acquired to understand and work with its applications to information processing. Familiarity with a few fundamental facts about finite-dimensional vector spaces over the complex numbers (summarized and reviewed in Appendix A) is the only real prerequisite.

One of the secondary readerships I have in mind consists of physicists who, like myself – I am a theorist who has worked in statistical physics, solid-state physics, low-temperature physics, and mathematical physics – know very little about computer science, but would like to learn about this extraordinary new application of their discipline. I stress, however, that my subject is quantum computer science, not quantum computer design. This is a book about quantum computational software – not hardware. The difficult question of how one might actually build a quantum computer is beyond its scope.

Another secondary readership is made up of those philosophers and physicists who – again like myself – are puzzled by so-called foundational issues: what the strange quantum formalism implies about the nature of the world that it so accurately describes. By applying quantum mechanics in an entirely new way – and especially by applying it to the processing of knowledge – quantum computation gives a new perspective on interpretational questions. While I rarely address such matters explicitly, for purely pedagogical reasons my presentation is suffused with a perspective on the quantum theory that is very close to the venerable but recently much reviled Copenhagen interpretation. Those with a taste for such things may be startled to see how well quantum computation resonates with the Copenhagen point of view. Indeed, it had been my plan to call this book *Copenhagen Computation* until the excellent people at Cambridge University Press and my computer-scientist friends persuaded me that virtually no members of my primary readership would then have had any idea what it was about.

Several years ago I mentioned to a very distinguished theoretical physicist that I spent the first four lectures of a course in quantum computation giving an introduction to quantum mechanics for mathematically literate people who knew nothing about quantum mechanics, and quite possibly little if anything about physics. His immediate response was that any application of quantum mechanics that can be taught after only a four-hour introduction to the subject cannot have serious intellectual content. After all, he remarked, it takes any physicist many years to develop a feeling for quantum mechanics.

It's a good point. Nevertheless computer scientists and mathematicians with no background in physics have been able quickly to learn enough quantum mechanics to understand and make major contributions to the theory of quantum computation. There are two main reasons for this.

First of all, a quantum computer – or, more accurately, the abstract quantum computer that one hopes someday to be able to embody in actual hardware – is an extremely simple example of a physical system. It is discrete, not continuous. It is made up out of a finite number of units, each of which is the simplest possible kind of quantum-mechanical system, a so-called two-state system, whose behavior, as we shall see, is highly constrained and easily specified. Much of the analytical complexity of learning quantum mechanics is connected with mastering the description of continuous (infinite-state) systems. By restricting attention to collections of two-state systems (or even d-state systems for finite d) one can avoid much suffering. Of course one also loses much wisdom, but hardly any of it – at least at this stage of the art – is relevant to the basic theory of quantum computation.

Second, and just as important, the most difficult part of learning quantum mechanics is to get a good feeling for how the formalism can be applied to actual phenomena. This almost invariably involves formulating oversimplified abstract models of real physical systems, to which the quantum formalism can then be applied. The best physicists have an extraordinary intuition for what features of the phenomena are essential and must be represented in a model, and what features are inessential and can be ignored. It takes years to develop such intuition. Some never do. The theory of quantum computation, however, is entirely concerned with an abstract model – the easy part of the problem.

To understand how to *build* a quantum computer, or even to study what physical systems are promising candidates for realizing such a device, you must indeed have many years of experience in quantum mechanics and its applications under your belt. But if you only want to know what such a device is capable in principle of doing once you have it, then there is no reason to get involved in the really difficult physics of the subject. Exactly the same thing holds for ordinary classical computers. One can be a masterful practitioner of computer science without having the foggiest notion of what a transistor is, not to mention how it works.

So while you should be warned that the subset of quantum mechanics you will acquire from this book is extremely focused and quite limited in its scope, you can also rest assured that it is neither oversimplified nor incomplete, when applied to the special task for which it is intended.

I might note that a third impediment to developing a good intuition for quantum physics is that in some ways the behavior implied by quantum mechanics is highly counterintuitive, if not downright weird. Glimpses of such strange behavior sometimes show up at the level of quantum computation. Indeed, for me one of the major appeals of quantum computation is that it affords a new conceptual arena for trying to come to a better understanding of quantum weirdness. When opportunities arise I will call attention to some of this strange behavior, rather than (as I easily could) letting it pass by unremarked upon and unnoticed.

The book evolved as notes for a course of 28 one-hour lectures on quantum computation that I gave six times between 2000 and 2006 to a diverse group of Cornell University undergraduates, graduate students, and faculty, in computer science, electrical engineering, mathematics, physics, and applied physics. With so broad an audience, little common knowledge could be assumed. My lecture notes, as well as my own understanding of the subject, repeatedly benefited from comments and questions in and after class, coming from a number of different perspectives. What made sense to one of my constituencies was often puzzling, absurd, or irritatingly simple-minded to others. This final form of my notes bears little resemblance to my earliest versions, having been improved by insightful remarks, suggestions, and complaints about everything from notation to number theory.

In addition to the 200 or so students who passed through P481-P681-CS483, I owe thanks to many others. Albert J. Sievers, then Director of Cornell's Laboratory of Atomic and Solid State Physics, started me thinking hard about quantum computation by asking me to put together a two-week set of introductory lectures for members of our laboratory, in the Fall of 1999. So many people showed up from all over the university that I decided it might be worth expanding this survey into a full course. I'm grateful to two Physics Department chairs, Peter Lepage and Saul Teukolsky, for letting me continue teaching that course for six straight years, and to the Computer Science Department chair, Charlie van Loan, for support, encouragement, and a steady stream of wonderful students. John Preskill, though he may not know it, taught me much of the subject from his superb online Caltech lecture notes. Charles Bennett first told me about quantum information processing, back when the term might not even have been coined, and he has always been available as a source of wisdom and clarification. Gilles Brassard has on many occasions supplied me with help from the computer-science side. Chris Fuchs has been an indispensable quantum-foundational critic and consultant. Bob Constable made me, initially against my will, a certified Cornell Information Scientist and introduced me to many members of that excellent community. But most of all, I owe thanks to David DiVincenzo, who collaborated with me on the 1999 two-week LASSP Autumn School and has acted repeatedly over the following years as a sanity check on my ideas, an indispensable source of references and historical information, a patient teacher, and an encouraging friend.

A note on references

Quantum Computer Science is a pedagogical introduction to the basic structure and procedures of the subject – a quantum-computational primer. It is not a historical survey of the development of the field. Many of these procedures are named after the people who first put them forth, but although I use their names, I do not cite the original papers unless they add something to my own exposition. This is because, not surprisingly, work done since the earliest papers has led to clearer expositions of those ideas. I learned the subject myself almost exclusively from secondary, tertiary, or even higher-order sources, and then reformulated it repeatedly in the course of teaching it for six years.

On the few occasions when I do cite a paper it is either because it completes an exposition that I have only sketched, or because the work has not yet become identified in the field with the name(s) of the author(s) and I wanted to make clear that it was not original with me.

Readers interested in hunting down earlier work in the field can begin (and in most cases conclude) their search at the quantum-physics subdivision of the Cornell (formerly Los Alamos) E-print Archive, `http://arxiv.org/archive/quant-ph`, where most of the important papers in the field have been and are still being posted.

Chapter 1

Cbits and Qbits

1.1 What is a quantum computer?

It is tempting to say that a quantum computer is one whose operation is governed by the laws of quantum mechanics. But since the laws of quantum mechanics govern the behavior of all physical phenomena, this temptation must be resisted. Your laptop operates under the laws of quantum mechanics, but it is not a quantum computer. A quantum computer is one whose operation exploits certain very special transformations of its internal state, whose description is the primary subject of this book. The laws of quantum mechanics allow these peculiar transformations to take place under very carefully controlled conditions.

In a quantum computer the physical systems that encode the individual logical bits must have no physical interactions whatever that are not under the complete control of the program. All other interactions, however irrelevant they might be in an ordinary computer – which we shall call *classical* – introduce potentially catastrophic disruptions into the operation of a quantum computer. Such damaging encounters can include interactions with the external environment, such as air molecules bouncing off the physical systems that represent bits, or the absorption of minute amounts of ambient radiant thermal energy. There can even be disruptive interactions between the computationally relevant features of the physical systems that represent bits and other features of those same systems that are associated with computationally irrelevant aspects of their internal structure. Such destructive interactions, between what matters for the computation and what does not, result in *decoherence*, which is fatal to a quantum computation.

To avoid decoherence individual bits cannot in general be encoded in physical systems of macroscopic size, because such systems (except under very special circumstances) cannot be isolated from their own irrelevant internal properties. Such isolation can be achieved if the bits are encoded in a small number of states of a system of atomic size, where extra internal features do not matter, either because they do not exist, or because they require unavailably high energies to come into play. Such atomic-scale systems must also be decoupled from their surroundings except for the completely controlled interactions that are associated with the computational process itself.

Two things keep the situation from being hopeless. First, because the separation between the discrete energy levels of a system on the atomic scale can be enormously larger than the separation between the levels of a large system, the dynamical isolation of an atomic system is easier to achieve. It can take a substantial kick to knock an atom out of its state of lowest energy. The second reason for hope is the discovery that errors induced by extraneous interactions can actually be corrected if they occur at a sufficiently low rate. While error correction is routine for bits represented by classical systems, quantum error correction is constrained by the formidable requirement that it be done without knowing either the original or the corrupted state of the physical systems that represent the bits. Remarkably, this turns out to be possible.

Although the situation is therefore not hopeless, the practical difficulties in the way of achieving useful quantum computation are enormous. Only a rash person would declare that there will be no useful quantum computers by the year 2050, but only a rash person would predict that there will be. Never mind. Whether or not it will ever become a practical technology, there is a beauty to the theory of quantum computation that gives it a powerful appeal as a lovely branch of mathematics, and as a strange generalization of the paradigm of classical computer science, which had completely escaped the attention of computer scientists until the 1980s. The new paradigm demonstrates that the theory of computation can depend profoundly on the physics of the devices that carry it out. Quantum computation is also a valuable source of examples that illustrate and illuminate, in novel ways, the mysterious phenomena that quantum behavior can give rise to.

For computer scientists the most striking thing about quantum computation is that a quantum computer can be vastly more efficient than anything ever imagined in the classical theory of computational complexity, for certain computational tasks of considerable practical interest. The time it takes the quantum computer to accomplish such tasks scales up much more slowly with the size of the input than it does in any classical computer. Much of this book is devoted to examining the most celebrated examples of this speed-up.

This exposition of quantum computation begins with an introduction to quantum mechanics, specially tailored for this particular application. The quantum-mechanics lessons are designed to give you, as efficiently as possible, the conceptual tools needed to delve into quantum computation. This is done by restating the rules of quantum mechanics, not as the remarkable revision of classical Newtonian mechanics required to account for the behavior of matter at the atomic and subatomic levels, but as a curious generalization of rules describing an ordinary classical digital computer. By focusing exclusively on how quantum mechanics enlarges the possibilities for the physical manipulation of digital information, it is possible to characterize how the quantum theory works in an elementary and quite concise way, which is nevertheless rigorous and complete for this special area of application.

While I assume no prior familiarity with quantum physics (or any other kind of physics), I do assume familiarity with elementary linear algebra and, in particular, with the theory of finite-dimensional vector spaces over the complex numbers. Appendix A summarizes the relevant linear algebra. It is worth examining even if you are well acquainted with the mathematics of such vector spaces, since it also provides a compact summary of the mathematically unconventional language – *Dirac notation* – in which linear algebra is couched in all treatments of quantum computation. Dirac notation is also developed, more informally, throughout the rest of this chapter.

1.2 Cbits and their states

We begin with an offbeat formulation of what an ordinary classical computer does. I frame the elementary remarks that follow in a language which may look artificial and cumbersome, but is designed to accommodate the richer variety of things that a computer can do if it takes full advantage of the possibilities made available by the quantum-mechanical behavior of its constituent parts. By introducing and applying the unfamiliar nomenclature and notation of quantum mechanics in a familiar classical context, I hope to make a little less strange its subsequent extension to the broader quantum setting.

A classical computer operates on strings of zeros and ones, such as 110010111011000, converting them into other such strings. Each position in such a string is called a *bit*, and it contains either a 0 or a 1. To represent such collections of bits the computer must contain a corresponding collection of physical systems, each of which can exist in two unambiguously distinguishable physical states, associated with the value (0 or 1) of the abstract bit that the physical system represents. Such a physical system could be, for example, a switch that could be open (0) or shut (1), or a magnet whose magnetization could be oriented in two different directions, "up" (0) or "down" (1).

It is a common practice in quantum computer science to use the same term "bit" to describe the two-state classical system that represents the value of the abstract bit. But this use of a single term to characterize both the abstract bit (0 or 1) and the physical system whose two states represent the two values is a potential source of confusion. To avoid such confusion, I shall use the term *Cbit* ("C" for "classical") to describe the two-state classical physical system and *Qbit* to describe its quantum generalization. This terminology is inspired by Paul Dirac's early use of *c-number* and *q-number* to describe classical quantities and their quantum-mechanical generalizations. "Cbit" and

"Qbit" are preferable to "c-bit" and "q-bit" because the terms themselves often appear in hyphenated constructions.

Unfortunately the preposterous spelling *qubit* currently holds sway for the quantum system. The term *qubit* was invented and first used in print by the otherwise admirable Benjamin Schumacher.[1] A brief history of the term can be found in the acknowledgments at the end of his paper. Although "qubit" honors the English (German, Italian, ...) rule that q should be followed by u, it ignores the equally powerful requirement that *qu* should be followed by a vowel. My guess is that "qubit" has gained acceptance because it visually resembles an obsolete English unit of distance, the homonymic *cubit*. To see its ungainliness with fresh eyes, it suffices to imagine that Dirac had written *qunumber* instead of *q-number*, or that one erased transparencies and cleaned one's ears with *Qutips*.

Because clear distinctions among bits, Cbits, and Qbits are crucial in the introduction to quantum computation that follows, I shall use this currently unfashionable terminology. If you are already addicted to the term *qubit*, please regard *Qbit* as a convenient abbreviation.

To prepare for the extension from Cbits to Qbits, I introduce what may well strike you as a degree of notational overkill in the discussion of Cbits that follows. We shall represent the state of each Cbit as a kind of box, depicted by the symbol $|\ \rangle$, into which we place the value, 0 or 1, represented by that state. Thus the two distinguishable states of a Cbit are represented by the symbols $|0\rangle$ and $|1\rangle$. It is the common practice to call the symbol $|0\rangle$ or $|1\rangle$ itself the *state* of the Cbit, thereby using the same term to refer to both the physical condition of the Cbit and the abstract symbol that represents that physical condition. There is nothing unusual in this. For example one commonly uses the term "position" to refer to the symbol x that represents the physical position of an object. I call this common, if little noted, practice to your attention only because in the quantum case "state" refers *only* to the symbol, there being *no* internal property of the Qbit that the symbol represents. The subtle relation between Qbits and their state symbol will emerge later in this chapter.

Along the same lines, we shall characterize the states of the five Cbits representing 11001, for example, by the symbol

$$|1\rangle|1\rangle|0\rangle|0\rangle|1\rangle, \tag{1.1}$$

and refer to this object as the *state* of all five Cbits. Thus a pair of Cbits can have (or "be in") any of the four possible states

$$|0\rangle|0\rangle,\ \ |0\rangle|1\rangle,\ \ |1\rangle|0\rangle,\ \ |1\rangle|1\rangle, \tag{1.2}$$

1 Benjamin Schumacher, "Quantum coding," *Physical Review* A **51**, 2738–2747 (1995).

three Cbits can be in any of the eight possible states

$$|0\rangle|0\rangle|0\rangle,\ |0\rangle|0\rangle|1\rangle,\ |0\rangle|1\rangle|0\rangle,\ |0\rangle|1\rangle|1\rangle,\ |1\rangle|0\rangle|0\rangle,\ |1\rangle|0\rangle|1\rangle,\ |1\rangle|1\rangle|0\rangle,\ |1\rangle|1\rangle|1\rangle, \tag{1.3}$$

and so on.

As (1.4) already makes evident, when there are many Cbits such products are often much easier to read if one encloses the whole string of zeros and ones in a single bigger box of the form $|\quad\rangle$ rather than having a separate box for each Cbit:

$$|000\rangle,\ |001\rangle,\ |010\rangle,\ |011\rangle,\ |100\rangle,\ |101\rangle,\ |110\rangle,\ |111\rangle. \tag{1.4}$$

We shall freely move between these two equivalent ways of expressing the state of several Cbits that represent a string of bits, boxing the whole string or boxing each individual bit. Whether the form (1.3) or (1.4) is to be preferred depends on the context.

There is also a third form, which is useful when we regard the zeros and ones as constituting the binary expansion of an integer. We can then replace the representations of the 3-Cbit states in (1.4) by the even shorter forms

$$|0\rangle,\ |1\rangle,\ |2\rangle,\ |3\rangle,\ |4\rangle,\ |5\rangle,\ |6\rangle,\ |7\rangle. \tag{1.5}$$

Note that, unlike the forms (1.3) and (1.4), the form (1.5) is ambiguous, unless we are told that these symbols express states of three Cbits. If we are not told, then there is no way of telling, for example, whether $|3\rangle$ represents the 2-Cbit state $|11\rangle$, the 3-Cbit state $|011\rangle$, or the 4-Cbit state $|0011\rangle$, etc. This ambiguity can be removed, when necessary, by adding a subscript making the number of Cbits explicit:

$$|0\rangle_3,\ |1\rangle_3,\ |2\rangle_3,\ |3\rangle_3,\ |4\rangle_3,\ |5\rangle_3,\ |6\rangle_3,\ |7\rangle_3. \tag{1.6}$$

Be warned, however, that, when there is no need to emphasize how many Cbits $|x\rangle$ represents, it can be useful to use such subscripts for other purposes. If, for example, Alice and Bob each possess a single Cbit it can be convenient to describe the state of Alice's Cbit (if it has the value 1) by $|1\rangle_a$, Bob's (if it has the value 0) by $|0\rangle_b$, and the joint state of the two by $|1\rangle_a|0\rangle_b$ or $|10\rangle_{ab}$.

Dirac introduced the $|\ \rangle$ notation (known as Dirac notation) in the early days of the quantum theory, as a useful way to write and manipulate *vectors*. For silly reasons he called such vectors *kets*, a terminology that has survived to this day. In Dirac notation you can put into the box $|\ \rangle$ anything that serves to specify what the vector is. If, for example, we were talking about displacement vectors in ordinary three-dimensional space, we could have a vector

$$|\text{5 horizontal centimeters northeast}\rangle. \tag{1.7}$$

In using Dirac notation to express the state of a Cbit, or a collection of Cbits, I'm suggesting that there might be some utility in thinking of the states as vectors. Is there? Well, in the case of Cbits, not very much, but maybe a little. We now explore this way of thinking about Cbit states, because when we come to the generalization to Qbits, it becomes absolutely essential to consider them to be vectors – so much so that the term *state* is often taken to be synonymous with *vector* (or, more precisely, "vector that represents the state").

We shall briefly explore what one can do with Cbits when one takes the two states $|0\rangle$ and $|1\rangle$ of a single Cbit to be represented by two *orthogonal unit vectors in a two-dimensional space.* While this is little more than a curious and unnecessarily elaborate way of describing Cbits, it is fundamental and unavoidable in dealing with Qbits. Playing unfamiliar and somewhat silly games with Cbits will enable you to become acquainted with much of the quantum-mechanical formalism in a familiar setting.

If you prefer your vectors to be expressed in terms of components, note that we can represent the two orthogonal states of a single Cbit, $|0\rangle$ and $|1\rangle$, as column vectors

$$|0\rangle = \begin{pmatrix} 1 \\ 0 \end{pmatrix}, \qquad |1\rangle = \begin{pmatrix} 0 \\ 1 \end{pmatrix}. \tag{1.8}$$

In the case of two Cbits the vector space is four-dimensional, with an orthonormal basis

$$|00\rangle,\ |01\rangle,\ |10\rangle,\ |11\rangle. \tag{1.9}$$

The alternative notation for this basis,

$$|0\rangle|0\rangle,\ |0\rangle|1\rangle,\ |1\rangle|0\rangle,\ |1\rangle|1\rangle, \tag{1.10}$$

is deliberately designed to suggest multiplication, since it is, in fact, a short-hand notation for the *tensor product* of the two single-Cbit 2-vectors, written in more formal mathematical notation as

$$|0\rangle \otimes |0\rangle,\ |0\rangle \otimes |1\rangle,\ |1\rangle \otimes |0\rangle,\ |1\rangle \otimes |1\rangle. \tag{1.11}$$

In terms of components, the tensor product $\mathbf{a} \otimes \mathbf{b}$ of an M-component vector $\mathbf{a}$ with components a_μ and an N-component vector $\mathbf{b}$ with components b_ν is the (MN)-component vector with components indexed by all the MN possible pairs of indices (μ, ν), whose (μ, ν)th component is just the product $a_\mu b_\nu$. A broader view can be found in the extended review of vector-space concepts in Appendix A. I shall freely move back and forth between the various ways (1.9)–(1.11) of writing the tensor product and their generalizations to multi-Cbit states, using in each case a form that makes the content clearest.

Once one agrees to regard the two 1-Cbit states as orthogonal unit vectors, the tensor product is indeed the natural way to represent multi-Cbit states, since it leads to the obvious multi-Cbit generalization of the representation (1.8) of 1-Cbit states as column vectors. If we express the states $|0\rangle$ and $|1\rangle$ of each single Cbit as column vectors, then we can get the column vector describing a multi-Cbit state by repeatedly applying the rule for the components of the tensor product of two vectors. The result is illustrated here for a three-fold tensor product:

$$\begin{pmatrix} x_0 \\ x_1 \end{pmatrix} \otimes \begin{pmatrix} y_0 \\ y_1 \end{pmatrix} \otimes \begin{pmatrix} z_0 \\ z_1 \end{pmatrix} = \begin{pmatrix} x_0 y_0 z_0 \\ x_0 y_0 z_1 \\ x_0 y_1 z_0 \\ x_0 y_1 z_1 \\ x_1 y_0 z_0 \\ x_1 y_0 z_1 \\ x_1 y_1 z_0 \\ x_1 y_1 z_1 \end{pmatrix}. \tag{1.12}$$

On applying this, for example, to the case $|5\rangle_3$, we have

$$|5\rangle_3 = |101\rangle = |1\rangle|0\rangle|1\rangle = \begin{pmatrix} 0 \\ 1 \end{pmatrix} \otimes \begin{pmatrix} 1 \\ 0 \end{pmatrix} \otimes \begin{pmatrix} 0 \\ 1 \end{pmatrix} = \begin{pmatrix} 0 \\ 0 \\ 0 \\ 0 \\ 0 \\ 1 \\ 0 \\ 0 \end{pmatrix}. \tag{1.13}$$

If we label the vertical components of the 8-vector on the right $0, 1, \ldots, 7$, from the top down, then the single nonzero component is the 1 in position 5 – precisely the position specified by the state vector in its form on the left of (1.13). This is indeed the obvious multi-Cbit generalization of the column-vector form (1.8) for 1-Cbit states.

This is quite general: the tensor-product structure of multi-Cbit states is just what one needs in order for the 2^n-dimensional column vector representing the state $|m\rangle_n$ to have all its entries zero except for a single 1 in the mth position down from the top.

One can turn this development upside down, taking as one's starting point the simple rule that an integer x in the range $0 \leq x < N$ is represented by one of N orthonormal vectors in an N-dimensional space. One can then pick a basis so that 0 is represented by an N-component column vector $|0\rangle$ that has 0 in every position except for a 1 in the top position, and x is to be represented by an N-component column vector $|x\rangle$ that has 0 in every position except for a 1 in the position x down from the top. It then follows from the nature of the tensor product that if $N = 2^n$ and x has the binary expansion $x = \sum_{j=0}^{n-1} x_j 2^j$, then the column vector $|x\rangle_n$ is the tensor product of the n 2-component column vectors $|x_j\rangle$:

$$|x\rangle_n = |x_{n-1}\rangle \otimes |x_{n-2}\rangle \otimes \cdots \otimes |x_1\rangle \otimes |x_0\rangle. \tag{1.14}$$

In dealing with n-Cbit states of the form (1.14) we shall identify each of the n 1-Cbit states, out of which they are composed, by giving the power of 2 associated with the individual bit that the Cbit represents. Thus the 1-Cbit state on the extreme right of (1.14) represents Cbit 0, the state immediately to its left represents Cbit 1, and so on.

This relation between tensor products of vectors and positional notation for integers is not confined to the binary system. Suppose, for example, one represents a decimal digit $x = 0, 1, \ldots, 9$ as a 10-component column vector $\mathbf{v}^{(x)}$ with all components 0 except for a 1, x positions down from the top. If the n-digit decimal number $X = \sum_{j=0}^{n-1} x_j 10^j$ is represented by the tensor product $\mathbf{V} = \mathbf{v}^{(x_{n-1})} \otimes \mathbf{v}^{(x_{n-2})} \otimes \cdots \otimes \mathbf{v}^{(1)} \otimes \mathbf{v}^{(0)}$, then $\mathbf{V}$ will be a 10^n-component column vector with all components 0 except for a 1, x positions down from the top.

Although the representation of Cbit states by column vectors clearly shows why tensor products give a natural description of multi-Cbit states, for almost all other purposes it is better and much simpler to forget about column vectors and components, and deal directly with the state vectors in their abstract forms (1.3)–(1.6).

1.3 Reversible operations on Cbits

Quantum computers do an important part of their magic through *reversible* operations, which transform the initial state of the Qbits into its final form using only processes whose action can be inverted. There is only a single *irreversible* component to the operation of a quantum computer, called *measurement*, which is the only way to extract useful information from the Qbits after their state has acquired its final form. Although measurement is a nontrivial and crucial part of any quantum computation, in a classical computer the extraction of information from the state of the Cbits is so conceptually straightforward that it is not viewed as an inherent part of the computational process, though it is, of course, a nontrivial concern for those who design digital displays or printers. Because the only computationally relevant operations on a classical computer that can be extended to operations on a quantum computer are reversible, only operations on Cbits that are reversible will be of interest to us here.

In a reversible operation every final state arises from a unique initial state. An example of an irreversible operation is ERASE, which forces a Cbit into the state $|0\rangle$ regardless of whether its initial state is $|0\rangle$ or $|1\rangle$. ERASE is irreversible in the sense that, given only the final state and the fact that it was the output of the operation ERASE, there is no way to recover the initial state.

The only nontrivial reversible operation we can apply to a single Cbit is the NOT operation, denoted by the symbol $\mathbf{X}$, which interchanges the two states $|0\rangle$ and $|1\rangle$:

$$\mathbf{X} : |x\rangle \to |\tilde{x}\rangle; \quad \tilde{1} = 0, \quad \tilde{0} = 1. \tag{1.15}$$

This is sometimes referred to as *flipping* the Cbit. NOT is reversible because it has an inverse: applying $\mathbf{X}$ a second time brings the state of the Cbit back to its original form:

$$\mathbf{X}^2 = \mathbf{1}, \tag{1.16}$$

where $\mathbf{1}$ is the unit (identity) operator. If we represent the two orthogonal states of the Cbit by the column vectors (1.8), then we can express NOT by a linear operator $\mathbf{X}$ on the two-dimensional vector space, whose action on the column vectors is given by the matrix

$$\mathbf{X} = \begin{pmatrix} 0 & 1 \\ 1 & 0 \end{pmatrix}. \tag{1.17}$$

So the two reversible things you can do to a single Cbit – leaving it alone and flipping it – correspond to the two linear operators $\mathbf{X}$ and $\mathbf{1}$,

$$\mathbf{1} = \begin{pmatrix} 1 & 0 \\ 0 & 1 \end{pmatrix}, \tag{1.18}$$

on its two-dimensional vector space.

A pedantic digression: since multiplication by the scalar 1 and action by the unit operator $\mathbf{1}$ achieve the same result, I shall sometimes follow the possibly irritating practice of physicists and not distinguish notationally between them. I shall take similar liberties with the scalar 0, the zero vector $\mathbf{0}$, and the zero operator $\mathbf{0}$.

Possibilities for reversible operations get richer when we go from a single Cbit to a pair of Cbits. The most general reversible operation on two Cbits is any permutation of their four possible states. There are 4! = 24 such operations. Perhaps the simplest nontrivial example is the *swap* (or *exchange*) operator $\mathbf{S}_{ij}$, which simply interchanges the states of Cbits i and j:

$$\mathbf{S}_{10}|xy\rangle = |yx\rangle. \tag{1.19}$$

Since the swap operator $\mathbf{S}_{10}$ interchanges $|01\rangle = |1\rangle_2$ and $|10\rangle = |2\rangle_2$, while leaving $|00\rangle = |0\rangle_2$ and $|11\rangle = |3\rangle_2$ fixed, its matrix in the basis $|0\rangle_2, |1\rangle_2, |2\rangle_2, |3\rangle_2$ is

$$\mathbf{S}_{10} = \mathbf{S}_{01} = \begin{pmatrix} 1 & 0 & 0 & 0 \\ 0 & 0 & 1 & 0 \\ 0 & 1 & 0 & 0 \\ 0 & 0 & 0 & 1 \end{pmatrix}. \tag{1.20}$$

The 2-Cbit operator whose extension to Qbits plays by far the most important role in quantum computation is the *controlled-NOT* or cNOT operator $\mathbf{C}_{ij}$. If the state of the ith Cbit (the *control Cbit*) is $|0\rangle$, $\mathbf{C}_{ij}$ leaves the state of the jth Cbit (the *target Cbit*) unchanged, but, if the state of the control Cbit is $|1\rangle$, $\mathbf{C}_{ij}$ applies the NOT operator $\mathbf{X}$ to the state of the target Cbit. In either case the state of the control Cbit is left unchanged.

We can summarize this compactly by writing

$$\mathbf{C}_{10}|x\rangle|y\rangle = |x\rangle|y \oplus x\rangle, \qquad \mathbf{C}_{01}|x\rangle|y\rangle = |x \oplus y\rangle|y\rangle, \tag{1.21}$$

where $\oplus$ denotes addition modulo 2:

$$y \oplus 0 = y, \qquad y \oplus 1 = \tilde{y} = 1 - y. \tag{1.22}$$

The modulo-2 sum $x \oplus y$ is also called the "exclusive OR" (or XOR) of x and y.

You can construct SWAP out of three cNOT operations:

$$\mathbf{S}_{ij} = \mathbf{C}_{ij}\mathbf{C}_{ji}\mathbf{C}_{ij}. \tag{1.23}$$

This can easily be verified by repeated applications of (1.21), noting that $x \oplus x = 0$. We note some other ways of showing it below.

To construct the matrix for the cNOT operation in the four-dimensional 2-Cbit space, note that if the control Cbit is on the left then cNOT leaves $|00\rangle = |0\rangle_2$ and $|01\rangle = |1\rangle_2$ fixed and exchanges $|10\rangle = |2\rangle_2$ and $|11\rangle = |3\rangle_2$. Therefore the $4 \otimes 4$ matrix representing $\mathbf{C}_{10}$ is just

$$\mathbf{C}_{10} = \begin{pmatrix} 1 & 0 & 0 & 0 \\ 0 & 1 & 0 & 0 \\ 0 & 0 & 0 & 1 \\ 0 & 0 & 1 & 0 \end{pmatrix}. \tag{1.24}$$

If the control Cbit is on the right, then the states $|01\rangle = |1\rangle_2$ and $|11\rangle = |3\rangle_2$ are interchanged, and $|00\rangle = |0\rangle_2$ and $|10\rangle = |2\rangle_2$ are fixed, so the matrix representing $\mathbf{C}_{01}$ is

$$\mathbf{C}_{01} = \begin{pmatrix} 1 & 0 & 0 & 0 \\ 0 & 0 & 0 & 1 \\ 0 & 0 & 1 & 0 \\ 0 & 1 & 0 & 0 \end{pmatrix}. \tag{1.25}$$

The construction (1.23) of $\mathbf{S}$ out of cNOT operators also follows from (1.20), (1.24), and (1.25), using matrix multiplication. As a practical matter, it is almost always more efficient to establish operator identities by dealing with them directly as operators, avoiding matrix representations.

A very common kind of 2-Cbit operator consists of the tensor product $\otimes$ of two 1-Cbit operators:

$$(\mathbf{a} \otimes \mathbf{b})|xy\rangle = (\mathbf{a} \otimes \mathbf{b})|x\rangle \otimes |y\rangle = \mathbf{a}|x\rangle \otimes \mathbf{b}|y\rangle, \tag{1.26}$$

from which it follows that

$$(\mathbf{a} \otimes \mathbf{b})(\mathbf{c} \otimes \mathbf{d}) = (\mathbf{ac}) \otimes (\mathbf{bd}). \tag{1.27}$$

This tensor-product notation for operators can become quite ungainly when one is dealing with a large number of Cbits and wants to write a 2-Cbit operator that affects only a particular pair of Cbits. If, for example, the 2-Cbit operator in (1.26) acts only on the second and fourth Cbits from the right in a 6-Cbit state, then the operator on the 6-Cbit state has to be written as

$$\mathbf{1}\otimes\mathbf{1}\otimes\mathbf{a}\otimes\mathbf{1}\otimes\mathbf{b}\otimes\mathbf{1}. \tag{1.28}$$

To avoid such typographical monstrosities, we simplify (1.28) to

$$\mathbf{1}\otimes\mathbf{1}\otimes\mathbf{a}\otimes\mathbf{1}\otimes\mathbf{b}\otimes\mathbf{1} = \mathbf{a}_3\mathbf{b}_1 = \mathbf{b}_1\mathbf{a}_3, \tag{1.29}$$

where the subscript indicates which Cbit the 1-Cbit operator acts on, and it is understood that those Cbit states whose subscripts do not appear remain unmodified – i.e. they are acted on by the unit operator. As noted above, we label each 1-Cbit state by the power of 2 it would represent if the n Cbits were representing an integer: the state on the extreme right is labeled 0, the one to its left, 1, etc. Since the order in which $\mathbf{a}$ and $\mathbf{b}$ are written is clearly immaterial if their subscripts specify different 1-Cbit states, the order in which one writes them in (1.29) doesn't matter: 1-Cbit operators that act on different 1-Cbit states commute.

Sometimes we deal with 1-Cbit operators that already have subscripts in their names; under such conditions it is more convenient to indicate which Cbit state the operator acts on by a superscript, enclosed in parentheses to avoid confusion with an exponent: thus $\mathbf{X}^{(2)}$ represents the 1-Cbit operator that flips the third Cbit state from the right, but $\mathbf{X}^2$ represents the square of the flip operator (i.e. the unit operator) without reference to which Cbit state it acts on.

To prepare for some of the manipulations we will be doing with operations on Qbits, we now examine a few examples of working with operators on Cbits.

1.4 Manipulating operations on Cbits

It is useful to introduce a 1-Cbit operator $\mathbf{n}$ that is simply the projection operator onto the state $|1\rangle$:

$$\mathbf{n}|x\rangle = x|x\rangle, \quad x = 0 \text{ or } 1. \tag{1.30}$$

Because $|0\rangle$ and $|1\rangle$ are eigenvectors of $\mathbf{n}$ with eigenvalues 0 and 1, $\mathbf{n}$ is called the 1-Cbit *number operator*. We also define the complementary operator,

$$\tilde{\mathbf{n}} = \mathbf{1} - \mathbf{n}, \tag{1.31}$$

which projects onto the state $|0\rangle$, so $|0\rangle$ and $|1\rangle$ are eigenvectors of $\tilde{\mathbf{n}}$ with eigenvalues 1 and 0. These operators have the matrix representations

$$\mathbf{n} = \begin{pmatrix} 0 & 0 \\ 0 & 1 \end{pmatrix}, \qquad \tilde{\mathbf{n}} = \begin{pmatrix} 1 & 0 \\ 0 & 0 \end{pmatrix}. \tag{1.32}$$

It follows directly from their definitions that

$$\mathbf{n}^2 = \mathbf{n}, \qquad \tilde{\mathbf{n}}^2 = \tilde{\mathbf{n}}, \qquad \mathbf{n}\tilde{\mathbf{n}} = \tilde{\mathbf{n}}\mathbf{n} = \mathbf{0}, \qquad \mathbf{n} + \tilde{\mathbf{n}} = \mathbf{1}. \tag{1.33}$$

We also have

$$\mathbf{n}\mathbf{X} = \mathbf{X}\tilde{\mathbf{n}}, \qquad \tilde{\mathbf{n}}\mathbf{X} = \mathbf{X}\mathbf{n}, \tag{1.34}$$

since flipping the state of a Cbit and then acting on it with $\mathbf{n}$ ($\tilde{\mathbf{n}}$) is the same as acting on the state with $\tilde{\mathbf{n}}$ ($\mathbf{n}$) and then flipping it. All the simple relations in (1.33) and (1.34) also follow, as they must, from the matrix representations (1.17) and (1.32) for $\mathbf{X}$, $\mathbf{n}$, and $\tilde{\mathbf{n}}$.

Although $\mathbf{n}$ has no interpretation as a physical operation on Cbits – replacing the state of a Cbit by the zero vector corresponds to no physical operation – it can be useful in deriving relations between operations that do have physical meaning. Since, for example, the SWAP operator $\mathbf{S}_{ij}$ acts as the identity if the states of the Cbits i and j are the same, and flips the numbers represented by both Cbits if their states are different, it can be written as

$$\mathbf{S}_{ij} = \mathbf{n}_i\mathbf{n}_j + \tilde{\mathbf{n}}_i\tilde{\mathbf{n}}_j + (\mathbf{X}_i\mathbf{X}_j)(\mathbf{n}_i\tilde{\mathbf{n}}_j + \tilde{\mathbf{n}}_i\mathbf{n}_j). \tag{1.35}$$

At the risk of belaboring the obvious, I note that (1.35) acts as the swap operator because if both Cbits are in the state $|1\rangle$ (so swapping their states does nothing) then only the first term in the sum acts (i.e. each of the other three terms gives 0) and multiplies the state by 1; if both Cbits are in the state $|0\rangle$, only the second term acts and again multiplies the state by 1; if Cbit i is in the state $|1\rangle$ and Cbit j is in the state $|0\rangle$, only the third term acts and the effect of flipping both Cbits is to swap their states; and if Cbit i is in the state $|0\rangle$ and Cbit j is in the state $|1\rangle$, only the fourth term acts and the effect of the two $\mathbf{X}$s is again to swap their states.

To help you become more at home with this notation, you are urged to prove from (1.35) that $\mathbf{S}_{ij}^2 = \mathbf{1}$, using only the relations in (1.33) and (1.34), the fact that $\mathbf{X}^2 = \mathbf{1}$, and the fact that 1-Cbit operators acting on different Cbits commute.

The construction (1.23) of SWAP out of cNOT operators can also be demonstrated using a more algebraic approach. Note first that $\mathbf{C}_{ij}$ can be expressed in terms of $\mathbf{n}$s and $\mathbf{X}$s by

$$\mathbf{C}_{ij} = \tilde{\mathbf{n}}_i + \mathbf{X}_j\mathbf{n}_i, \tag{1.36}$$

since if the state of Cbit i is $|0\rangle$ only the first term acts, which leaves the states of both Cbits unchanged, but if the state of Cbit i is $|1\rangle$ only the second term acts, which leaves the state of Cbit i unchanged, while $\mathbf{X}_j$

flips Cbit j. If you substitute expressions of the form (1.36) for each of the three terms in (1.23), then you can show by purely algebraic manipulations that four of the eight terms into which the products expand vanish and the remaining four can be rearranged to give the swap operator (1.35).

An operator that has no direct role to play in classical computation, but which is as important as the NOT operator $\mathbf{X}$ in quantum computation, is the operator $\mathbf{Z}$ defined by

$$\mathbf{Z} = \tilde{\mathbf{n}} - \mathbf{n} = \begin{pmatrix} 1 & 0 \\ 0 & -1 \end{pmatrix}. \tag{1.37}$$

It follows from (1.34) (or from the matrix representations (1.17) and (1.37)) that $\mathbf{X}$ *anticommutes* with $\mathbf{Z}$:

$$\mathbf{ZX} = -\mathbf{XZ}. \tag{1.38}$$

Since $\tilde{\mathbf{n}} + \mathbf{n} = \mathbf{1}$, we can use (1.37) to express the 1-Cbit projection operators $\tilde{\mathbf{n}}$ and $\mathbf{n}$ in terms of $\mathbf{1}$ and $\mathbf{Z}$:

$$\mathbf{n} = \tfrac{1}{2}(\mathbf{1} - \mathbf{Z}), \qquad \tilde{\mathbf{n}} = \tfrac{1}{2}(\mathbf{1} + \mathbf{Z}). \tag{1.39}$$

Using this we can rewrite the cNOT operator (1.36) in terms of $\mathbf{X}$ and $\mathbf{Z}$ operators:

$$\begin{aligned} \mathbf{C}_{ij} &= \tfrac{1}{2}\big(\mathbf{1} + \mathbf{Z}_i\big) + \tfrac{1}{2}\mathbf{X}_j\big(\mathbf{1} - \mathbf{Z}_i\big) \\ &= \tfrac{1}{2}\big(\mathbf{1} + \mathbf{X}_j\big) + \tfrac{1}{2}\mathbf{Z}_i\big(\mathbf{1} - \mathbf{X}_j\big). \end{aligned} \tag{1.40}$$

The second form follows from the first because $\mathbf{X}_j$ and $\mathbf{Z}_i$ commute when $i \neq j$. Note that, if we were to interchange $\mathbf{X}$ and $\mathbf{Z}$ in the second line of (1.40), we would get back the expression directly above it except for the interchange of i and j. So interchanging the $\mathbf{X}$ and $\mathbf{Z}$ operators has the effect of switching which Cbit is the control and which is the target, changing $\mathbf{C}_{ij}$ into $\mathbf{C}_{ji}$. An operator that can produce just this effect is the *Hadamard transformation* (also sometimes called the *Walsh–Hadamard transformation*),

$$\mathbf{H} = \frac{1}{\sqrt{2}}(\mathbf{X} + \mathbf{Z}) = \frac{1}{\sqrt{2}}\begin{pmatrix} 1 & 1 \\ 1 & -1 \end{pmatrix}. \tag{1.41}$$

This is another operator of fundamental importance in quantum computation.[2]

---

2 Physicists should note here an unfortunate clash between the notations of quantum computer science and physics. Quantum physicists invariably use H to denote the Hamiltonian function (in classical mechanics) or Hamiltonian operator (in quantum mechanics). Fortunately Hamiltonian operators, although of crucial importance in the design of quantum computers, play a very limited role in the general theory of quantum computation, being completely overshadowed by the unitary transformations that they generate. So physicists can go along with the computer-science notation without getting into serious trouble.

Since $\mathbf{X}^2 = \mathbf{Z}^2 = \mathbf{1}$ and $\mathbf{XZ} = -\mathbf{ZX}$, one easily shows from the definition (1.41) of $\mathbf{H}$ in terms of $\mathbf{X}$ and $\mathbf{Z}$ that

$$\mathbf{H}^2 = \mathbf{1} \tag{1.42}$$

and that

$$\mathbf{HXH} = \mathbf{Z}, \qquad \mathbf{HZH} = \mathbf{X}. \tag{1.43}$$

This shows how $\mathbf{H}$ can be used to interchange the $\mathbf{X}$ and $\mathbf{Z}$ operators in $\mathbf{C}_{ji}$: it follows from (1.43), together with (1.40) and (1.42), that

$$\mathbf{C}_{ji} = \big(\mathbf{H}_i\mathbf{H}_j\big)\mathbf{C}_{ij}\big(\mathbf{H}_i\mathbf{H}_j\big). \tag{1.44}$$

We shall see that this simple relation can be put to some quite remarkable uses in a quantum computer. While one can achieve this interchange on a classical computer using the SWAP operation, $\mathbf{C}_{ji} = \mathbf{S}_{ij}\mathbf{C}_{ij}\mathbf{S}_{ij}$, the crucial difference between $\mathbf{S}_{ij}$ and $\mathbf{H}_i\mathbf{H}_j$ is that the latter is a product of two 1-Cbit operators, while the former is not.

Of course, the action of $\mathbf{H}$ on the state of a Cbit that follows from (1.41),

$$\mathbf{H}|0\rangle = \tfrac{1}{\sqrt{2}}(|0\rangle + |1\rangle), \qquad \mathbf{H}|1\rangle = \tfrac{1}{\sqrt{2}}(|0\rangle - |1\rangle), \tag{1.45}$$

describes no meaningful transformation of Cbits. Nevertheless, when combined with other operations, as on the right side of (1.44), the Hadamard operations result in the perfectly sensible operation given on the left side. In a quantum computer the action of $\mathbf{H}$ on 1-Qbit states turns out to be not only meaningful but also easily implemented, and the possibility of interchanging control and target Qbits using only 1-Qbit operators in the manner shown in (1.44) turns out to have some striking consequences.

The use of Hadamards to interchange the control and target Qbits of a cNOT operation is sufficiently important in quantum computation to merit a second derivation of (1.44), which further illustrates the way in which one uses the operator formalism. In strict analogy to the definition of cNOT (see (1.21) and the preceding paragraph) we can define a controlled-Z operation, $\mathbf{C}^Z_{ij}$, which leaves the state of the target Cbit j unchanged if the state of the control Cbit i is $|0\rangle$, and operates on the target Cbit with $\mathbf{Z}$ if the state of the control Cbit is $|1\rangle$. As a result $\mathbf{C}^Z_{10}|xy\rangle$ acts as the identity on $|xy\rangle$ unless both x and y are 1, in which case it simply takes $|11\rangle$ into $-|11\rangle$. This behavior is completely symmetric in the two Cbits, so

$$\mathbf{C}^Z_{ij} = \mathbf{C}^Z_{ji}. \tag{1.46}$$

It is a straightforward consequence of (1.42) and (1.43) that sandwiching the target Cbit of a cNOT between Hadamards converts it to a $\mathbf{C}^Z$:

$$\mathbf{H}_j\mathbf{C}_{ij}\mathbf{H}_j = \mathbf{C}^Z_{ij}, \qquad \mathbf{H}_i\mathbf{C}_{ji}\mathbf{H}_i = \mathbf{C}^Z_{ji}. \tag{1.47}$$

In view of (1.46), we then have

$$\mathbf{H}_j\mathbf{C}_{ij}\mathbf{H}_j = \mathbf{H}_i\mathbf{C}_{ji}\mathbf{H}_i, \tag{1.48}$$

which is equivalent to (1.44), since $\mathbf{H}^2 = \mathbf{1}$.

As a final exercise in treating operations on Cbits as linear operations on vectors, we construct an alternative form for the swap operator. If we use (1.39) to reexpress each $\mathbf{n}$ and $\tilde{\mathbf{n}}$ appearing in the swap operator (1.35) in terms of $\mathbf{Z}$, we find that

$$\mathbf{S}_{ij} = \tfrac{1}{2}(\mathbf{1} + \mathbf{Z}_i\mathbf{Z}_j) + \tfrac{1}{2}(\mathbf{X}_i\mathbf{X}_j)(\mathbf{1} - \mathbf{Z}_i\mathbf{Z}_j). \tag{1.49}$$

If we define

$$\mathbf{Y} = i\mathbf{X}\mathbf{Z} = \begin{pmatrix} 0 & -i \\ i & 0 \end{pmatrix} \qquad (i = \sqrt{-1}), \tag{1.50}$$

we get the more compact form

$$\mathbf{S}_{ij} = \tfrac{1}{2}(\mathbf{1} + \mathbf{X}_i\mathbf{X}_j + \mathbf{Y}_i\mathbf{Y}_j + \mathbf{Z}_i\mathbf{Z}_j). \tag{1.51}$$

For three quarters of a century physicists have enjoyed grouping the matrix representations of the three operators $\mathbf{X}$, $\mathbf{Y}$, and $\mathbf{Z}$ into a "3-vector" $\vec{\sigma}$ whose "components" are $2 \otimes 2$ matrices:

$$\sigma_x = \begin{pmatrix} 0 & 1 \\ 1 & 0 \end{pmatrix}, \qquad \sigma_y = \begin{pmatrix} 0 & -i \\ i & 0 \end{pmatrix}, \qquad \sigma_z = \begin{pmatrix} 1 & 0 \\ 0 & -1 \end{pmatrix}. \tag{1.52}$$

The swap operator then becomes[3]

$$\mathbf{S}_{ij} = \tfrac{1}{2}\left(\mathbf{1} + \vec{\sigma}^{(i)} \cdot \vec{\sigma}^{(j)}\right), \tag{1.53}$$

where "·" represents the ordinary three-dimensional scalar product:

$$\vec{\sigma}^{(i)} \cdot \vec{\sigma}^{(j)} = \sigma^{(i)}_x\sigma^{(j)}_x + \sigma^{(i)}_y\sigma^{(j)}_y + \sigma^{(i)}_z\sigma^{(j)}_z. \tag{1.54}$$

The three components of $\vec{\sigma}$ have many properties that are unchanged under cyclic permutations of $x, y,$ and z. All three are Hermitian.[4] All square to unity,

$$\sigma_x^2 = \sigma_y^2 = \sigma_z^2 = \mathbf{1}. \tag{1.55}$$

3 Physicists might enjoy the simplicity of this "computational" derivation of the form of the exchange operator, compared with the conventional quantum-mechanical derivation, which invokes the full apparatus of angular-momentum theory.

4 The elements of a Hermitian matrix A satisfy $A_{ji} = A^*_{ij}$, where * denotes complex conjugation. A fuller statement in a broader context can be found in Appendix A.

They all anticommute in pairs and the product of any two of them is simply related to the third:

$$\begin{aligned}\boldsymbol{\sigma}_x\boldsymbol{\sigma}_y &= -\boldsymbol{\sigma}_y\boldsymbol{\sigma}_x = i\boldsymbol{\sigma}_z,\\ \boldsymbol{\sigma}_y\boldsymbol{\sigma}_z &= -\boldsymbol{\sigma}_z\boldsymbol{\sigma}_y = i\boldsymbol{\sigma}_x,\\ \boldsymbol{\sigma}_z\boldsymbol{\sigma}_x &= -\boldsymbol{\sigma}_x\boldsymbol{\sigma}_z = i\boldsymbol{\sigma}_y.\end{aligned} \tag{1.56}$$

The three relations (1.56) differ only by cyclic permutations of x, y, and z.

All the relations in (1.55) and (1.56) can be summarized in a single compact and useful identity. Let $\vec{a}$ and $\vec{b}$ be two 3-vectors with components a_x, a_y, a_z and b_x, b_y, b_z that are ordinary real numbers. (They can also be complex numbers, but in most useful applications they are real.) Then one easily confirms that all the relations in (1.55) and (1.56) imply and are implied by the single identity

$$(\vec{a}\cdot\vec{\boldsymbol{\sigma}})(\vec{b}\cdot\vec{\boldsymbol{\sigma}}) = (\vec{a}\cdot\vec{b})\mathbf{1} + i(\vec{a}\times\vec{b})\cdot\vec{\boldsymbol{\sigma}}, \tag{1.57}$$

where $\vec{a}\times\vec{b}$ denotes the vector product (or "cross product") of $\vec{a}$ and $\vec{b}$,

$$\begin{aligned}(\vec{a}\times\vec{b})_x &= a_yb_z - a_zb_y,\\ (\vec{a}\times\vec{b})_y &= a_zb_x - a_xb_z,\\ (\vec{a}\times\vec{b})_z &= a_xb_y - a_yb_x.\end{aligned} \tag{1.58}$$

Together with the unit matrix $\mathbf{1}$, the matrices $\boldsymbol{\sigma}_x$, $\boldsymbol{\sigma}_y$, and $\boldsymbol{\sigma}_z$ form a basis for the four-dimensional algebra of two-dimensional matrices of complex numbers: any such matrix is a unique linear combination of these four with complex coefficients. Because the four are all Hermitian, any two-dimensional Hermitian matrix A of complex numbers must be a *real* linear combination of the four, and therefore of the form

$$A = a_0\mathbf{1} + \vec{a}\cdot\vec{\boldsymbol{\sigma}}, \tag{1.59}$$

where a_0 and the components of the 3-vector $\vec{a}$ are all real numbers.

The matrices $\boldsymbol{\sigma}_x$, $\boldsymbol{\sigma}_y$, and $\boldsymbol{\sigma}_z$ were introduced in the early days of quantum mechanics by Wolfgang Pauli, to describe the angular momentum associated with the spin of an electron. They have many other useful purposes, being simply related to the quaternions invented by Hamilton to deal efficiently with the composition of three-dimensional rotations.[5] It is pleasing to find them here, buried in the interior of the operator that simply swaps two classical bits. We shall have extensive occasion to use Pauli's 1-Qbit operators when we come to the subject of

5 Hamilton's quaternions i, j, k are represented by $i\boldsymbol{\sigma}_x, i\boldsymbol{\sigma}_y, i\boldsymbol{\sigma}_z$. The beautiful and useful connection between Pauli matrices and three-dimensional rotations discovered by Hamilton is developed in Appendix B.

quantum error correction. Some of their properties, developed further in Appendix B, prove to be quite useful in treating Qbits, to which we now turn.

1.5 Qbits and their states

The state of a Cbit is a pretty miserable specimen of a two-dimensional vector. The only vectors with any classical meaning in the whole two-dimensional vector space are the two orthonormal vectors $|0\rangle$ and $|1\rangle$, since those are the only two states a Cbit can have. Happily, nature has provided us with physical systems, Qbits, described by states that do not suffer from this limitation. The state $|\psi\rangle$ associated with a Qbit can be any unit vector in the two-dimensional vector space spanned by $|0\rangle$ and $|1\rangle$ over the complex numbers. The general state of a Qbit is

$$|\psi\rangle = \alpha_0|0\rangle + \alpha_1|1\rangle = \begin{pmatrix} \alpha_0 \\ \alpha_1 \end{pmatrix}, \tag{1.60}$$

where α_0 and α_1 are two complex numbers constrained only by the requirement that $|\psi\rangle$, like $|0\rangle$ and $|1\rangle$, should be a unit vector in the complex vector space – i.e. only by the normalization condition

$$|\alpha_0|^2 + |\alpha_1|^2 = 1. \tag{1.61}$$

The state $|\psi\rangle$ is said to be a *superposition* of the states $|0\rangle$ and $|1\rangle$ with *amplitudes* α_0 and α_1. If one of α_0 and α_1 is 0 and the other is 1 – i.e. the special case in which the state of the Qbit is one of the two classical states $|0\rangle$ or $|1\rangle$ – it can be convenient to retain the language appropriate to Cbits, speaking of the Qbit "having the value" 0 or 1. More correctly, however, one is entitled to say only that the state of the Qbit is $|0\rangle$ or $|1\rangle$. Qbits, in contrast to Cbits, cannot be said to "have values." They have – or, more correctly, *are described by*, or, better still, are *associated with* – states. We shall often sacrifice correctness for ease of expression. Some reasons for this apparently pedantic terminological hair splitting will emerge below.

Just as the general state of a single Qbit is any normalized superposition (1.60) of the two possible classical states, the general state $|\Psi\rangle$ that nature allows us to associate with two Qbits is any normalized superposition of the four orthogonal classical states,

$$|\Psi\rangle = \alpha_{00}|00\rangle + \alpha_{01}|01\rangle + \alpha_{10}|10\rangle + \alpha_{11}|11\rangle = \begin{pmatrix} \alpha_{00} \\ \alpha_{01} \\ \alpha_{10} \\ \alpha_{11} \end{pmatrix}, \tag{1.62}$$

with the complex amplitudes being constrained only by the normalization condition

$$|\alpha_{00}|^2 + |\alpha_{01}|^2 + |\alpha_{10}|^2 + |\alpha_{11}|^2 = 1. \tag{1.63}$$

This generalizes in the obvious way to n Qbits, whose general state can be any superposition of the 2^n different classical states, with amplitudes whose squared magnitudes sum to unity:

$$|\Psi\rangle = \sum_{0 \leq x < 2^n} \alpha_x |x\rangle_n, \tag{1.64}$$

$$\sum_{0 \leq x < 2^n} |\alpha_x|^2 = 1. \tag{1.65}$$

In the context of quantum computation, the set of 2^n classical states – all the possible tensor products of n individual Qbit states $|0\rangle$ and $|1\rangle$ – is called the *computational basis*. For most purposes *classical basis* is a more appropriate term. I shall use the two interchangeably. The states that characterize n Cbits – the classical-basis states – are an extremely limited subset of the states of n Qbits, which can be any (normalized) superposition with complex coefficients of these classical-basis states.

If we have two Qbits, one in the state $|\psi\rangle = \alpha_0|0\rangle + \alpha_1|1\rangle$ and the other in the state $|\phi\rangle = \beta_0|0\rangle + \beta_1|1\rangle$, then the state $|\Psi\rangle$ of the pair, in a straightforward generalization of the rule for multi-Cbit states, is taken to be the tensor product of the individual states,

$$\begin{aligned} |\Psi\rangle &= |\psi\rangle \otimes |\phi\rangle = \big(\alpha_0|0\rangle + \alpha_1|1\rangle\big) \otimes \big(\beta_0|0\rangle + \beta_1|1\rangle\big) \\ &= \alpha_0\beta_0|00\rangle + \alpha_0\beta_1|01\rangle + \alpha_1\beta_0|10\rangle + \alpha_1\beta_1|11\rangle \\ &= \begin{pmatrix} \alpha_0\beta_0 \\ \alpha_0\beta_1 \\ \alpha_1\beta_0 \\ \alpha_1\beta_1 \end{pmatrix}. \end{aligned} \tag{1.66}$$

Note that a general 2-Qbit state (1.62) is of the special form (1.66) if and only if $\alpha_{00}\alpha_{11} = \alpha_{01}\alpha_{10}$. Since the four amplitudes in (1.62) are constrained only by the normalization condition (1.63), this relation need not hold, and the general 2-Qbit state, unlike the general state of two Cbits, is *not* a product (1.66) of two 1-Qbit states. The same is true for states of n Qbits. Unlike Cbits, whose general state can only be one of the 2^n products of $|0\rangle$s and $|1\rangle$s, a general state of n Qbits is a superposition of these 2^n product states and cannot, in general, be expressed as a product of any set of 1-Qbit states. Individual Qbits making up a multi-Qbit system, in contrast to individual Cbits, cannot always be characterized as having individual states of their own.[6]

Such nonproduct states of two or more Qbits are called *entangled* states. The term is a translation of Schrödinger's *verschränkt*, which I

6 More precisely, they do not always have what are called *pure states* of their own. It is often convenient to give a statistical description of an individual Qbit (or a group of Qbits) in terms of what is called a *density matrix* or *mixed state*. If one wishes to emphasize that one is not talking about a mixed state, one uses the term "pure state." In this book the term "state" always means "pure state."

am told is rendered more accurately as "entwined" or "enfolded." But Schrödinger himself used the English word "entangled," and may even have used it before coining the German term. When the state of several Qbits is entangled, they can sometimes behave in some very strange ways. An example of such peculiar behavior is discussed in Appendix D. Aside from its intrinsic interest, the appendix provides some further exercise in the analytical manipulation of Qbits.

1.6 Reversible operations on Qbits

The only nontrivial reversible operation a classical computer can perform on a single Cbit is the NOT operation $\mathbf{X}$. Nature has been far more versatile in what it allows us to do to a Qbit. The reversible operations that a quantum computer can perform upon a single Qbit are represented by the action on the state of the Qbit of any *linear* transformation that takes unit vectors into unit vectors. Such transformations $\mathbf{u}$ are called *unitary* and satisfy the condition[7]

$$\mathbf{u}\mathbf{u}^\dagger = \mathbf{u}^\dagger\mathbf{u} = 1. \tag{1.67}$$

Since any unitary transformation has a unitary inverse, such actions of a quantum computer on a Qbit are reversible. The reason why reversibility is crucial for the effective functioning of a quantum computer will emerge in Chapter 2.

The most general reversible n-Cbit operation in a classical computer is a permutation of the $(2^n)!$ different classical-basis states. The most general reversible operation that a quantum computer can perform upon n Qbits is represented by the action on their state of any linear transformation that takes unit vectors into unit vectors – i.e. any 2^n-dimensional unitary transformation $\mathbf{U}$, satisfying

$$\mathbf{U}\mathbf{U}^\dagger = \mathbf{U}^\dagger\mathbf{U} = 1. \tag{1.68}$$

Any reversible operation on n Cbits – i.e. any permutation $\mathbf{P}$ of the 2^n Cbit states – can be associated with a unitary operation $\mathbf{U}$ on n Qbits. One defines the action of $\mathbf{U}$ on the classical-basis states of the Qbit to be identical to the operation of $\mathbf{P}$ on the corresponding classical states of the Cbit. Since the classical basis *is* a basis, $\mathbf{U}$ can be extended to arbitrary n-Qbit states by requiring it to be linear. Since the action of $\mathbf{U}$ on the classical-basis states is to permute them, its effect on any superposition of such states $\sum \alpha_x|x\rangle_n$ is to permute the amplitudes α_x. Such a permutation preserves the value of $\sum |\alpha_x|^2$, so $\mathbf{U}$ takes unit vectors into unit vectors. Being norm-preserving and linear, $\mathbf{U}$ is indeed unitary.

7 These and other facts about linear operators on vector spaces over the complex numbers are also reviewed and summarized in Appendix A.

Many important unitary operations on Qbits that we shall be examining below are defined in this way, as permutations of the classical-basis states, which are implicitly understood to be extended by linearity to all Qbit states. In particular, the transformations NOT, SWAP, and cNOT on Cbits are immediately defined in this way for Qbits as well. But the available unitary transformations on Qbits are, of course, much more general than straightforward extensions of classical operations. We have already encountered two such examples, the operator **Z** and the Hadamard transformation **H**. Both of these take the classical-basis states of a Qbit into another orthonormal basis, so their linear extensions to all Qbit states are necessarily unitary.

In designing quantum algorithms, the class of allowed unitary transformations is almost always restricted to ones that can be built entirely out of products of unitary transformations that act on only one Qbit at a time, called *1-Qbit gates*, or that act on just a pair of Qbits, called *2-Qbit gates*. This restriction is imposed because the technical problems of making higher-order quantum gates are even more formidable than the (already difficult) problems of constructing reliable 1- and 2-Qbit gates.

It turns out that this is not a fundamental limitation, since arbitrary unitary transformations can be approximated to an arbitrary degree of precision by sufficiently many 1- and 2-Qbit gates. We shall not prove this general result,[8] because all of the quantum algorithms to be developed here will be explicitly built up entirely out of 1- and 2-Qbit gates. One very important illustration of the sufficiency of 1- and 2-Qbit gates will emerge in Chapter 2. For a reversible classical computer, it can be shown that at least one 3-Cbit gate is needed to build up general logical operations. But, in a quantum computer, we shall find, remarkably – and importantly for the feasibility of practical quantum computation – that the quantum extension of this 3-Cbit gate can be constructed out of a small number of 1- and 2-Qbit gates.

While unitarity is generally taken to be the hallmark of the transformations nature allows us to perform on quantum states, what is really remarkable about the transformations of Qbit states is their *linearity* (which is, of course, one aspect of their unitarity). It is easy to dream up simple classical models for a Qbit, particularly if one restricts its states to real linear combinations of the two computational basis states. It is not hard to invent classical models for NOT and Hadamard 1-Qbit gates that act linearly on all the 1-Qbit states of the model Qbit. But I know of no classical model that can extend a cNOT on the four computational basis states of two Cbits to an operation that acts

8 The argument is given by David P. DiVincenzo, "Two-bit gates are universal for quantum computation," *Physical Review* A **51**, 1015–1022 (1995), `http://arxiv.org/abs/quant-ph/9407022`.

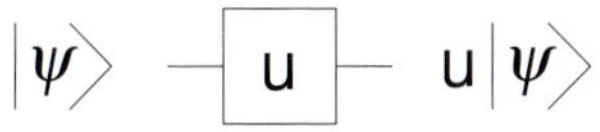

Fig 1.1 A circuit diagram representing the action on a single Qbit of the 1-Qbit gate $\mathbf{u}$. Initially the Qbit is described by the input state $|\psi\rangle$ on the left. The thin line (wire) represents the subsequent history of the Qbit. After emerging from the box representing $\mathbf{u}$, the Qbit is described on the right by the final state $\mathbf{u}|\psi\rangle$.

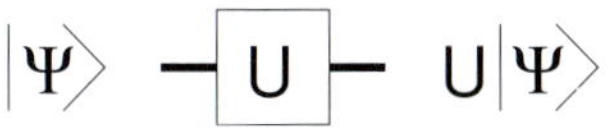

Fig 1.2 A circuit diagram representing the action on n Qbits of the n-Qbit gate $\mathbf{U}$. Initially the Qbits ares described by the input state $|\Psi\rangle$ on the left. The thick line (bar) represents the subsequent history of the Qbits. After emerging from the box representing $\mathbf{U}$, the Qbits are described on the right by the final state $\mathbf{U}|\Psi\rangle$.

linearly on all the states of two model Qbits. It is a remarkable and highly nontrivial fact about the physical world that nature does allow us, with much ingenuity and hard work, to fabricate unitary cNOT gates for a pair of genuine quantum Qbits.

1.7 Circuit diagrams

It is the practice in quantum computer science to represent the action of a sequence of gates acting on n Qbits by a circuit diagram. The initial state of the Qbits appears on the left, the final state on the right, and the gates themselves in the central part of the figure. Figure 1.1 shows the simplest possible such diagram: a Qbit initially in the state $|\psi\rangle$ is acted on by a 1-Qbit gate $\mathbf{u}$, with the result that the Qbit is assigned the new state $\mathbf{u}|\psi\rangle$. Figure 1.2 shows the analogous diagram for an n-Qbit gate $\mathbf{U}$ and an n-Qbit initial state $|\Psi\rangle$. The line that goes into and out of the box representing the unitary transformation – which becomes useful when one starts chaining together a sequence of gates – is sometimes called a *wire* in the case of a single Qbit, and the thicker line (which represents n wires) associated with an n-Qbit gate is sometimes called a *bar*.

Figure 1.3 reveals a peculiar feature of these circuit diagrams that it is important to be aware of. The diagrams are read from left to right (as one reads ordinary prose in European languages). Part (a) portrays a circuit that acts first with $\mathbf{V}$ and then with $\mathbf{U}$ on the initial state $|\Psi\rangle$. The result is the state $\mathbf{UV}|\Psi\rangle$, because it is the convention, in writing equations for linear operators on vector spaces, that the operation appears to the

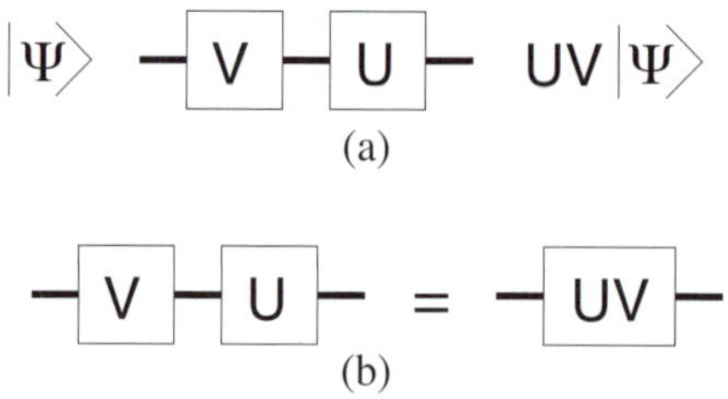

Fig 1.3 (a) A circuit diagram representing the action on n Qbits of two n-Qbit gates. Initially the Qbits are described by the input state $|\Psi\rangle$ on the left. They are acted upon first by the gate **V** and then by the gate **U**, emerging on the right in the final state $\mathbf{UV}|\Psi\rangle$. Note that the order in which the Qbits encounter unitary gates in the figure is opposite to the order in which the corresponding symbols are written in the symbol for the final state on the right. (b) This emphasizes the unfortunate convention that, because gates on the left act before gates on the right in a circuit diagram, a circuit showing **V** on the left and **U** on the right represents the operation conventionally denoted by **UV**.

left of the state on which it acts. Thus the sequence of symbols $|\Psi\rangle$, **V**, and **U** on the left of the circuit diagram in (a) is reversed from the sequence in which they appear in the mathematical representation of the state that is produced on the right. Part (b) shows the consequences of this for the part of the circuit diagram containing just the gates: a diagram in which a gate **V** (on the left) is followed by a gate **U** on the right describes the unitary transformation **UV**.

One should be wary of the possibility for confusion arising from the fact that operators (and states) in circuit diagrams always appear in the diagrams in the opposite sequence from the order in which they appear on the page in the corresponding equations. While several of the most important diagrams we shall encounter are left–right symmetric, many are not, so one should be on guard against getting things backwards when translating equations into circuit diagrams and vice versa.

In lecturing on quantum computation I tried for several years to reverse the computer-science convention, putting the initial state on the right of the circuit diagram and letting the gates on the right act first. This has the great advantage of making the diagram look like the equation it represents. It has, however, a major disadvantage, even setting aside the fact that it flies in the face of well established convention. It requires one to write on the blackboard in the wrong direction, from right to left, whenever one wishes to produce a circuit diagram. Guessing how far to the right one should start is hard to do if the diagram is a lengthy one, and for this reason I gave up after a few years and reverted to the conventional form. A better alternative would be for physicists to start writing their equations with the states on the left (represented by bra vectors rather than ket vectors[9]) and with linear operators appearing to the right of the states on which they act. But this would require abandoning a tradition that goes back three quarters of a century. So we are stuck with a clash of cultures, and must simply keep in mind that confusion can arise if one forgets the elementary fact represented in Figure 1.3(b).

There is little utility to circuit diagrams of the simple form in Figures 1.1–1.3, but they are important as building blocks out of which

9 See Appendix A for the distinction between bras and kets.

larger circuit diagrams are constructed. As the number of operations increases, the diagrams enable one to see at a glance the action of a sequence of 1- and 2-Qbit unitary gates on a collection of many Qbits in a way that is far more transparent and much more easily remembered than the corresponding formulae. Indeed, many calculations that involve rather lengthy equations can be simply accomplished by manipulating circuit diagrams, as we shall see.

When the state vectors entering or leaving a wire or bar in a circuit diagram are computational-basis states like $|x\rangle$, one sometimes omits the symbol $|\ \rangle$ and simply writes x.

1.8 Measurement gates and the Born rule

To give the state of a single Cbit you need only one bit of information: whether the state of the Cbit is $|0\rangle$ or $|1\rangle$. But to specify the state (1.60) of a single Qbit to an arbitrarily high degree of precision, you need arbitrarily many bits of information, since you must specify two complex numbers α and β subject only to the normalization constraint (1.61). Because Qbits not only have a much richer set of states than Cbits, but also can be acted on by a correspondingly richer set of transformations, it might appear obvious that a quantum computer would be vastly more powerful than a classical computer. *But there is a major catch!*

The catch is this: if you have n Cbits, each representing either 0 or 1, you can find out the state of each just by looking. There is nothing problematic about learning the state of a Cbit, and hence learning the result of any calculation you may have built up out of operations on those Cbits. Furthermore – and this is taken for granted in any discussion of a classical computer – the state of Cbits is not altered by the process of reading them. The act of acquiring the information from Cbits is not disruptive. You can read the Cbits at any stage of a computation without messing up subsequent stages.

In stark contrast, if you have n Qbits in a superposition (1.64) of computational basis states, there is nothing whatever you can do to them to extract from those Qbits the vast amount of information contained in the amplitudes α_x. You cannot read out the values of those amplitudes, and therefore you cannot find out what the state is. The state of n Qbits is not associated with any ascertainable property of those Qbits, as it is for Cbits.

There is only one way to extract information from n Qbits in a given state. It is called *making a measurement*.[10] Making a measurement

10 Physicists will note – others need pay no attention to this remark – that what follows is more accurately characterized as "making a (von Neumann) measurement in the computational (classical) basis." There are other ways

consists of performing a certain test on each Qbit, the outcome of which is either 0 or 1. The particular collection of zeros and ones produced by the test is not in general determined by the state $|\Psi\rangle$ of the Qbits; the state determines only the *probability* of the possible outcomes, according to the following rule: the probability of getting a particular result – say 01100, if you have five Qbits – is given by the squared magnitude of the amplitude of the state $|01100\rangle$ in the expansion of the state $|\Psi\rangle$ of the Qbits in the 2^5 computational basis states. More generally, if the state of n Qbits is

$$|\Psi\rangle_n = \sum_{0 \le x < 2^n} \alpha_x |x\rangle_n, \tag{1.69}$$

then the probability that the zeros and ones resulting from measurements of all the Qbits will give the binary expansion of the integer x is

$$p(x) = |\alpha_x|^2. \tag{1.70}$$

This basic rule for how information can be extracted from a quantum state was first enunciated by Max Born, and is known as the *Born rule*. It provides the link between amplitudes and the numbers you can actually read out when you test – i.e. measure – the Qbits. The squared magnitudes of the amplitudes give the probabilities of outcomes of measurements. Normalization conditions like (1.65) are just the requirements that the probabilities for all of the 2^n mutually exclusive outcomes add up to 1.

The process of measurement is carried out by a piece of hardware with a digital display, known as an n-Qbit *measurement gate*. Such an n-Qbit measurement gate is depicted schematically in Figure 1.4. In contrast to unitary gates, which have a unique output state for each input state, the state of the Qbits emerging from a measurement gate is only *statistically* determined by the state of the input Qbits. In further contrast to unitary gates, the action of a measurement gate cannot be undone: given the final state $|x\rangle$, there is no way of reconstructing the initial state $|\Psi\rangle$. Measurement is irreversible. Nor is the action of a measurement gate in any sense linear.

To the extent that it suggests that some preexisting property is being revealed, "measurement" is a dangerously misleading term, but it is

---

to make such a measurement, but they can all be reduced to measurements in the computational basis if an appropriate unitary transformation is applied to the n-Qbit state of the computer just before carrying out the measurement. In this book the term "measurement" always means measurement in the computational basis. Measurements in other bases will always be treated as measurements in the computational basis preceded by suitable unitary transformations. There are also more general forms of measurement than von Neumann measurements, going under the unpleasant acronym *POVM* (for "positive operator-valued measure"). We shall make no explicit use of POVMs.

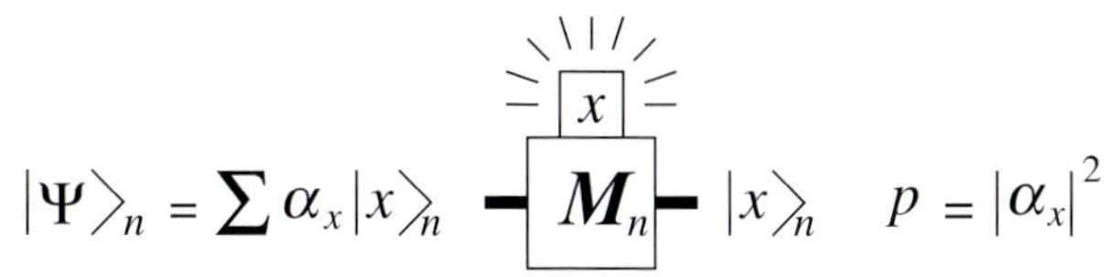

Fig 1.4 A circuit diagram representing an n-Qbit measurement gate. The Qbits are initially described by the n-Qbit state

$$|\Psi\rangle_n = \sum_{0 \le x < 2^n} \alpha_x |x\rangle_n,$$

on the left. After the measurement gate M_n has acted, with probability $p = |\alpha_x|^2$ it indicates an integer x, $0 \le x < 2^n$, and the Qbits are subsequently described by the state $|x_n\rangle$ on the right.

hallowed by three quarters of a century of use by quantum physicists, and impossible to avoid in treatments of quantum computation. One should avoid being misled by such spurious connotations of "measurement," though it confused many physicists in the early days of quantum mechanics and may well continue to confuse some to this day. In quantum computation "measurement" means nothing more or less than applying and reading the display of an appropriate measurement gate, whose action is fully specified by the Born rule, as described above, and expanded upon below. While measurement in quantum mechanics is not at all like measuring somebody's weight, it does have some resemblance to measuring Alice's IQ, which, one can argue, reveals no preexisting numerical property of Alice, but only what happens when she is subjected to an IQ test.

The simplest statement of the Born rule is for a single Qbit. If the state of the Qbit is the superposition (1.60) of the states $|0\rangle$ and $|1\rangle$ with amplitudes α_0 and α_1 then the result of the measurement is 0 with probability $|\alpha_0|^2$ and 1 with probability $|\alpha_1|^2$. This measurement is carried out by a 1-Qbit measurement gate, as illustrated in Figure 1.5. We shall see below that n-Qbit measurement gates can be realized by applying 1-Qbit measurement gates to each of the n Qbits. The process of measurement can thus be reduced to applying multiple copies of a *single elementary piece of hardware: the 1-Qbit measurement gate.*

In addition to displaying an n-bit integer with probabilities determined by the amplitudes, there is a second very important aspect of the action of measurement gates: if n Qbits, initially described by a state $|\Psi\rangle$, are sent through an n-Qbit measurement gate, and the display of the measurement gate indicates the integer x, then one must associate with the Qbits emerging from that measurement gate the classical-basis state $|x\rangle_n$, as shown in Figures 1.4 and 1.5. This means that all traces of the amplitudes α_x characterizing the input state have vanished from the output state. The only role they have played in the measurement is to determine the probability of a particular output.

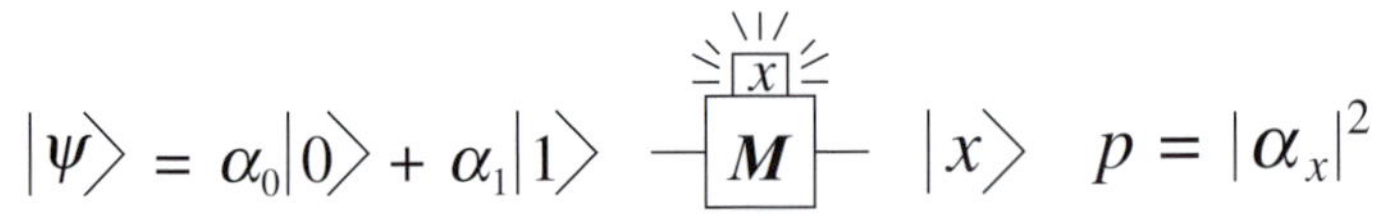

Fig 1.5 A special case of Figure 1.4: a 1-Qbit measurement gate. The reading x of the gate is either 0 or 1.

If the state of the input Qbits is one of the classical-basis states $|x\rangle_n$, then according to the Born rule the probability that the measurement gate will read x and the output state will remain $|x\rangle_n$ is 1. But for superpositions (1.69) with more than a single nonzero amplitude α_x, the output state is not determined. Being a single one of the classical basis states $|x\rangle_n$, the output state no longer carries any information about the amplitudes characterizing the initial state, other than certifying that the particular amplitude α_x was not zero, and, in all likelihood, was not exceedingly small.

So once you send n Qbits through an n-Qbit measurement gate, you remove the possibility of extracting any further information about their original state $|\Psi\rangle$. After such a measurement of five Qbits, if the result is 01100, then the post-measurement state associated with the Qbits is no longer $|\Psi\rangle$, but $|01100\rangle$. The original state $|\Psi\rangle$, with all the rich information potentially available in its amplitudes, is irretrievably lost. Qbits emerging from a measurement gate that indicates the outcome x are characterized by the state $|x\rangle$, regardless of what their pre-measurement state may have been.

This change of state attendant upon a measurement is often referred to as a *reduction* or *collapse* of the state. One says that the pre-measurement state *reduces* or *collapses* to the post-measurement state, as a consequence of the measurement. This should not be taken to imply (though, alas, it often is) that the Qbits themselves suffer a catastrophic "reduction" or "collapse." It is important to keep in mind, in this context, that the state of n Qbits is nothing more than an abstract symbol, used, via the Born rule, to calculate probabilities of measurement outcomes. As has already been noted, there is no internal property of the Qbits that corresponds to their state.

You might well wonder how one can learn anything at all of computational interest under these wretched conditions. The artistry of quantum computation consists of producing, through a cunningly constructed unitary transformation, a superposition in which most of the amplitudes α_x are zero or extremely close to zero, with useful information being carried by *any* of the values of x that have an appreciable probability of being indicated by the measurement. It is thus important to be seeking information that, once possessed, can easily be confirmed, perhaps with an ordinary (classical) computer (e.g. the factors of a large number), so that one is not misled by rare and irrelevant low-probability outcomes. How this is actually accomplished in various cases of interest will be one of our major preoccupations.

It is important to note and immediately reject a possible misunderstanding of the Born rule. One might be tempted to infer from the rule that for a Qbit to be in a superposition, such as the state $|\psi\rangle = \alpha_0|0\rangle + \alpha_1|1\rangle$, means nothing more than that the "actual state" of the Qbit is either $|0\rangle$ with probability $|\alpha_0|^2$ or $|1\rangle$ with probability $|\alpha_1|^2$. Such an assertion goes beyond the rule, of course, which merely asserts that if one subjects a Qbit in the state $|\psi\rangle$ to an appropriate test – a measurement – then the outcome of the test will be 0 or 1 with those probabilities and the post-measurement state of the Qbit can correspondingly be taken to be $|0\rangle$ or $|1\rangle$. This does not imply that prior to the test the Qbit already carried the value revealed by the test and was already described by the corresponding classical-basis state, since, among other possibilities, the action of the test itself might well play a role in bringing forth the outcome.

In fact, it is easy to produce examples that demonstrate that the Qbit, prior to the test, *could not* have been in either of the states $|0\rangle$ and $|1\rangle$. We can see this with the help of the Hadamard transformation (1.41). We have defined the action of the 1-Qbit operators **H**, **X**, and **Z** only on the computational-basis states $|0\rangle$ and $|1\rangle$, but, as noted above, we can extend their action to arbitrary linear combinations of these states by requiring the extensions to be linear operators. Since the states $|0\rangle$ and $|1\rangle$ form a basis, this determines the action of **H**, **X**, and **Z** on any 1-Qbit state.

Because it is linear and norm-preserving, **H** is unitary, and is therefore the kind of operation a quantum computer can apply to the state of a Qbit: a 1-Qbit gate. The result of the operation of a Hadamard gate is to change the state $|\phi\rangle$ of a Qbit to $\mathbf{H}|\phi\rangle$. Suppose, now, that we apply **H** to a Qbit that is initially in the state

$$|\phi\rangle = \tfrac{1}{\sqrt{2}}(|0\rangle + |1\rangle). \tag{1.71}$$

It follows from (1.45) that the result is just

$$\mathbf{H}|\phi\rangle = |0\rangle. \tag{1.72}$$

So according to the Born rule, if we measure a Qbit described by the state $\mathbf{H}|\phi\rangle$, the result will be 0 with probability 1.

But suppose that a Qbit in the state $|\phi\rangle$ were indeed either in the state $|0\rangle$ with probability $\frac{1}{2}$ or in the state $|1\rangle$ with probability $\frac{1}{2}$. In either case, according to (1.45), the subsequent action of **H** would produce a state – either $(1/\sqrt{2})(|0\rangle + |1\rangle)$ or $(1/\sqrt{2})(|0\rangle - |1\rangle)$ – that under measurement yielded 0 or 1 with equal probability. This contradicts the fact just extracted directly from (1.72) that the result of making a measurement on a Qbit in the state $\mathbf{H}|\phi\rangle$ is invariably 0.

So a Qbit in a quantum superposition of $|0\rangle$ and $|1\rangle$ cannot be viewed as being either in the state $|0\rangle$ or in the state $|1\rangle$ with certain probabilities. Such a state represents something quite different. Although the Qbit reveals only a 0 or a 1 when you query it with a measurement gate, prior to the query its state is not in general either $|0\rangle$ or $|1\rangle$, but a superposition of the form (1.60). Such a superposition is as natural and irreducible a description of a Qbit as $|0\rangle$ and $|1\rangle$ are. This point is expanded on in Appendix C.

If the states of n Qbits are restricted to computational-basis states then the process of measurement is just like the classical process of "learning the value" of x without altering the state. Thus a quantum computer can be made to simulate a reversible classical computer by allowing only computational-basis states as input, and using only unitary gates that take computational-basis states into computational-basis states.

The Born rule, relating the amplitudes α_x in the expansion (1.64) of a general n-Qbit state $|\Psi\rangle$ to the probabilities of measuring x, is often stated in terms of inner products or projection operators.[11] The probability of a measurement giving the result x $(0 \le x < 2^n)$ is

$$p_\Psi(x) = |\,\alpha_x|^2 = |\langle x|\Psi\rangle|^2. \tag{1.73}$$

It can also be usefully expressed in terms of projection operators:

$$p_\Psi(x) = \langle x|\Psi\rangle\langle\Psi|x\rangle = \langle x|\mathbf{P}_\Psi|x\rangle \tag{1.74}$$

or

$$p_\Psi(x) = \langle\Psi|x\rangle\langle x|\Psi\rangle = \langle\Psi|\mathbf{P}_x|\Psi\rangle, \tag{1.75}$$

where $\mathbf{P}_\Psi = |\Psi\rangle\langle\Psi|$ is the projection operator on the state $|\Psi\rangle$, and $\mathbf{P}_x = |x\rangle\langle x|$ is the projection operator on the state $|x\rangle$.

1.9 The generalized Born rule

There is a stronger version of the Born rule, which plays an important role in quantum computation, even though, surprisingly, it is rarely explicitly mentioned in most standard quantum-mechanics texts. We shall call it the *generalized Born rule.* This stronger form applies when one measures only a single one of $n + 1$ Qbits, by sending it through a standard 1-Qbit measurement gate.

To formulate the generalized Born rule, note that any state of all $n + 1$ Qbits can be represented in the form

$$|\Psi\rangle_{n+1} = \alpha_0|0\rangle|\Phi_0\rangle_n + \alpha_1|1\rangle|\Phi_1\rangle_n, \qquad |\alpha_0|^2 + |\alpha_1|^2 = 1, \tag{1.76}$$

11 The Dirac notation for inner products and projection operators is described in Appendix A.

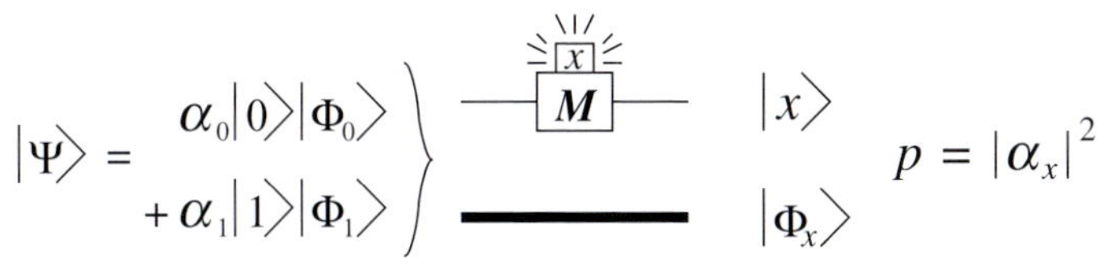

Fig 1.6 The action of a 1-Qbit measurement gate on a single one of $n+1$ Qbits, according to the generalized Born rule. The initial state (on the left) is a general $(n+1)$-Qbit state, expressed in the form $|\Psi\rangle_{n+1} = a_0|0\rangle|\Phi_0\rangle_n + a_1|1\rangle|\Phi_1\rangle_n$. Only the single Qbit on the left of this expression is subjected to a measurement gate.

where $|\Phi_0\rangle_n$ and $|\Phi_1\rangle_n$ are normalized (but not necessarily orthogonal). This follows directly from the general form,

$$|\Psi\rangle_{n+1} = \sum_{x=0}^{2^{n+1}-1} \gamma(x)|x\rangle_{n+1}, \qquad \sum_{x=0}^{2^{n+1}-1} |\gamma(x)|^2 = 1. \tag{1.77}$$

The states $|\Phi_0\rangle_n$ and $|\Phi_1\rangle_n$ are given by

$$|\Phi_0\rangle_n = (1/\alpha_0)\sum_{x=0}^{2^n-1} \gamma(x)|x\rangle_n, \qquad |\Phi_1\rangle_n = (1/\alpha_1)\sum_{x=0}^{2^n-1} \gamma(2^n+x)|x\rangle_n, \tag{1.78}$$

where

$$\alpha_0^2 = \sum_{x=0}^{2^n-1} |\gamma(x)|^2, \qquad \alpha_1^2 = \sum_{x=0}^{2^n-1} |\gamma(2^n+x)|^2. \tag{1.79}$$

(The α_0 and α_1 in (1.78) and (1.79) are real numbers, but can be multiplied by arbitrary phase factors if $|\Phi_0\rangle_n$ and $|\Phi_1\rangle_n$ are multiplied by the inverse phase factors.)

The generalized Born rule asserts that if one measures only the single Qbit whose state symbol is explicitly separated out from the others in the $(n+1)$-Qbit state (1.76), then the 1-Qbit measurement gate will indicate x (0 or 1) with probability $|\alpha_x|^2$, after which the $(n+1)$-Qbit state can be taken to be the product state $|x\rangle|\Phi_x\rangle_n$. (The rule holds for the measurement of any single Qbit – there is nothing special about the Qbit whose state symbol appears on the left in the $(n+1)$-Qbit state symbol.) This action of a 1-Qbit measurement gate on an $(n+1)$-Qbit state is depicted schematically in Figure 1.6.

If the Qbit on which the 1-Qbit gate acts is initially unentangled with the remaining n Qbits, then the action of the gate on the measured Qbit is just that specified by the ordinary Born rule, and the unmeasured Qbits play no role at all, remaining in their original state throughout the process. This is evident from the above statement of the generalized Born rule, specialized to the case in which the two states $|\Phi_0\rangle_n$ and $|\Phi_1\rangle_n$ are identical. It is illustrated in Figure 1.7.

If one applies the generalized Born rule n times to successive 1-Qbit measurements of each of n Qbits, initially in the general n-Qbit state (1.69), one can show by a straightforward argument, given in Appendix E, that the final state of the n Qbits is x with probability $|\alpha_x|^2$, where x is the n-bit integer whose bits are given by the readings

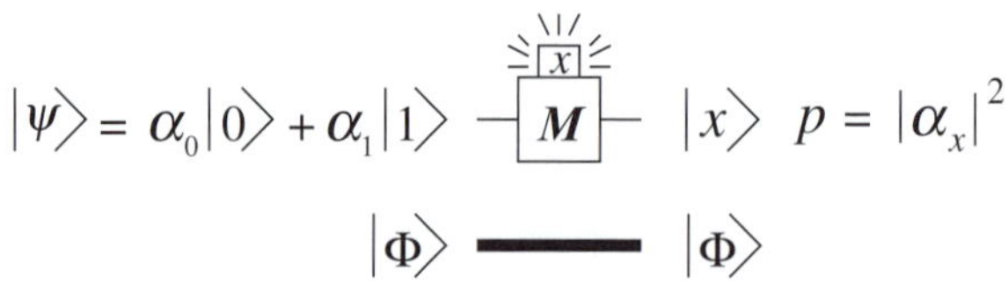

Fig 1.7 A simplification of Figure 1.6 when $|\Phi_0\rangle = |\Phi_1\rangle = |\Phi\rangle$. In this case the initial state on the left is just the product state $|\Psi\rangle_n = |\psi\rangle|\Phi\rangle = \big(a_0|0\rangle + a_1|1\rangle\big)|\Phi\rangle$, and the final state of the unmeasured Qbits continues to be $|\Phi\rangle$ regardless of the value of x indicated by the 1-Qbit measurement gate. The unmeasured Qbits are unentangled with the measured Qbit and described by the state $|\Phi\rangle$ throughout the process. The 1-Qbit measurement gate acts on the measured Qbit exactly as it does in Figure 1.5 when no other Qbits are present, and the generalized Born rule of Figure 1.6 reduces to the ordinary Born rule.

on the n 1-Qbit measurement gates. This is nothing but the ordinary Born rule, with the n 1-Qbit measurement gates playing the role of the single n-Qbit measurement gate. There is thus, as remarked upon above, only a single primitive piece of measurement hardware: the 1-Qbit measurement gate. The construction of an n-Qbit measurement gate out of n 1-Qbit measurement gates is depicted in Figure 1.8.

An even more general version of the Born rule follows from the generalized Born rule itself. The general state of $m + n$ Qbits can be written as

$$|\Psi\rangle_{m+n} = \sum_{x=0}^{2^m} \alpha_x |x\rangle_m |\Phi_x\rangle_n, \tag{1.80}$$

where $\sum_x |\alpha_x|^2 = 1$ and the states $|\Phi_x\rangle_n$ are normalized, but not necessarily orthogonal. By applying the generalized Born rule m times to m Qbits in an $(m+n)$-Qbit state, one establishes the rule that if just the m Qbits on the left of (1.80) are measured, then with probability $|\alpha_x|^2$ the result will be x, and after the measurement the state of all $m + n$ Qbits will be the product state

$$|x\rangle_m |\Phi_x\rangle_n \tag{1.81}$$

in which the m measured Qbits are in the state $|x\rangle_m$ and the n unmeasured ones are in the state $|\Phi_x\rangle_n$.

1.10 Measurement gates and state preparation

In addition to providing an output at the end of a computation, measurement gates also play a crucial role (which is not often emphasized) at the beginning. Since there is no way to determine the state of a given collection of Qbits – indeed, in general such a collection might be entangled with other Qbits and therefore not even have a state of its own – how can one produce a set of Qbits in a definite state for the gates of a quantum computer to transform into another computationally useful state?

The answer is by measurement. If one takes n Qbits off the shelf, and subjects them to an n-Qbit measurement gate that registers x, then the Qbits emerging from that gate are assigned the classical-basis state $|x\rangle_n$. If one then applies the 1-Qbit operation **X** to each Qbit that registered a 1 in the measurement, doing nothing to the Qbits that

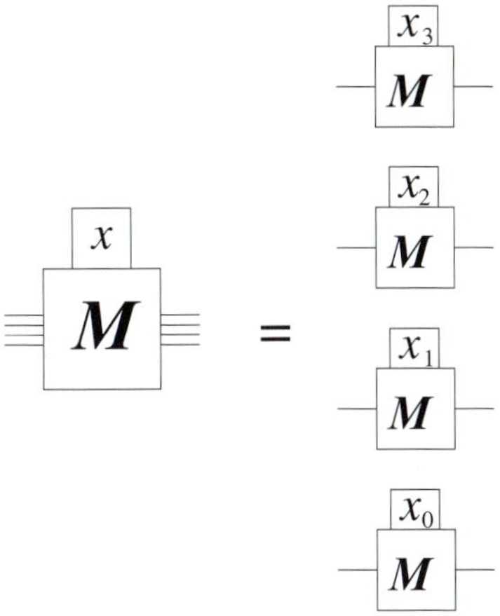

Fig 1.8 Constructing a 4-Qbit measurement gate out of four 1-Qbit measurement gates. The integer x has the binary expansion $x_3x_2x_1x_0$.

registered 0, the resulting set of Qbits will be described by the state $|0\rangle_n$. It is this state that most quantum-computational algorithms take as their input. Such a use of a measurement gate to produce a Qbit described by the state $|0\rangle$ is shown in Figure 1.9.

Measurement gates therefore play *two* roles in a quantum computation. They get the Qbits ready for the subsequent action of the computer, and they extract from the Qbits a digital output after the computer has acted. The initial action of the measurement gates is called *state preparation*, since the Qbits emerging from the process can be characterized by a definite state. The association of unitary operators with the gates that subsequently act on the Qbits permits one to update that initial state assignment into the corresponding unitary transformation of the initial state, thereby making it possible to calculate, using the Born rule, the probabilities of the outcomes of the final measurement gates.

This role of measurement gates in state preparation follows from the Born rule if the Qbits that are to be prepared already have a state of their own, even though that state might not be known to the user of the quantum computer. It also follows from the generalized Born rule if the Qbits already share an entangled state – again, not necessarily known to the user – with additional (unmeasured) Qbits. But one cannot deduce from the Born rules that measurement gates serve to prepare states for Qbits "off the shelf," whose past history nobody knows anything about. In such cases the use of measurement gates to assign a state to the Qbits is a reasonable and plausible extension of the Born rules. It is consistent with them, but goes beyond them.

For particular physical realizations of Qbits, there may be other ways to produce the standard initial state $|0\rangle_n$. Suppose, for example, that each Qbit is an atom, the state $|0\rangle$ is the lowest-energy state (the *ground state*) of the atom, and the state $|1\rangle$ is the atomic state of next-lowest

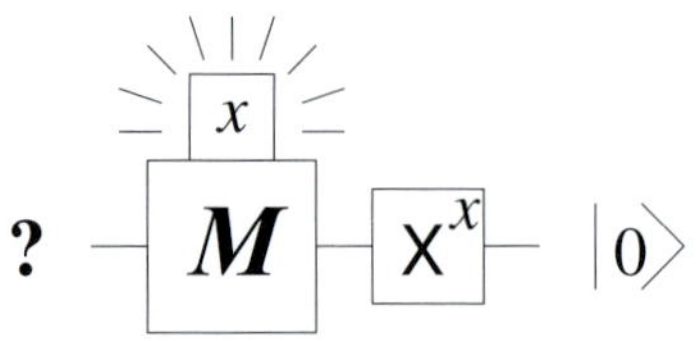

Fig 1.9 Using a 1-Qbit measurement gate to prepare an off-the-shelf Qbit so that its associated state is $|0\rangle$. The input on the left is a Qbit in an unknown condition – i.e. nothing is known of its past history. After the measurement gate is applied, the NOT gate **X** is or is not applied, depending on whether the measurement gate indicates 1 or 0. The Qbit that emerges (on the right) is described by the state $|0\rangle$.

energy (the *first excited state*). Then one can produce the state $|0\rangle_n$ by cooling n such atoms to an appropriately low temperature (determined by the energy difference between the two states – the smaller that energy, the lower the temperature must be).

From the conceptual point of view, state preparation by the use of measurement gates is the simplest way. An acceptable physical candidate for a Qbit must be a system for which measurement gates are readily available. Otherwise there would be no way of extracting information from the computation, however well the unitary gates did their job. So the hardware for state preparation by measurement is already there. Whether one chooses to use it or other (e.g. cryogenic) methods to initialize the Qbits to the state $|0\rangle_n$ is a practical matter that need not concern us here. It is enough to know that it can always be done with measurement gates.

1.11 Constructing arbitrary 1- and 2-Qbit states

The art of quantum computation is to construct circuits out of 1- and 2-Qbit gates that produce final states capable of revealing useful information, when measured. The expectation is that 1-Qbit gates will be comparatively easy to construct. Two-Qbit gates that are not mere tensor products of 1-Qbit gates are likely to be substantially more difficult to make. Attention has focused strongly on the cNOT gate, and gates that can be constructed from it in combination with 1-Qbit unitaries. All of the circuits we shall be examining can, in fact, be reduced to combinations of 1-Qbit gates and 2-Qbit cNOT gates. Given the difficulty in making cNOT gates, it is generally considered desirable to keep their number as small as possible. As an illustration of such constructions, we now examine how to assign arbitrary states to one or two Qbits, starting with the standard 1-Qbit state $|0\rangle$ or the standard 2-Qbit state $|00\rangle$. (Both of these standard states can be produced with the help of measurement gates, as described in Section 1.10.)

The situation for 1-Qbit states is quite simple. Let $|\psi\rangle$ be any 1-Qbit state, and let $|\phi\rangle$ be the orthogonal state (unique to within an overall phase), satisfying $\langle\phi|\psi\rangle = 0$. Since $|0\rangle$ and $|1\rangle$ are linearly independent, there is a unique linear transformation taking them into $|\psi\rangle$ and $|\phi\rangle$. But, since $|\psi\rangle$ and $|\phi\rangle$ are an orthonormal pair (as are $|0\rangle$ and $|1\rangle$), this linear transformation is easily verified to preserve the norm of arbitrary states, so it is a unitary transformation $\mathbf{u}$. Thus, for any $|\psi\rangle$ there is a 1-Qbit unitary gate $\mathbf{u}$ that takes $|0\rangle$ into $|\psi\rangle$:

$$|\psi\rangle = \mathbf{u}|0\rangle. \tag{1.82}$$

Things are more complicated for 2-Qbit states. An unentangled 2-Qbit state, being the product of two 1-Qbit states, can be constructed out of $|00\rangle$ by the application of 1-Qbit unitaries to each of the two Qbits. But a general 2-Qbit state is entangled, and its production requires a 2-Qbit gate that is not just a tensor product of 1-Qbit unitaries. Interestingly, a single cNOT gate, combined with 1-Qbit unitaries, is enough to do the trick.

To see this, note that the general 2-Qbit state,

$$|\Psi\rangle = \alpha_{00}|00\rangle + \alpha_{01}|01\rangle + \alpha_{10}|10\rangle + \alpha_{11}|11\rangle, \tag{1.83}$$

is of the form

$$|\Psi\rangle = |0\rangle \otimes |\psi\rangle + |1\rangle \otimes |\phi\rangle, \tag{1.84}$$

where $|\psi\rangle = \alpha_{00}|0\rangle + \alpha_{01}|1\rangle$ and $|\phi\rangle = \alpha_{10}|0\rangle + \alpha_{11}|1\rangle$. Apply $\mathbf{u} \otimes \mathbf{1}$ to $|\Psi\rangle$, where $\mathbf{u}$ is a linear transformation, whose action on the computational basis is of the form

$$\mathbf{u}|0\rangle = a|0\rangle + b|1\rangle, \qquad \mathbf{u}|1\rangle = -b^*|0\rangle + a^*|1\rangle; \qquad |a|^2 + |b|^2 = 1. \tag{1.85}$$

The transformation $\mathbf{u}$ is unitary because it preserves the orthogonality and normalization of the basis $|0\rangle$, $|1\rangle$.

We have

$$\begin{aligned}(\mathbf{u} \otimes \mathbf{1})|\Psi\rangle &= \big(a|0\rangle + b|1\rangle\big) \otimes |\psi\rangle + \big(-b^*|0\rangle + a^*|1\rangle\big) \otimes |\phi\rangle \\ &= |0\rangle \otimes |\psi'\rangle + |1\rangle \otimes |\phi'\rangle,\end{aligned} \tag{1.86}$$

where

$$|\psi'\rangle = a|\psi\rangle - b^*|\phi\rangle, \qquad |\phi'\rangle = b|\psi\rangle + a^*|\phi\rangle. \tag{1.87}$$

We would like to choose the complex numbers a and b to make $|\phi'\rangle$ and $|\psi'\rangle$ orthogonal. The inner product $\langle\phi'|\psi'\rangle$ is

$$\langle\phi'|\psi'\rangle = a^2\langle\phi|\psi\rangle - b^{*2}\langle\psi|\phi\rangle + ab^*\big(\langle\psi|\psi\rangle - \langle\phi|\phi\rangle\big). \tag{1.88}$$

If $\langle\phi|\psi\rangle \neq 0$, then setting $\langle\phi'|\psi'\rangle$ to 0 gives a quadratic equation for a/b^*, which has two complex solutions. If a in (1.85) is any nonzero complex number then either solution determines b, which, with a, gives a 1-Qbit unitary $\mathbf{u}$ for which

$$(\mathbf{u} \otimes \mathbf{1})|\Psi\rangle = |0\rangle \otimes |\psi'\rangle + |1\rangle \otimes |\phi'\rangle \tag{1.89}$$

where $|\psi'\rangle$ and $|\phi'\rangle$ are orthogonal. If $\langle\phi|\psi\rangle = 0$ then (1.84) is already of this form with $\mathbf{u} = \mathbf{1}$.

We can pick positive real numbers λ and μ so that $|\psi''\rangle = |\psi'\rangle/\lambda$ and $|\phi''\rangle = |\phi'\rangle/\mu$ are unit vectors, making $|\psi''\rangle$ and $|\phi''\rangle$ an orthonormal pair. They are therefore related to $|0\rangle$ and $|1\rangle$ by a unitary transformation $\mathbf{v}$:

$$|\psi''\rangle = \mathbf{v}|0\rangle, \qquad |\phi''\rangle = \mathbf{v}|1\rangle. \tag{1.90}$$

Equation (1.89) then gives[12]

$$|\Psi\rangle = \big(\mathbf{u}^\dagger \otimes \mathbf{v}\big)\big(\lambda|0\rangle \otimes |0\rangle + \mu|1\rangle \otimes |1\rangle\big). \tag{1.91}$$

We can write this as

$$|\Psi\rangle = \big(\mathbf{u}^\dagger \otimes \mathbf{v}\big)\mathbf{C}_{10}\big(\lambda|0\rangle + \mu|1\rangle\big) \otimes |0\rangle. \tag{1.92}$$

Since $|\Psi\rangle$ is a unit vector and unitary transformations preserve unit vectors, it follows from (1.91) that $\lambda|0\rangle + \mu|1\rangle$ is a unit vector. It can therefore be obtained from $|0\rangle$ by a unitary transformation $\mathbf{w}$. So

$$|\Psi\rangle = \big(\mathbf{u}^\dagger \otimes \mathbf{v}\big)\mathbf{C}_{10}\big(\mathbf{w} \otimes \mathbf{1}\big)\big(|0\rangle \otimes |0\rangle\big) = \mathbf{u}_1^\dagger \mathbf{v}_0 \mathbf{C}_{10} \mathbf{w}_1 |00\rangle. \tag{1.93}$$

We have thus established that a general 1-Qbit state $|\Psi\rangle$ can be constructed out of three 1-Qbit unitaries and a single cNOT gate, acting on the standard state $|00\rangle$. This is an early example of the usefulness of cNOT gates.

1.12 Summary: Qbits versus Cbits

Table 1.1 gives a concise comparison of the elementary properties of Cbits and Qbits. The table uses the term "Bit," with an upper-case B, to mean "Qbit or Cbit," which should be distinguished from "bit," with a lower-case b, which means "0 or 1." Alice (in the fifth line of the table) is anybody who knows the relevant history of the Qbits – their initial state preparation and the unitary gates that have subsequently acted on them.

12 This form for a general vector in a space of 2×2 dimensions is a special case of a more general result for $d \times d$ dimensions known as the *polar* (or Schmidt) *decomposition theorem*.

Table 1.1. A summary of the features of Qbits, contrasted to the analogous features of Cbits

	Cbits	**Qbits**
States of n Bits	$\lvert x\rangle_n,\quad 0 \le x < 2^n$	$\sum \alpha_x \lvert x\rangle_n, \sum \lvert\alpha_x\rvert^2 = 1$
Subsets of n Bits	Always have states	Generally have no states
Reversible operations on states	Permutations	Unitary transformations
Can state be learned from Bits?	Yes	No
To learn state of Bits	Examine them	Go ask Alice
To get information from Bits	Just look at them	Measure them
Information acquired	x	x with probability $\lvert\alpha_x\rvert^2$
State after information acquired	Same: still $\lvert x\rangle$	Different: now $\lvert x\rangle$

Chapter 2

General features and some simple examples

2.1 The general computational process

A suitably programmed quantum computer should act on a number x to produce another number $f(x)$ for some specified function f. Appropriately interpreted, with an accuracy that increases with increasing k, we can treat such numbers as non-negative integers less than 2^k. Each integer is represented in the quantum computer by the corresponding computational-basis state of k Qbits.

If we specify the numbers x as n-bit integers and the numbers $f(x)$ as m-bit integers, then we shall need at least $n + m$ Qbits: a set of n-Qbits, called the *input register*, to represent x, and another set of m Qbits, called the *output register*, to represent $f(x)$. Qbits being a scarce commodity, you might wonder why we need separate registers for input and output. One important reason is that if $f(x)$ assigns the same value to different values of x, as many interesting functions do, then the computation cannot be inverted if its only effect is to transform the contents of a single register from x to $f(x)$. Having separate registers for input and output is standard practice in the classical theory of reversible computation. Since quantum computers must operate reversibly to perform their magic (except for measurement gates), they are generally designed to operate with both input and output registers. We shall find that this dual-register architecture can also be usefully exploited by a quantum computer in some strikingly nonclassical ways.

The computational process will generally require many Qbits besides the $n + m$ in the input and output registers, but we shall ignore these additional Qbits for now, viewing a computation of f as doing nothing more than applying a unitary transformation, $\mathbf{U}_f$, to the $n + m$ Qbits of the input and output registers. We take up the fundamental question of why the additional Qbits can be ignored in Section 2.3, only noting for now that it is the reversibility of the computation that makes this possible.

We define the transformation $\mathbf{U}_f$ by specifying it as a reversible transformation taking computational-basis states into computational-basis states. As noted in Section 1.6, the linear extension of such a classically meaningful transformation to arbitrary complex superpositions of computational-basis states is necessarily unitary. The standard quantum-computational protocol, which we shall use repeatedly, defines the action of $\mathbf{U}_f$ on the computational-basis states $|x\rangle_n|y\rangle_m$ of the $n+m$ Qbits making up the input and output registers as follows:

$$\mathbf{U}_f\big(|x\rangle_n|y\rangle_m\big) = |x\rangle_n|y\oplus f(x)\rangle_m, \tag{2.1}$$

where $\oplus$ indicates modulo-2 bitwise addition (without carrying) or, if you prefer, the bitwise exclusive OR. If x and y are m-bit integers whose jth bits are x_j and y_j, then $x\oplus y$ is the m-bit integer whose jth bit is $x_j\oplus y_j$. Thus $1101\oplus 0111 = 1010$. This is a straightforward generalization of the single-bit $\oplus$ defined in Section 1.3.

If the initial value represented by the output register is $y=0$ then we have

$$\mathbf{U}_f\big(|x\rangle_n|0\rangle_m\big) = |x\rangle_n|f(x)\rangle_m \tag{2.2}$$

and we do indeed end up with $f(x)$ in the output register. Regardless of the initial value of y, the input register remains in its initial state $|x\rangle_n$.

The transformation (2.1) is clearly invertible. Indeed, $\mathbf{U}_f$ is its own inverse:

$$\begin{aligned}\mathbf{U}_f\mathbf{U}_f\big(|x\rangle|y\rangle\big) &= \mathbf{U}_f\big(|x\rangle|y\oplus f(x)\rangle\big)\\ &= |x\rangle|y\oplus f(x)\oplus f(x)\rangle = |x\rangle|y\rangle,\end{aligned} \tag{2.3}$$

since $z\oplus z=0$ for any z. (From this point on I shall use subscripts that specify the numbers of Qbits only when it is important to emphasize what those numbers are.)

The form (2.2) inspires the most important trick of the quantum-computational repertoire. If we apply to each Qbit in the 2-Qbit state $|0\rangle|0\rangle$ the 1-Qbit Hadamard transformation $\mathbf{H}$ (Equation (1.45)), then we get

$$\begin{aligned}\big(\mathbf{H}\otimes\mathbf{H}\big)\big(|0\rangle\otimes|0\rangle\big) &= \mathbf{H}_1\mathbf{H}_0|0\rangle|0\rangle = \big(\mathbf{H}|0\rangle\big)\big(\mathbf{H}|0\rangle\big)\\ &= \tfrac{1}{\sqrt{2}}\big(|0\rangle+|1\rangle\big)\tfrac{1}{\sqrt{2}}\big(|0\rangle+|1\rangle\big)\\ &= \tfrac{1}{2}\big(|0\rangle|0\rangle+|0\rangle|1\rangle+|1\rangle|0\rangle+|1\rangle|1\rangle\big)\\ &= \tfrac{1}{2}\big(|0\rangle_2+|1\rangle_2+|2\rangle_2+|3\rangle_2\big).\end{aligned} \tag{2.4}$$

This clearly generalizes to the n-fold tensor product of n Hadamards, applied to the n-Qbit state $|0\rangle_n$:

$$\mathbf{H}^{\otimes n}|0\rangle_n = \frac{1}{2^{n/2}}\sum_{0\le x<2^n}|x\rangle_n, \tag{2.5}$$

where

$$\mathbf{H}^{\otimes n} = \mathbf{H}\otimes\mathbf{H}\otimes\cdots\otimes\mathbf{H},\quad n \text{ times}. \tag{2.6}$$

So if the initial state of the input register is $|0\rangle_n$ and we apply an n-fold Hadamard transformation to that register, its state becomes an equally weighted superposition of all possible n-Qbit inputs. If we then apply $\mathbf{U}_f$ to that superposition, with 0 initially in the output register, then by linearity we get from (2.5) and (2.2)

$$\mathbf{U}_f(\mathbf{H}^{\otimes n} \otimes \mathbf{1}_m)\big(|0\rangle_n|0\rangle_m\big) = \frac{1}{2^{n/2}} \sum_{0 \le x < 2^n} \mathbf{U}_f\big(|x\rangle_n|0\rangle_m\big)$$
$$= \frac{1}{2^{n/2}} \sum_{0 \le x < 2^n} |x\rangle_n|f(x)\rangle_m. \qquad (2.7)$$

This contains an important part of the magic that underlies quantum computation. If before letting $\mathbf{U}_f$ act, we merely apply a Hadamard transformation to every Qbit of the input register, initially in the standard state $|0\rangle_n$, the result of the computation is described by a state whose structure cannot be explicitly specified without knowing the result of all 2^n evaluations of the function f. So if we have a mere hundred Qbits in the input register, initially all in the state $|0\rangle_{100}$ (and m more in the ouput register), if a hundred Hadamard gates act on the input register before the application of $\mathbf{U}_f$, then the form of the final state contains the results of $2^{100} \approx 10^{30}$ evaluations of the function f. A billion billion trillion evaluations! This apparent miracle is called *quantum parallelism.*

But a major part of the miracle is only apparent. One cannot say that the result of the calculation *is* 2^n evaluations of f, though some practitioners of quantum computation are rather careless about making such a claim. All one can say is that those evaluations characterize the *form* of the state that describes the output of the computation. One knows what the state *is* only if one already knows the numerical values of all those 2^n evaluations of f. Before drawing extravagant practical, or even only metaphysical, conclusions from quantum parallelism, it is essential to remember that when you have a collection of Qbits in a definite but unknown state, *there is no way to find out what that state is.*

If there *were* a way to learn the state of such a set of Qbits, then everyone could join in the rhapsodic chorus. (Typical verses: "Where were all those calculations done? In parallel universes!" "The possibility of quantum computation has established the existence of the multiverse." "Quantum computation achieves its power by dividing the computational task among huge numbers of parallel worlds.") But there is no way to learn the state. The only way to extract any information from Qbits is to subject them to a measurement.

When we send all $n + m$ Qbits through measurement gates, the Born rule tells us that if the state of the registers has the form (2.7), then with equal probability the result of measuring the Qbits in the input register will be any one of the values of x less than 2^n, while the result of measuring the Qbits in the ouput register will be the value of f for that particular value of x. So by measuring the Qbits we can learn a single value of f as well as learning a single (random)

x_0 at which f has that value. After the measurement the state of the registers reduces to $|x_0\rangle|f(x_0)\rangle$ and we are no longer able to learn anything about the values of f for any other values of x. So although we can learn something from the output of the "parallel computation," it is nothing more than what we would have learned had we simply run the computation starting with a classical state $|x\rangle$ in the input register, with the value of x chosen randomly. That, of course, could have been done with an ordinary classical computer.

To be sure, a hint of a miracle remains – hardly more than the smile of the Cheshire cat – in the fact that in the quantum case the random selection of the x, for which $f(x)$ can be learned, is made only *after* the computation has been carried out. (To assert that the selection was made *before* the computation was done is to make the same error as asserting that a Qbit described by a superposition of the states $|0\rangle$ and $|1\rangle$ is actually in one or the other of them, as discussed in Section 1.8.) This is a characteristic instance of what journalists like to call "quantum weirdness," in that (a) it is indeed vexing to contemplate the fact that the choice of the value of x for which f can be learned is made only after – quite possibly long after – the computation has been finished, but (b) since that choice is inherently random – beyond anyone's power to control in any way whatever – it does not matter for any practical purpose whether the selection was made astonishingly after or boringly before the calculation was executed.

If, of course, there were an easy way to make copies of the output state prior to making the measurement, without running the whole computation over again, then one could, with high probability, learn the values of f for several different (random) values of x. But such copying is prohibited by an elementary result called the "no-cloning theorem," which states that there is no such duplication procedure: there is no unitary transformation that can take the state $|\psi\rangle_n|0\rangle_n$ into the state $|\psi\rangle_n|\psi\rangle_n$ for arbitrary $|\psi\rangle_n$.

The no-cloning theorem is an immediate consequence of linearity. If

$$\mathbf{U}\big(|\psi\rangle|0\rangle\big) = |\psi\rangle|\psi\rangle \quad \text{and} \quad \mathbf{U}\big(|\phi\rangle|0\rangle\big) = |\phi\rangle|\phi\rangle, \tag{2.8}$$

then it follows from linearity that

$$\mathbf{U}\big(a|\psi\rangle + b|\phi\rangle\big)|0\rangle = a\mathbf{U}|\psi\rangle|0\rangle + b\mathbf{U}|\phi\rangle|0\rangle = a|\psi\rangle|\psi\rangle + b|\phi\rangle|\phi\rangle. \tag{2.9}$$

But if $\mathbf{U}$ cloned arbitrary inputs, we would have

$$\begin{aligned}\mathbf{U}\big(a|\psi\rangle + b|\phi\rangle\big)|0\rangle &= \big(a|\psi\rangle + b|\phi\rangle\big)\big(a|\psi\rangle + b|\phi\rangle\big)\\ &= a^2|\psi\rangle|\psi\rangle + b^2|\phi\rangle|\phi\rangle + ab|\psi\rangle|\phi\rangle + ab|\phi\rangle|\psi\rangle,\end{aligned} \tag{2.10}$$

which differs from (2.9) unless one of a and b is zero. Surprisingly, this very simple theorem was not proved until half a century after the discovery of quantum mechanics, presumably because it took that long for it to occur to somebody that it was an interesting proposition to formulate.

Of course, the ability to clone to a reasonable degree of approximation would be quite useful. But this is also impossible. Suppose that $\mathbf{U}$ approximately cloned both $|\phi\rangle$ and $|\psi\rangle$:

$$\mathbf{U}\big(|\psi\rangle|0\rangle\big) \approx |\psi\rangle|\psi\rangle \text{ and } \mathbf{U}\big(|\phi\rangle|0\rangle\big) \approx |\phi\rangle|\phi\rangle. \tag{2.11}$$

Then since unitary transformations preserve inner products, since the inner product of a tensor product of states is the (ordinary) product of their inner products, and since $\langle 0|0\rangle = 1$, it follows from (2.11) that

$$\langle\phi|\psi\rangle \approx \langle\phi|\psi\rangle^2. \tag{2.12}$$

But this requires $\langle\phi|\psi\rangle$ to be either close to 1 or close to 0. Hence a unitary transformation can come close to cloning both of two states $|\psi\rangle$ and $|\phi\rangle$ only if the states are very nearly the same, or very close to being orthogonal. In all other cases at least one of the two states will be badly copied.

If this were the full story, nobody but a few philosophers would be interested in quantum computation. The National Security Agency of the United States of America is interested because there are more clever things one can do. Typically these involve applying additional unitary gates to one or both of the input and output registers before and/or after applying $\mathbf{U}_f$, sometimes intermingled with intermediate measurement gates acting on subsets of the Qbits. All these additional gates are cunningly chosen so that when one finally does measure all the Qbits, one extracts useful information about *relations* between the values of f for several different values of x, which a classical computer could get only by making several independent evaluations. The price one inevitably pays for this relational information is the loss of the possibility of learning the actual value $f(x)$ for any individual x. This tradeoff of one kind of information for another is typical of quantum computation, and typical of quantum physics in general, where it is called the *uncertainty principle*. The principle was first enunciated by Werner Heisenberg in the context of mechanical information – the position of a particle versus its momentum.

So it is wrong and deeply misleading to say that in the process that assigns the state (2.7) to the Qbits, the quantum computer has evaluated the function $f(x)$ for all x in the range $0 \leq x < 2^n$. Such assertions are based on the mistaken view that the quantum state encodes a property inherent in the Qbits. The state encodes only the possibilities available for the extraction of information from those Qbits. You should keep this in mind as we examine some of the specific ways in which this nevertheless permits a quantum computer to perform tricks that no classical computer can accomplish.

2.2 Deutsch's problem

Deutsch's problem is the simplest example of a quantum tradeoff that sacrifices particular information to acquire relational information. A crude version of it appeared in a 1985 paper by David Deutsch that, together with a 1982 paper by Richard Feynman, launched the whole field. In that early version the trick could be executed successfully only half the time. It took a while for people to realize that the trick could be accomplished every single time. Here is how it works.

Let both input and output registers each contain only one Qbit, so we are exploring functions f that take a single bit into a single bit. There are two rather different ways to think about such functions.

(1) The first way is to note that there are just four such functions, as shown in Table 2.1. Suppose that we are given a black box that calculates one of these four functions in the usual quantum-computational format, by performing the unitary transformation

$$\mathbf{U}_f\big(|x\rangle|y\rangle\big) = |x\rangle|y \oplus f(x)\rangle, \tag{2.13}$$

where the state on the left is that of the 1-Qbit input register (i), and the state on the right is that of the 1-Qbit output register (o). Using the forms in Table 2.1 and the explicit structure (2.13) of $\mathbf{U}_f$, you can easily confirm that

$$\mathbf{U}_{f_0} = \mathbf{1}, \qquad \mathbf{U}_{f_1} = \mathbf{C}_{io}, \qquad \mathbf{U}_{f_2} = \mathbf{C}_{io}\mathbf{X}_o, \qquad \mathbf{U}_{f_3} = \mathbf{X}_o, \tag{2.14}$$

where $\mathbf{1}$ is the (2-Qbit) unit operator, $\mathbf{C}_{io}$ is the controlled-NOT with the input Qbit as control and the output as target, and $\mathbf{X}_o$ acts as NOT on the output register. These possibilities are illustrated in the circuit diagram of Figure 2.1.

Suppose that we are given a black box that executes $\mathbf{U}_f$ for one of the four functions, but are not told which of the four operations (2.14) the box carries out. We can, of course, find out by letting the black box

Table 2.1. The four distinct functions $f_j(x)$ that take one bit into one bit

	$x = 0$	$x = 1$
f_0	0	0
f_1	0	1
f_2	1	0
f_3	1	1

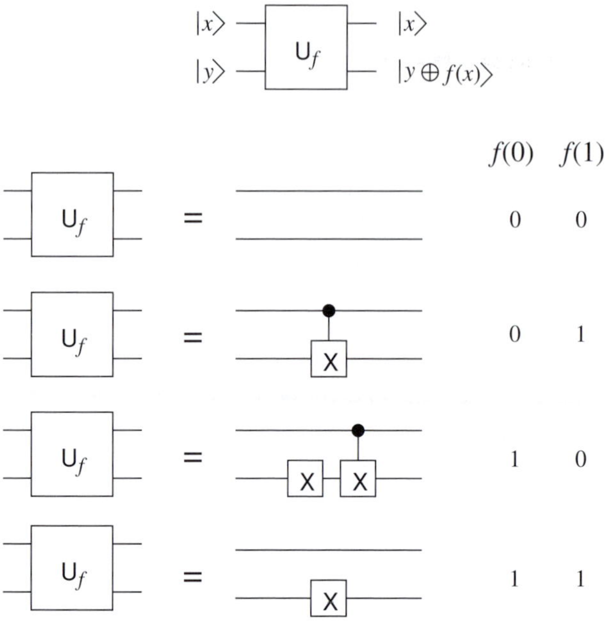

Fig 2.1 A way to construct, with elementary gates, each of the black boxes $\mathbf{U}_f$ that realize the four possible functions f that appear in Deutsch's problem. In case 00 f is identically 0 and it is evident from the general form at the top of the figure that $\mathbf{U}_f$ acts as the identity. In case 01 $f(x) = x$, so $\mathbf{U}_f$ acts as cNOT, with the input register as the control Qbit. In case 10 f interchanges 0 and 1, so $\mathbf{U}_f$ applies NOT to the target Qbit if and only if the computational-basis state of the control Qbit is $|0\rangle$. This is equivalent to combining a cNOT with an unconditional NOT on the target Qbit. In case 11 f is identically 1, and the effect of $\mathbf{U}_f$ is just to apply NOT to the output register, whatever the state of the input register. Note the diagrammatic convention for controlled operations: the control Qbit is represented by the wire with the black dot on it; the target Qbit is connected to the control by a vertical line ending in a box containing the controlled operation. An alternative representation for cNOT appears in Figure 2.7.

act twice – first on $|0\rangle|0\rangle$ and then on $|1\rangle|0\rangle$. But suppose that we can only let the box act once. What can we learn about f?

In a classical computer, where we are effectively restricted to letting the black box act on Qbits in one of the four computational-basis states, we can learn either the value of $f(0)$ (if we let $\mathbf{U}_f$ act on either $|0\rangle|0\rangle$ or $|0\rangle|1\rangle$) or the value of $f(1)$ (if we let $\mathbf{U}_f$ act on either $|1\rangle|0\rangle$ or $|1\rangle|1\rangle$). If we choose to learn the value of $f(0)$, then we can restrict f to being either f_0 or f_1 (if $f(0) = 0$) or to being either f_2 or f_3 (if $f(0) = 1$). If we choose to learn the value of $f(1)$, then we can restrict f to being either f_0 or f_2 (if $f(1) = 0$) or to being either f_1 or f_3 (if $f(1) = 1$).

Suppose, however, that we want to learn whether f is constant ($f(0) = f(1)$, satisfied by f_0 and f_3) or not constant ($f(0) \neq f(1)$, satisfied by f_1 and f_2). We then have no choice with a classical computer but to evaluate both $f(0)$ and $f(1)$ and compare them. In this way we determine whether or not f is constant, but we have to extract complete information about f to do so. We have to run $\mathbf{U}_f$ twice.

Remarkably, it turns out that with a quantum computer we do not have to run $\mathbf{U}_f$ twice to determine whether or not f is constant. We can do this in a single run. Interestingly, when we do this we learn nothing whatever about the individual values of $f(0)$ and $f(1)$, but we are nevertheless able to answer the question about their relative values: whether or not they are the same. Thus we get less information than we get in answering the question with a classical computer, but by renouncing the possibility of acquiring that part of the information which is irrelevant to the question we wish to answer, we can get the answer with only a *single* application of the black box.

(2) There is a second way to look at Deutsch's problem, which gives it nontrivial mathematical content. One can think of x as specifying a choice of two different inputs to an elaborate subroutine that requires many additional Qbits, and one can think of $f(x)$ as characterizing a two-valued property of the output of that subroutine. For example $f(x)$ might be the value of the millionth bit in the binary expansion of $\sqrt{2+x}$ so that $f(0)$ is the millionth bit in the expansion of $\sqrt{2}$ while $f(1)$ is the millionth bit of $\sqrt{3}$. In this case the input register feeds data into the subroutine and the subroutine reports back to the output register.

In the course of the calculation the input and output registers will in general become entangled with the additional Qbits used by the subroutine. If the entanglement persists to the end of the calculation, the input and output registers will have no final states of their own, and it will be impossible to describe the computational process as the simple unitary transformation (2.1). We shall see in Section 2.3, however, that it is possible to set things up so that at the end of the computation the additional Qbits required for the subroutine are no longer entangled with the input and output registers, so that the additional Qbits can indeed be ignored. The simple linear transformation (2.1) then correctly characterizes the net effect of the computation on those two registers.

Under interpretation (1) of Deutsch's problem, answering the question of whether f is or is not constant amounts to learning something about the nature of the black box that executes $\mathbf{U}_f$ without actually opening it up and looking inside. Under interpretation (2) it becomes the nontrivial question of whether the millionth bits of $\sqrt{2}$ and $\sqrt{3}$ agree or disagree. Under either interpretation, to answer the question with a classical computer we can do no better than to run the black box twice, with both 0 and 1 as inputs, and compare the two outputs.

In the quantum case we could try the standard trick, preparing the input register in the superposition $(1/\sqrt{2})\big(|0\rangle + |1\rangle\big)$. After a single application of $\mathbf{U}_f$ the final state of the 1-Qbit input and output registers would then be

$$\mathbf{U}_f(\mathbf{H}\otimes\mathbf{1})\big(|0\rangle|0\rangle\big) = \tfrac{1}{\sqrt{2}}|0\rangle|f(0)\rangle + \tfrac{1}{\sqrt{2}}|1\rangle|f(1)\rangle, \tag{2.15}$$

as described in (2.7). If we then measured the input and ouput registers we could learn, under case (2), the millionth bit of either $\sqrt{2}$ or $\sqrt{3}$, as well as learning which we had learned. The choice of which we did learn would be random. This procedure offers no improvement on the classical situation.

It was first noticed that, without making any further use of $\mathbf{U}_f$, there are additional unitary transformations one can apply to the state (2.15) before carrying out the measurement that enable you half the time to state with assurance whether or not $f(0) = f(1)$. (This imperfect solution to Deutsch's problem has some interesting features, which we explore further in Appendix F.) Some time later, it was realized that you can *always* answer the question if you apply appropriate unitary transformations *before* as well as after the single application of $\mathbf{U}_f$. Here is how the trick is done.

To get the output (2.15) we took the input to $\mathbf{U}_f$ to be the state

$$(\mathbf{H}\otimes\mathbf{1})\big(|0\rangle|0\rangle\big). \tag{2.16}$$

Instead of doing this, we again start with both input and output registers in the state $|0\rangle$, but then we apply the NOT operation $\mathbf{X}$ to both registers, followed by an application of the Hadamard transform to both. Since $\mathbf{X}|0\rangle = |1\rangle$ and $\mathbf{H}|1\rangle = (1/\sqrt{2})(|0\rangle - |1\rangle)$, the input to $\mathbf{U}_f$ is now described by the state

$$\begin{aligned}(\mathbf{H}\otimes\mathbf{H})(\mathbf{X}\otimes\mathbf{X})\big(|0\rangle|0\rangle\big) &= (\mathbf{H}\otimes\mathbf{H})\big(|1\rangle|1\rangle\big)\\ &= \Big(\tfrac{1}{\sqrt{2}}|0\rangle - \tfrac{1}{\sqrt{2}}|1\rangle\Big)\Big(\tfrac{1}{\sqrt{2}}|0\rangle - \tfrac{1}{\sqrt{2}}|1\rangle\Big)\\ &= \tfrac{1}{2}\big(|0\rangle|0\rangle - |1\rangle|0\rangle - |0\rangle|1\rangle + |1\rangle|1\rangle\big).\end{aligned} \tag{2.17}$$

If we take the state (2.17) as input to $\mathbf{U}_f$, then by linearity the resulting state is

$$\tfrac{1}{2}\big(\mathbf{U}_f(|0\rangle|0\rangle) - \mathbf{U}_f(|1\rangle|0\rangle) - \mathbf{U}_f(|0\rangle|1\rangle) + \mathbf{U}_f(|1\rangle|1\rangle)\big). \tag{2.18}$$

It follows from the explicit form (2.13) of the action of $\mathbf{U}_f$ on the computational-basis states that this is simply

$$\tfrac{1}{2}\big(|0\rangle|f(0)\rangle - |1\rangle|f(1)\rangle - |0\rangle|\tilde{f}(0)\rangle + |1\rangle|\tilde{f}(1)\rangle\big), \tag{2.19}$$

where, as earlier, $\tilde{x} = 1 \oplus x$ so that $\tilde{1} = 0$ and $\tilde{0} = 1$, and $\tilde{f}(x) = 1 \oplus f(x)$. So if $f(0) = f(1)$ the ouput state (2.19) is

$$\tfrac{1}{2}\big(|0\rangle - |1\rangle\big)\big(|f(0)\rangle - |\tilde{f}(0)\rangle\big),\quad f(0) = f(1), \tag{2.20}$$

but if $f(0) \neq f(1)$ then $f(1) = \tilde{f}(0)$, $\tilde{f}(1) = f(0)$, and the output state (2.19) becomes

$$\tfrac{1}{2}\big(|0\rangle + |1\rangle\big)\big(|f(0)\rangle - |\tilde{f}(0)\rangle\big),\quad f(0) \neq f(1). \tag{2.21}$$

If, finally, we apply a Hadamard transformation to the input register, these become

$$|1\rangle\tfrac{1}{\sqrt{2}}\big(|f(0)\rangle - |\tilde{f}(0)\rangle\big),\quad f(0) = f(1), \tag{2.22}$$

$$|0\rangle\tfrac{1}{\sqrt{2}}\big(|f(0)\rangle - |\tilde{f}(0)\rangle\big),\quad f(0) \neq f(1). \tag{2.23}$$

On putting together all the operations in a form we can compare with the more straightforward computation (2.15), we have

$$(\mathbf{H}\otimes\mathbf{1})\mathbf{U}_f(\mathbf{H}\otimes\mathbf{H})(\mathbf{X}\otimes\mathbf{X})\big(|0\rangle|0\rangle\big)$$
$$=\begin{cases}|1\rangle\frac{1}{\sqrt{2}}\Big(|f(0)\rangle-|\tilde{f}(0)\rangle\Big), & f(0)=f(1),\\ |0\rangle\frac{1}{\sqrt{2}}\Big(|f(0)\rangle-|\tilde{f}(0)\rangle\Big), & f(0)\neq f(1).\end{cases} \tag{2.24}$$

Thus the state of the input register has ended up as $|1\rangle$ or $|0\rangle$ depending on whether or not $f(0)=f(1)$, so by measuring the *input* register we can indeed answer the question of whether $f(0)$ and $f(1)$ are or are not the same!

Notice that in either case the output register is left in the state $(1/\sqrt{2})(|f(0)\rangle-|\tilde{f}(0)\rangle)$. Because the two terms in the superposition have amplitudes with exactly the same magnitude, if one measures the output register the result is equally likely to be $f(0)$ or $\tilde{f}(0)$, and one learns absolutely nothing about the actual value of $f(0)$. The output register contains no useful information at all.

Another way to put it is that the final state of the output register is $\pm(1/\sqrt{2})\big(|0\rangle-|1\rangle\big)$ depending on whether $f(0)=0$ or $f(0)=1$. Since a change in the overall sign of a state (or, more generally, the presence of an overall complex factor of modulus 1) has no effect on the statistical distribution of measurement outcomes, there is no way to distinguish between these two cases.

Thus the price one has paid to learn whether $f(0)$ and $f(1)$ are or are not the same is the loss of any information whatever about the actual value of either of them. One has still eliminated only two of the four possible forms for the function f. What the quantum computer gives us is the ability to make this particular discrimination with just a single invocation of the black box. No classical computer can do this.

There is a rather neat circuit-theoretic way of seeing why this trick enables one to learn whether or not $f(0)=f(1)$ in just one application of $\mathbf{U}_f$, without going through any of the above algebraic manipulations. This quite different way of looking at Deutsch's problem is illustrated in Figures 2.1–2.3. The basic idea is that for each of the four possible choices for the function f, the 2-Qbit unitary transformation $\mathbf{U}_f$ behaves in exactly the same way as the equivalent circuit constructed out of a NOT and/or a cNOT gate pictured in Figure 2.1. Consequently applying Hadamard gates to each Qbit, both before and after the application of $\mathbf{U}_f$, must produce exactly the same result as it would if the Hadamards were applied to the equivalent circuits in Figure 2.1. Using the elementary identities in Figure 2.2, one easily demonstrates that those results are as shown in Figure 2.3. But Figure 2.3 shows explicitly that when $\mathbf{U}_f$ is so sandwiched between Hadamards, the input register ends up in the state $|0\rangle$ if $f(0)=f(1)$ and in the state $|1\rangle$ if $f(0)\neq f(1)$.

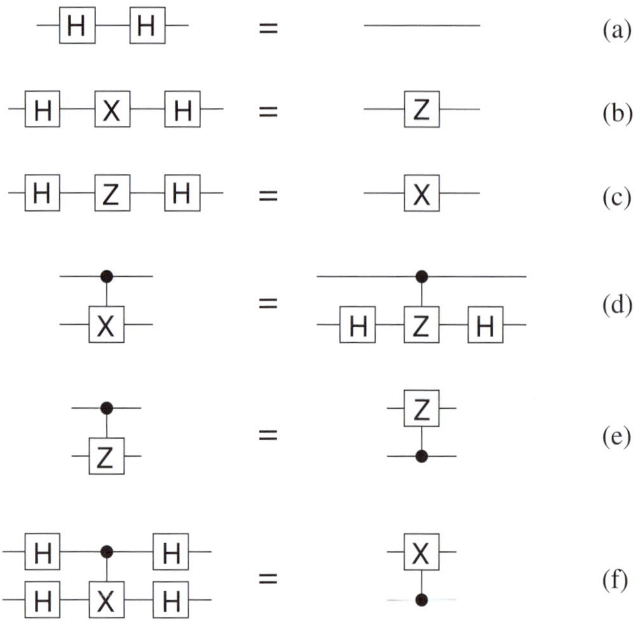

Fig 2.2 Some elementary circuit identities. (a) $\mathbf{H}^2 = \mathbf{1}$. (b) $\mathbf{HXH} = \mathbf{Z}$. (c) A consequence of (a) and (b). (d) A consequence of (a) and (c). (e) The action of the controlled-Z gate does not depend on which Qbit is control and which is target, since it acts as the identity on each of the states $|00\rangle$, $|01\rangle$, and $|10\rangle$ and multiplies the state $|11\rangle$ by -1. (f) This follows from (d), (a), and (e).

When one thinks of applying this to learn whether the millionth bits of $\sqrt{2}$ and $\sqrt{3}$ are the same or different, as in the second interpretation of Deutsch's problem, it is quite startling that one can do this with no more effort (except for a simple modification of the initial and final states) than one uses to calculate the millionth bit of either $\sqrt{2}$ or $\sqrt{3}$. In this case, however, there is an irritating catch, which we note at the end of Section 2.3.

2.3 Why additional Qbits needn't mess things up

Now that we have a specific example of a quantum computation to keep in mind, we can address an important and very general issue mentioned in Section 2.1. The computational process generally requires the use of many Qbits besides the $n + m$ in the input and output registers. In the second interpretation of Deutsch's problem, it may need a great many more. The action of the computer is then described by a unitary transformation $\mathbf{W}_f$ that acts on the space associated with *all* the Qbits: those in the input and output registers, together with the r additional Qbits used in calculating the function f. Only under very special circumstances will this global unitary transformation $\mathbf{W}_f$ on all $n + m + r$ Qbits induce a transformation on the input and output registers that can be be described by a unitary transformation $\mathbf{U}_f$ that acts only on those two registers, as in (2.1). In general the input and output registers will become entangled with the states of the additional r Qbits, and cannot even be assigned a state.

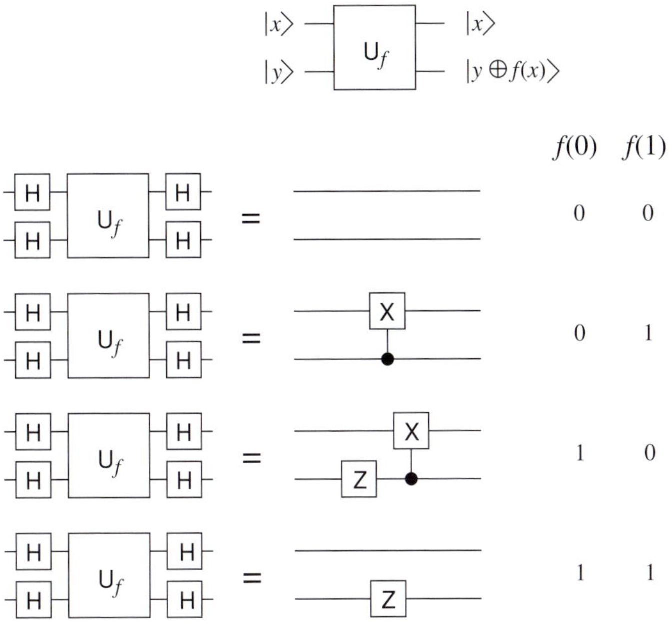

Fig 2.3 We can get the action of $\mathbf{U}_f$, when it is preceded and followed by Hadamards on both Qbits, by applying the appropriate identities of Figure 2.2 to the diagrams of Figure 2.1. Case 00 is unchanged because of Figure 2.2(a). In case 01 the target and control Qbits of the cNOT are interchanged because of Figure 2.2(f). The form in case 10 follows from the corresponding form in Figure 2.1 because of Figures 2.2(f) and 2.2(b). The form in case 11 follows from Figures 2.2(a) and 2.2(b). If the initial state of the output register (lower wire) is $|1\rangle$ and the initial state of the input register (upper wire) is either of the two computational-basis states, then the initial state of the input register will be unchanged in cases 00 and 11, and flipped in cases 01 and 10, so by measuring the input register after the action of $(\mathbf{H} \otimes \mathbf{H})\mathbf{U}_f(\mathbf{H} \otimes \mathbf{H})$ one can determine whether or not $f(0) = f(1)$.

But if the action of the computer on all $n + m + r$ Qbits has a very special form, then the input and output registers can indeed end up with a state, related to their initial states through the desired unitary transformation $\mathbf{U}_f$. Let the additional r Qbits start off in some standard initial state $|\psi\rangle_r$, so that the initial state of input register, output register, and additional Qbits is

$$|\Psi\rangle_{n+m+r} = |x\rangle_n |y\rangle_m |\psi\rangle_r. \tag{2.25}$$

Although the r additional Qbits may well become entangled with those in the input and output registers in the course of the calculation – they will have to if they are to serve any useful purpose – we require that when the calculation is finished the final state of the computer must be of the form

$$\mathbf{W}_f |\Psi\rangle_{n+m+r} = |x\rangle_n |y \oplus f(x)\rangle_m |\phi\rangle_r, \tag{2.26}$$

where the additional r Qbits not only are unentangled with the input and output registers, but also have a state $|\phi\rangle_r$ that is independent of the initial state of the input and output registers.

Because $\mathbf{W}_f$ is linear on the whole $(n + m + r)$-Qbit subspace, and because $|\psi\rangle_r$ and $|\phi\rangle_r$ are independent of the initial computational-basis state of the input and output registers, it follows that if the input and output registers are initially assigned any superposition of computational-basis states, then $\mathbf{W}_f$ leaves them with a definite final state, which is related to their initial state by precisely the unitary transformation $\mathbf{U}_f$ of (2.1).

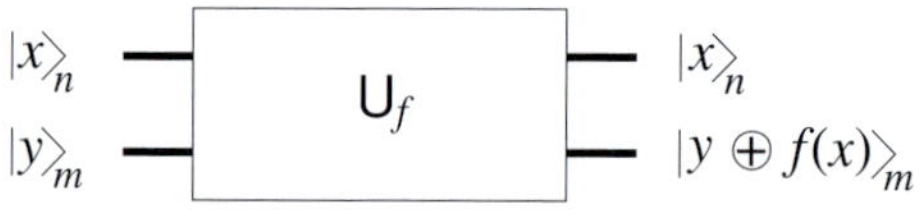

Fig 2.4 A schematic representation of the standard unitary transformation $\mathbf{U}_f$ for evaluating a function f taking a number $0 \le x < 2^n$ into a number $0 \le f(x) < 2^m$. The heavy horizontal lines (bars) represent multiple-Qbit inputs. In order for the computation to be reversible even when f is not one-to-one, two multi-Qbit registers must be used.

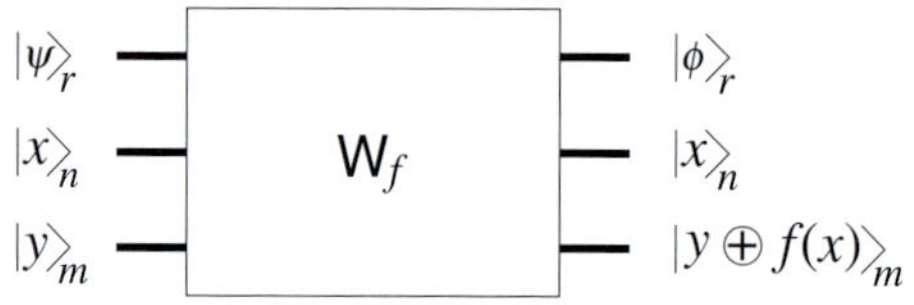

Fig 2.5 A more realistic picture of the computation represented in Figure 2.4. Many additional Qbits may be needed to carry out the calculation. These are represented by an r-Qbit bar in addition to the n- and m-Qbit bars representing the input and output registers. The computation is actually executed by a unitary transformation $\mathbf{W}_f$ that acts on the larger space of all $n + m + r$ Qbits. The representation of Figure 2.4 is correct only if the action of this larger unitary transformation $\mathbf{W}_f$ on the input and output registers alone can be represented by a unitary transformation $\mathbf{U}_f$. This will be the case if the action of $\mathbf{W}_f$ on the residual r Qbits is to take them from an initial pure state $|\psi\rangle_r$ to a final pure state $|\phi\rangle_r$ that is independent of the initial contents of the input and output registers.

Therefore we can indeed use (2.1), ignoring complications associated with the additional r Qbits needed to compute the function f, if both the initial and the final states of the additional Qbits are independent of the initial states of the input and output registers. Independence of the initial states can be arranged by initializing the additional r Qbits to some standard state, for example $|\Psi\rangle_r = |0\rangle_r$. A standard final state $|\phi\rangle_r$ of the r Qbits, which is, in fact, identical to their initial state $|\psi\rangle_r$, can be produced by taking appropriate advantage of the fact that unitary transformations are *reversible*.

We do the trick in three stages.

(1) Begin the computation by applying a unitary transformation $\mathbf{V}$ that acts only on the n-Qbit input register and the r additional Qbits,

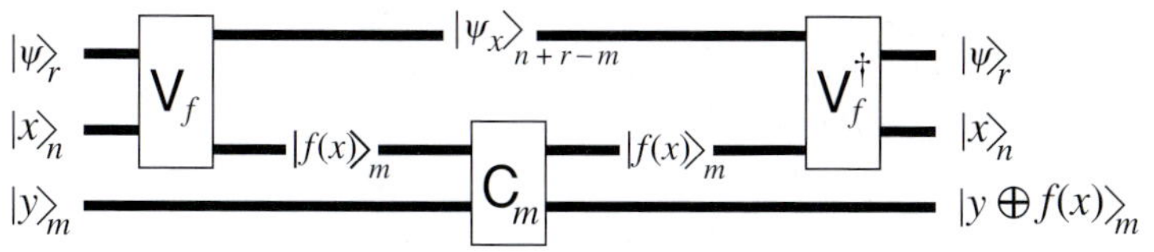

Fig 2.6 A more detailed view of the structure of the unitary transformation $\mathbf{W}_f$ of Figure 2.5. Algebraically, $\mathbf{W}_f = \mathbf{V}_f^\dagger \mathbf{C}_m \mathbf{V}_f$. First a unitary transformation $\mathbf{V}_f$ acts only on the n-Qbit input register and r additional Qbits, acting as the identity on the output register. This transformation takes the $n + r$ Qbits into a state in which an m-Qbit subset represents the result of the calculation, $f(x)$. Second, m controlled-NOT transformations (described in more detail in Figure 2.7) act only on the m Qbits representing $f(x)$ and the m Qbits of the output register, leaving the former m unchanged but changing the number represented by the output register from y to $y \oplus f(x)$. Finally, the inverse $\mathbf{V}_f^\dagger$ of $\mathbf{V}_f$ is applied to the $n + r$ Qbits on the top two bars, to restore them to their (unentangled) initial states.

doing nothing to the output register. Because there is no action on the output register, the $n + r$ Qbits on which $\mathbf{V}$ acts continue to have a state of their own. If the initial state of the input register is $|x\rangle_n$ the unitary transformation $\mathbf{V}$ is designed, using standard tricks of reversible classical computation (about which we shall have more to say in Section 2.6) to construct $f(x)$ in an appropriate m-Qbit subset of the $n + r$ Qbits, given x in the input register.

(2) Next change the y initially in the output register to $y \oplus f(x)$, as (2.1) or (2.26) specifies, without altering the state of the $n + r$ other Qbits. This can be done with m cNOT gates that combine to make up a unitary transformation $\mathbf{C}_m$. The m control Qbits are those among the $n + r$ that represent the result of the computation $f(x)$; the m target Qbits are the ones in the corresponding positions of the output register.

(3) Since the state of the $n + r$ Qbits is not altered by the application of $\mathbf{C}_m$, we can finally apply to them the inverse transformation $\mathbf{V}^\dagger$ to restore them to their original state. We have thus produced the required unitary transformation $\mathbf{W}$ in (2.26), with the final state $|\phi\rangle_r$ of the r additional Qbits being identical to their initial state $|\psi\rangle_r$. This whole construction is illustrated by the circuit diagrams of Figures 2.4–2.7.

The need for this, or some equivalent procedure, negates some of the hype one sometimes encounters in discussions of Deutsch's problem. It is sometimes said that by using a quantum computer one can learn whether or not $f(x) = f(y)$ in no more time than it takes to perform a single evaluation of f. This is true only under the first, arithmetically uninteresting, interpretation of Deutsch's problem. If, however, one is thinking of f as a function of mathematical interest evaluated by an elaborate subroutine, then to evaluate f for a single value of x there is no need to undo the effect of the unitary transform $\mathbf{V}$ on the additional registers. But for the trick that determines whether or not $f(x) = f(y)$ it is absolutely essential to apply $\mathbf{V}^\dagger$ to undo the effect of $\mathbf{V}$. This doubles the time of the computation.

This misrepresentation of the situation is not entirely dishonorable, however, since in almost all other examples the speed-up is by considerably more than a factor of two, and the necessary doubling of computational time is an insignificant price to pay. We turn immediately to an elementary example.

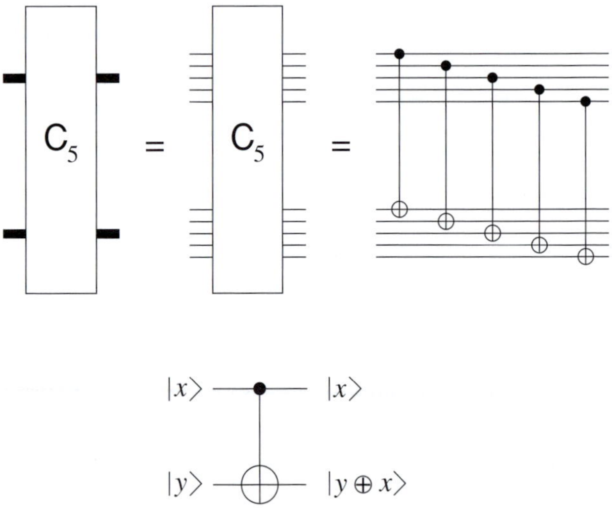

Fig 2.7 A more detailed picture of the $\mathbf{C}_m$ unitary transformation in Figure 2.6, for the case $m = 5$. Each of the input and output bars contains five Qbits, represented by sets of five thin lines (wires). Five different 2-Qbit controlled-NOT gates link the five upper wires representing $f(x)$ to the five lower wires representing the corresponding positions in the output register. The action of a single such cNOT gate is shown in the lower part of the figure. Note the alternative convention for a cNOT gate: the black dot on the wire representing the control Qbit is connected by a vertical line to an open circle on the wire representing the target Qbit. The other convention (used above in Figures 2.1–2.3) replaces the open circle by a square box containing the NOT operator $\mathbf{X}$ that may act on the target Qbit. The advantages of the circle representation are that it suggests the symbol $\oplus$ that represents the XOR operation, and that it is easier to draw quickly on a blackboard. The advantages of using $\mathbf{X}$ are that it makes the algebraic relations more evident when NOT operations $\mathbf{X}$, $\mathbf{Z}$ operations, or controlled-Z operations also appear, and that it follows the form used for all other controlled unitaries.

2.4 The Bernstein–Vazirani problem

Like many of the examples discovered before Shor's factoring algorithm, this has a somewhat artificial character. Its significance lies not in the intrinsic arithmetical interest of the problem, but in the fact that it can be solved dramatically and unambiguously faster on a quantum computer.

Let a be an unknown non-negative integer less than 2^n. Let $f(x)$ take any other such integer x into the modulo-2 sum of the products of corresponding bits of a and x, which we denote by $a \cdot x$ (in recognition of the fact that it is a kind of bitwise modulo-2 inner product):

$$a \cdot x = a_0x_0 \oplus a_1x_1 \oplus a_2x_2 \cdots . \tag{2.27}$$

Suppose that we have a subroutine that evaluates $f(x) = a \cdot x$. How many times do we have to call that subroutine to determine the value of the integer a? Here and in all subsequent examples, we shall assume that any Qbits acted on by such subroutines, except for the Qbits of the input and output registers, are returned to their initial state at the end of the computation, as discussed in Section 2.3.

The mth bit of a is $a \cdot 2^m$, since the binary expansion of 2^m has 1 in position m and 0 in all the other positions. So with a classical computer we can learn the n bits of a by applying f to the n values $x = 2^m$, $0 \leq m < n$. This, or any other classical method one can think of, requires n different invocations of the subroutine. But with a quantum computer a *single* invocation is enough to determine a completely, regardless of how big n is!

I first describe the conventional way of seeing how this can be done, and then describe a much simpler way to understand the process. The conventional way exploits a trick (implicitly exploited in our solution to Deutsch's problem) that is useful in dealing with functions like f that act on n Qbits with output to a single Qbit. If the 1-Qbit output register is initially prepared in the state $\mathbf{HX}|0\rangle = \mathbf{H}|1\rangle = (1/\sqrt{2})\big(|0\rangle - |1\rangle\big)$ then, since $\mathbf{U}_f$ applied to the computational basis state $|x\rangle_n|y\rangle_1$ flips the value y of the output register if and only if $f(x) = 1$, we have

$$\mathbf{U}_f|x\rangle_n \tfrac{1}{\sqrt{2}}\big(|0\rangle - |1\rangle\big) = (-1)^{f(x)}|x\rangle_n \tfrac{1}{\sqrt{2}}\big(|0\rangle - |1\rangle\big). \tag{2.28}$$

So by taking the state of the 1-Qbit output register to be $(1/\sqrt{2})\big(|0\rangle - |1\rangle\big)$, we convert a bit flip to an overall change of sign. This becomes useful because of a second trick, which exploits a generalization of the action (2.5) of $\mathbf{H}^{\otimes n}$ on $|0\rangle_n$.

The action of $\mathbf{H}$ on a single Qbit can be compactly summarized as

$$\mathbf{H}|x\rangle_1 = \frac{1}{\sqrt{2}}\big(|0\rangle + (-1)^x|1\rangle\big) = \frac{1}{\sqrt{2}}\sum_{y=0}^{1}(-1)^{xy}|y\rangle. \tag{2.29}$$

If we apply $\mathbf{H}^{\otimes n}$ to an n-Qbit computational-basis state $|x\rangle_n$ we can therefore express the result as

$$\begin{aligned}\mathbf{H}^{\otimes n}|x\rangle_n &= \frac{1}{2^{n/2}}\sum_{y_{n-1}=0}^{1}\cdots\sum_{y_0=0}^{1}(-1)^{\sum_{j=0}^{n-1}x_j y_j}|y_{n-1}\rangle\cdots|y_0\rangle \\ &= \frac{1}{2^{n/2}}\sum_{y=0}^{2^n-1}(-1)^{x\cdot y}|y\rangle_n,\end{aligned} \tag{2.30}$$

where the product $x \cdot y$ is the one defined in (2.27). (Because -1 is raised to the power $\sum x_j y_j$, all that matters about the sum is its value modulo 2.)

So if we start with the n-Qbit input register in the standard initial state $\mathbf{H}^{\otimes n}|0\rangle$, put the 1-Qbit output register into the state $\mathbf{H}|1\rangle$, apply

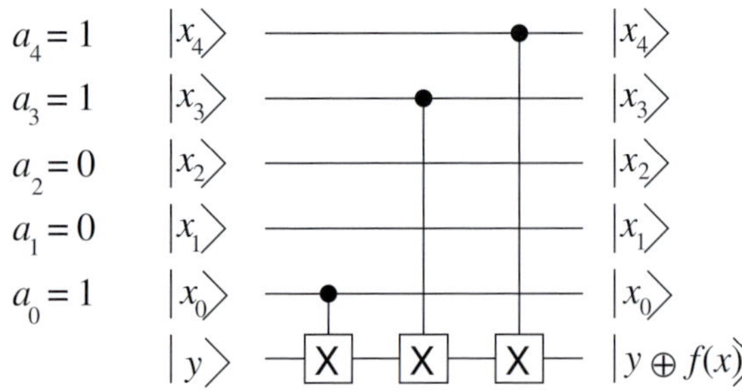

Fig 2.8 An illustration of a circuit that implements the unitary subroutine $\mathbf{U}_f$ taking n-Qbit input and 1-Qbit output registers, initially in the state $|x\rangle_n|y\rangle_1$, into $|x\rangle_n|y \oplus f(x)\rangle_1$, where $f(x) = a \cdot x = \sum_{j=0}^{n-1} a_j x_j$ (mod 2). The Bernstein–Vazirani problem asks us to determine all the bits of a with a single invocation of the subroutine. In the illustration $n = 5$ and $a = 25 = 11001$. For $j = 0, \ldots, n-1$, each of the cNOT gates adds 1 (mod 2) to the output register if and only if $a_j x_j = 1$. In addition to their normal labeling with the 1-Qbit states they represent, the wires of the input register are labeled with the bits of a, to make it clear which (those associated with $a_j = 1$) act as control bits for a cNOT targeted on the output register.

$\mathbf{U}_f$, and then again apply $\mathbf{H}^{\otimes n}$ to the input register, we get

$$
\begin{aligned}
&(\mathbf{H}^{\otimes n} \otimes \mathbf{1})\mathbf{U}_f(\mathbf{H}^{\otimes n} \otimes \mathbf{H})|0\rangle_n|1\rangle_1 \\
&\quad = (\mathbf{H}^{\otimes n} \otimes \mathbf{1})\mathbf{U}_f\left(\frac{1}{2^{n/2}}\sum_{x=0}^{2^n-1}|x\rangle\right)\frac{1}{\sqrt{2}}\big(|0\rangle - |1\rangle\big) \\
&\quad = \frac{1}{2^{n/2}}\left(\mathbf{H}^{\otimes n}\sum_{x=0}^{2^n-1}(-1)^{f(x)}|x\rangle\right)\frac{1}{\sqrt{2}}\big(|0\rangle - |1\rangle\big) \\
&\quad = \frac{1}{2^n}\sum_{x=0}^{2^n-1}\sum_{y=0}^{2^n-1}(-1)^{f(x)+x\cdot y}|y\rangle\frac{1}{\sqrt{2}}\big(|0\rangle - |1\rangle\big).
\end{aligned}
\tag{2.31}
$$

We do the sum over x first. If the function $f(x)$ is $a \cdot x$ then this sum produces the factor

$$
\sum_{x=0}^{2^n-1}(-1)^{(a\cdot x)}(-1)^{(y\cdot x)} = \prod_{j=1}^{n}\sum_{x_j=0}^{1}(-1)^{(a_j+y_j)x_j}. \tag{2.32}
$$

At least one term in the product vanishes unless every bit y_j of y is equal to the corresponding bit a_j of a – i.e. unless $y = a$. Therefore the entire computational process (2.31) reduces to

$$
\mathbf{H}^{\otimes(n+1)}\mathbf{U}_f\mathbf{H}^{\otimes(n+1)}|0\rangle_n|1\rangle_1 = |a\rangle_n|1\rangle_1, \tag{2.33}
$$

where I have applied a final $\mathbf{H}$ to the 1-Qbit output register to make the final expression look a little neater and more symmetric. (I have also restored subscripts to the state symbols for the n-Qbit input register and 1-Qbit output register.)

So by putting the input and output registers into the appropriate initial states, after a single invocation of the subroutine followed by an application of $\mathbf{H}^{\otimes n}$ to the input register, the state of the input register becomes $|a\rangle$. As promised, all n bits of the number a can now be determined by measuring the input register, even though we have called the subroutine only once!

There is a second, complementary way to look at the Bernstein–Vazirani problem that bypasses all of the preceding analysis, making it evident why (2.33) holds, by examining a few circuit diagrams. The

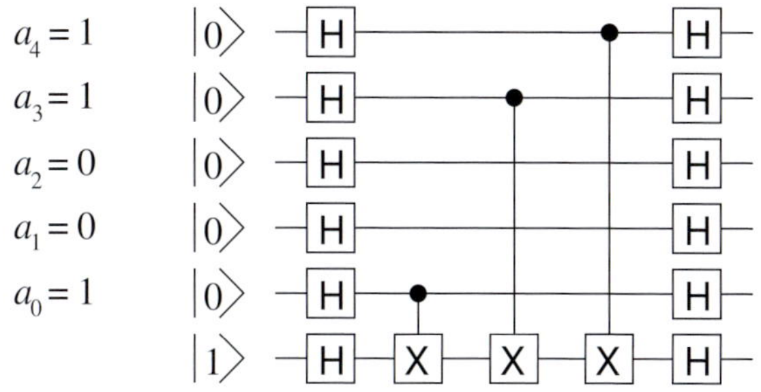

Fig 2.9 The solution to the Bernstein–Vazirani problem is to start with the input register in the state $|0\rangle_n$ and the output register in the state $|1\rangle_1$ and apply Hadamard transforms to all $n+1$ registers before applying $\mathbf{U}_f$. Another $n+1$ Hadamards are applied after $\mathbf{U}_f$ has acted. The cNOT gates reproduce the action of $\mathbf{U}_f$, as shown in Figure 2.8. The conventional analysis deduces the final state by calculating the effect of the Hadamards on the initial state of the Qbits and on the state subsequently produced by the action of $\mathbf{U}_f$. A much easier way to understand what is going on is to examine the effect of the Hadamards on the collection of cNOT gates equivalent to $\mathbf{U}_f$. This is shown in Figure 2.10.

idea is to note, just as we did for the black box of Deutsch's problem in (2.14), that the actions of the black boxes that implement $\mathbf{U}_f$ for the different available choices of f are identical to the actions of some simple circuits.

When $f(x) = a \cdot x$, the action of $\mathbf{U}_f$ on the computational basis is to flip the 1-Qbit output register once, whenever a bit of x and the corresponding bit of a are both 1. When the state of the input register is $|x\rangle_n$ this action can be performed by a collection of cNOT gates all targeted on the output register. There is one cNOT for each nonzero bit of a, controlled by the Qbit representing the corresponding bit of x. The combined effect of these cNOT gates on every computational-basis state is precisely that of $\mathbf{U}_f$. Therefore the effect of any other transformations preceding and/or following $\mathbf{U}_f$ can be understood by examining their effect on this equivalent collection of cNOT gates, even though $\mathbf{U}_f$ may actually be implemented in a completely different way.

The encoding of a in the disposition of the equivalent cNOT gates is illustrated in Figure 2.8. The application (2.33) of $\mathbf{H}$ to every Qbit in the input and output registers both before and after the application of $\mathbf{U}_f$, pictured in Figure 2.9, converts every cNOT gate in the equivalent representation of $\mathbf{U}_f$ from $\mathbf{C}_{ij}$ to $(\mathbf{H}_i\mathbf{H}_j)\mathbf{C}_{ij}(\mathbf{H}_i\mathbf{H}_j) = \mathbf{C}_{ji}$, as pictured in Figure 2.10 (see also Equation (1.44).) After this reversal of target and control Qbits, the output register controls every one of the cNOT gates, and since the state of the output register is $|1\rangle$, every one of the NOT operators acts. That action flips just those Qbits of the input register for which the corresponding bit of a is 1. Since the input register starts in the state $|0\rangle_n$, this changes the state of each Qbit of the input register to $|1\rangle$, if and only if it corresponds to a nonzero bit of a. As a result, the state of the input register changes from $|0\rangle_n$ to $|a\rangle_n$, just as (2.33) asserts.

Note how different these two explanations are. The first applies $\mathbf{U}_f$ to the quantum superposition of all possible inputs, and then applies operations that lead to perfect destructive interference of all states in the superposition except for the one in which the input register is in the state $|a\rangle$. The second suggests a specific mechanism for representing the subroutine that executes $\mathbf{U}_f$ and then shows that sandwiching such a

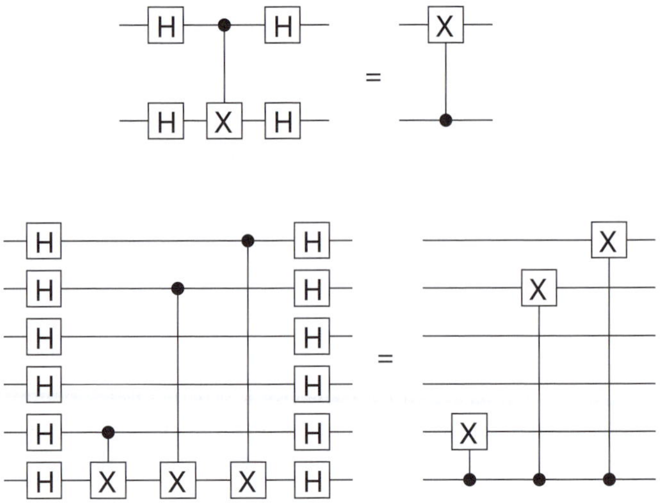

Fig 2.10 Sandwiching a cNOT gate between Hadamards that act on the control and target Qbits has the effect of interchanging control and target, as shown at the top of the figure. (See Equation (1.44) or Figure 2.2.) Consequently the action of all the Hadamards in Figure 2.9 on the cNOT gates between them is simply to interchange the control and target Qbits, as shown in the lower part of the figure. In establishing this one uses the fact that $\mathbf{H}^2 = \mathbf{1}$, so that the **H** gates on wires that are not control or target Qbits combine to give **1**, and pairs of Hadamards can be introduced between every **X** on the lowest wire, converting **HXXXH** into (**HXH**)(**HXH**)(**HXH**). After the action of the Hadamards the cNOT gates are controlled by the output register, so if the output register is in the state $|1\rangle$ then all the **X** act on their input-register targets. If the initial state of the input register is $|0\rangle_n$ then the effect of each **X** is to change to $|1\rangle$ the state of each Qbit associated with a bit of a that is 1. This converts the state of the input register to $|a\rangle_n$.

mechanism between Hadamards automatically imprints a on the input register.

Interestingly, quantum mechanics appears in the second method only because it allows the reversal of the control and target Qbits of a cNOT operation solely by means of 1-Qbit (Hadamard) gates. One can also reverse control and target bits of a cNOT classically, but this requires the use of 2-Qbit SWAP gates, rather than 1-Qbit Hadamards. You can confirm for yourself that this circuit-theoretic solution to the Bernstein–Vazirani problem no longer works if one tries to replace all the Hadamard gates by any arrangement of SWAP gates.

2.5 Simon's problem

Simon's problem, like the Bernstein–Vazirani problem, has an n-bit nonzero number a built into the action of a subroutine $\mathbf{U}_f$, and the aim is to learn the value of a with as few invocations of the subroutine as possible. In the Bernstein–Vazirani problem a classical computer must call the subroutine n times to determine the value of a, while a quantum computer need call the subroutine only once. The number of calls grows linearly with n in the classical case, while being independent of n in the quantum case. In Simon's problem the speed-up with a quantum computer is substantially more dramatic. With a classical computer the number of times one must call the subroutine grows exponentially in n, but with a quantum computer it grows only linearly.

This spectacular speed-up involves a probabilistic element characteristic of many quantum computations. The characterization of how the number of calls of the subroutine scales with the number of bits in a applies not to calculating a directly, but to learning it with probability very close to 1.

The subroutine $\mathbf{U}_f$ in Simon's problem evaluates a function f on n bits that is two to one – i.e. it is a function from n to $n - 1$ bits. It is constructed so that $f(x) = f(y)$ if and only if the n-bit integers x and y are related by $x = y \oplus a$ or, equivalently and more symmetrically, $x \oplus y = a$, where $\oplus$ again denotes bitwise modulo-2 addition. One can think of this as a period-finding problem. One is told that f is periodic under bitwise modulo-2 addition,

$$f(x \oplus a) = f(x) \tag{2.34}$$

for all x, and the problem is to find the period a. Simon's problem is thus a precursor of Shor's much subtler and spectacularly more useful period-finding algorithm – the heart of his factoring procedure – where one finds the unknown period a of a function that is periodic under ordinary addition: $f(x + a) = f(x)$.

To find the value of a in (2.34) with a classical computer all you can do is feed the subroutine different $x_1, x_2, x_3, \ldots$, listing the resulting values of f until you stumble on an x_j that yields one of the previously computed values $f(x_i)$. You then know that $a = x_j \oplus x_i$. At any stage of the process prior to success, if you have picked m different values of x, then all you know is that $a \neq x_i \oplus x_j$ for all pairs of previously selected values of x. You have therefore eliminated at most $\frac{1}{2}m(m - 1)$ values of a. (You would have eliminated fewer values of a if you were careless enough to pick an x equal to $x_i \oplus x_j \oplus x_k$ for three values of x already selected.) Since there are $2^n - 1$ possibilities for a, your chances of success will not be appreciable while $\frac{1}{2}m(m - 1)$ remains small compared with 2^n. You are unlikely to succeed until m becomes of the order of $2^{n/2}$, so the number of times the subroutine has to be run to give an appreciable probability of determining a grows with the number of bits n as $2^{n/2}$ – i.e. exponentially. If a has 100 bits a classical computer would have to run the subroutine about $2^{50} \approx 10^{15}$ times to have a significant chance of determining a. At ten million calls per second it would take about three years.

In contrast, a quantum computer can determine a with high probability (say less than one chance in a million of failing) by running the subroutine not very much more than n times – e.g. with about 120 invocations of the subroutine if a has 100 bits. This remarkable feat can be accomplished with the following strategy.

We return to the standard procedure and apply the unitary transformation $\mathbf{U}_f$ only after the state of the input register has been transformed into the uniformly weighted superposition (2.5) of all possible inputs by the application of $\mathbf{H}^{\otimes n}$, so that the effect of $\mathbf{U}_f$ is to assign to the input and output registers the entangled state

$$\frac{1}{2^{n/2}} \sum_{x=0}^{2^n-1} |x\rangle|f(x)\rangle. \tag{2.35}$$

If we now subject only the output register to a measurement, then the measurement gate is equally likely to indicate each of the 2^{n-1} different values of f. Since each value of f appears in two terms in (2.35) that have the same amplitudes, the generalized Born rule tells us that the input register will be left in the state

$$\frac{1}{\sqrt{2}}\big(|x_0\rangle + |x_0 \oplus a\rangle\big) \tag{2.36}$$

for that value of x_0 for which $f(x_0)$ agrees with the random value of f given by the measurement.

At first glance this looks like great progress. We have produced a superposition of just two computational-basis states, associated with two n-bit integers, that differ (in the sense of $\oplus$) by a. If we knew those two integers their bitwise modulo 2 sum would be a. But unfortunately, as already noted, when a register is in a given quantum state there is in general no way to learn what that state is. To be sure, if we could clone the state, then by measuring a mere ten copies of it in the computational basis we could with a probability of about 0.998 learn both x_0 and $x_0 \oplus a$ and therefore a itself. But unfortunately, as we have also noted earlier, one cannot clone an unknown quantum state. Nor does it help to run the algorithm many times, since we are overwhelmingly likely to get states of the form (2.36) for different random values of x_0. By subjecting (2.36) to a direct measurement all we can learn is either x_0 – a random number, or $x_0 \oplus a$ – another random number. The number a that we would like to know appears only in the *relation* between two random numbers, only one of which we can learn.

Nevertheless, as in Deutsch's problem, if we renounce the possibility of learning either number (which alone is of no interest at all), then by applying some further operations before measuring we can extract some useful partial information about their relationship – in this case their modulo-2 sum a. With the input register in the state (2.36), we apply the n-fold Hadamard transformation $\mathbf{H}^{\otimes n}$. Equation (2.30) then gives

$$\mathbf{H}^{\otimes n}\frac{1}{\sqrt{2}}\big(|x_0\rangle + |x_0 \oplus a\rangle\big) = \frac{1}{2^{(n+1)/2}}\sum_{y=0}^{2^n-1}\big((-1)^{x_0\cdot y} + (-1)^{(x_0\oplus a)\cdot y}\big)|y\rangle. \tag{2.37}$$

Since $(-1)^{(x_0\oplus a)\cdot y} = (-1)^{x_0\cdot y}(-1)^{a\cdot y}$, the coefficient of $|y\rangle$ in (2.37) is 0 if $a \cdot y = 1$ and $2(-1)^{x_0\cdot y}$ if $a \cdot y = 0$. Therefore (2.37) becomes

$$\frac{1}{2^{(n-1)/2}}\sum_{a\cdot y=0}(-1)^{x_0\cdot y}|y\rangle, \tag{2.38}$$

where the sum is now restricted to those y for which the modulo-2 bitwise inner product $a \cdot y$ is 0 rather than 1. So if we now measure the input register, we learn (with equal probability) any of the values of y

for which $a \cdot y = 0$ – i.e. for which

$$\sum_{i=0}^{n-1} y_i a_i = 0 \pmod 2, \tag{2.39}$$

where a_i and y_i are corresponding bits in the binary expansions of a and y.

This completes our description of the quantum computation: with each invocation of $\mathbf{U}_f$ we learn a random y satisfying $a \cdot y = 0$. What remains is the purely mathematical demonstration that this information enables us to determine a with high probability with not many more than n invocations. To see that this is plausible, note first that with just a single invocation of $\mathbf{U}_f$, unless we are unlucky enough to get $y = 0$ (which happens with the very small probability $1/2^{n-1}$), we learn a nonzero value of y, and therefore a nontrivial subset of the n bits of a whose modulo-2 sum vanishes. One of those bits is thus entirely determined by the others in the subset, so we have cut the number of possible choices for a in half, from $2^n - 1$ (the -1 reflecting the fact that we are told that $a \neq 0$) to $2^{n-1} - 1$. In one invocation of the subroutine we can, with very high probability, eliminate half the candidates for a! (Contrast this to the classical case, in which a single invocation of $\mathbf{U}_f$ can tell us nothing whatever about a.)

If we now repeat the whole procedure, then with very high probability the new value of y that we learn will be neither 0 nor the same as the value we learned the first time. We will therefore learn a new nontrivial relation among the bits of a, which enables us to reduce the number of candidates by another factor of 2, eliminating three quarters of the possibilities available for a with two invocations of the subroutine. (Compare this to the classical situation in which only a single value of a can be removed with two invocations.)

If every time we repeat the procedure we have a good chance of reducing the number of choices for a by another factor of 2, then with n invocations of the subroutine we might well expect to have a significant chance of learning a. This intuition is made precise in Appendix G, where some slightly subtle but purely mathematical analysis shows that with $n + x$ invocations of $\mathbf{U}_f$ the probability q of acquiring enough information to determine a is

$$q = \left(1 - \frac{1}{2^{n+x}}\right)\left(1 - \frac{1}{2^{n+x-1}}\right)\cdots\left(1 - \frac{1}{2^{x+2}}\right) > 1 - \frac{1}{2^{x+1}}. \tag{2.40}$$

Thus the odds are more than a million to one that with $n + 20$ invocations of $\mathbf{U}_f$ we will learn a, no matter how large n may be.

The intrusion of some mildly arcane arithmetic arguments, to confirm that the output of the quantum computer does indeed provide the needed information in the advertised number of runs, is characteristic of many quantum-computational algorithms. The action of the quantum computer itself is rather straightforward, but we must engage in more strenuous mathematical exertions to show that the outcome of the quantum computation does indeed enable us to accomplish the required task.

2.6 Constructing Toffoli gates

As noted in Section 1.6, constraints on what is physically feasible limit us to unitary transformations that can be built entirely out of 1- and 2-Qbit gates. It is assumed that 1-Qbit unitary gates will be relatively straightforward to make, though even this can be challenging for many of the physical systems proposed for Qbits. Making 2-Qbit gates presents an even tougher challenge to the quantum-computational engineer, since they will require one to manipulate with precision the physical interaction between the two Qbits. Making an inherently 3-Qbit gate goes beyond present hopes.

It has been known since before the arrival of quantum computation that to build up all arithmetical operations on a reversible *classical* computer it is necessary (and sufficient) to use at least one classically irreducible 3-Qbit gate – for example controlled-controlled-NOT (cc-NOT) gates, known as *Toffoli gates*. Such 3-Qbit gates cannot be built up out of 1- and 2-Cbit gates. This would appear to be bad news for the prospects of practical quantum computation.

Remarkably, however, the linear extension of the Toffoli gate to Qbits *can* be built up out of 1-Qbit unitary gates acting in suitable combination with 2-Qbit cNOT gates. The quantum extension of this classically irreducible 3-Cbit gate can be realized with a rather small number of 1- and 2-Qbit gates.

The 3-Qbit Toffoli gate T acts on the computational basis to flip the state of the third (target) Qbit if and only if the states of *both* of the first two (control) Qbits) are 1:

$$\mathsf{T}|x\rangle|y\rangle|z\rangle = |x\rangle|y\rangle|z \oplus xy\rangle. \tag{2.41}$$

Since T is its own inverse, it is clearly reversible, and therefore its linear extension from the classical basis to arbitrary 3-Qbit states is unitary, by the general argument in Section 1.6.

The Toffoli gate enables one to calculate the logical AND of two bits (i.e. their product) since $\mathsf{T}|x\rangle|y\rangle|0\rangle = |x\rangle|y\rangle|xy\rangle$. Since all Boolean operations can be built up out of AND and NOT, and since all of arithmetic can be constructed out of Boolean operations, with Toffoli gates one can build up all of classical computation through reversible operations. (One can even produce NOT with a Toffoli gate: $\mathsf{T}|1\rangle|1\rangle|x\rangle = |1\rangle|1\rangle|\overline{x}\rangle$, but this would be a ridiculously hard way to implement NOT on a quantum computer.)

There are (at least) two rather different ways to construct a ccNOT gate T out of cNOT gates and 1-Qbit unitaries. The first way to be found requires eight cNOT gates. Later a more efficient construction was discovered that requires only six cNOT gates. Nobody has found a construction with fewer than six cNOT gates, but I do not know of a proof that six are required. I describe both constructions, since they take advantage of and therefore illustrate several useful quantum-computational tricks.

The construction of a Toffoli gate from eight cNOT gates is based on three ingredients. (a) For any 1-Qbit unitary $\mathbf{U}$ one defines the 2-Qbit controlled-U gate $\mathbf{C}_{10}^{U}$ as one that acts on the computational basis as the identity if the state of Qbit 1 (the control Qbit) is $|0\rangle$ and acts on Qbit 0 (the target Qbit) as $\mathbf{U}$ if the state of the control Qbit is $|1\rangle$:

$$\mathbf{C}_{10}^{U}|x_1x_0\rangle = \mathbf{U}_0^{x_1}|x_1x_0\rangle. \tag{2.42}$$

(The cNOT operation $\mathbf{C}$ is thus a $\mathbf{C}^X$ operation, but so important a one as to make it the default form when no U is specified.) We shall show that a controlled-U gate for arbitrary $\mathbf{U}$ can be built out of two cNOT gates and 1-Qbit unitaries. (b) We shall show that a 3-Qbit doubly-controlled-U^2 gate, which takes $|x_2x_1x_0\rangle$ into $(\mathbf{U}_0^{2x_2x_1})|x_2x_1x_0\rangle$, can be constructed out of two controlled-U gates, one controlled-$U^\dagger$ gate, and two additional cNOT gates, making a total of eight cNOT gates. (c) We shall show that there is a unitary square-root-of-NOT gate, $\sqrt{\mathbf{X}}$. Taking $\mathbf{U}$ in (b) to be $\sqrt{\mathbf{X}}$ gives the desired Toffoli gate. We now elaborate on each part of the construction.

(a) Let $\mathbf{V}$ and $\mathbf{W}$ be two arbitrary 1-Qbit unitary transformations and consider the product

$$\mathbf{V}_0\mathbf{C}_{10}\mathbf{V}_0^\dagger\mathbf{W}_0\mathbf{C}_{10}\mathbf{W}_0^\dagger. \tag{2.43}$$

One easily confirms that (2.43) acts on the computational basis as $\mathbf{C}_{10}^{U}$ with

$$\mathbf{U} = (\mathbf{V}\mathbf{X}\mathbf{V}^\dagger)(\mathbf{W}\mathbf{X}\mathbf{W}^\dagger) = \big(\mathbf{V}(\vec{x}\cdot\vec{\sigma})\mathbf{V}^\dagger\big)\big(\mathbf{W}(\vec{x}\cdot\vec{\sigma})\mathbf{W}^\dagger\big). \tag{2.44}$$

As shown in Appendix B, one can pick $\mathbf{V}$ and $\mathbf{W}$ so that

$$\begin{aligned}&\big(\mathbf{V}(\vec{x}\cdot\vec{\sigma})\mathbf{V}^\dagger\big)\big(\mathbf{W}(\vec{x}\cdot\vec{\sigma})\mathbf{W}^\dagger\big)\\ &\quad= (\vec{a}\cdot\vec{\sigma})(\vec{b}\cdot\vec{\sigma}) = \vec{a}\cdot\vec{b} + i(\vec{a}\otimes\vec{b})\cdot\vec{\sigma},\end{aligned} \tag{2.45}$$

for arbitrary unit vectors $\vec{a}$ and $\vec{b}$. Appendix B also establishes that any 1-Qbit unitary transformation has, to within a multiplicative numerical phase factor $e^{i\alpha}$, the form

$$\mathbf{u}(\vec{n},\theta) = \exp(i\tfrac{1}{2}\theta\,\vec{n}\cdot\vec{\sigma}) = \cos(\tfrac{1}{2}\theta)\,\mathbf{1} + i\sin(\tfrac{1}{2}\theta)\,\vec{n}\cdot\vec{\sigma}. \tag{2.46}$$

If $\mathbf{a}$ and $\mathbf{b}$ are in the plane perpendicular to $\mathbf{n}$ and the angle between them is $\frac{1}{2}\theta$, then $\mathbf{U} = \mathbf{u}(\mathbf{n},\theta)$. The 1-Qbit unitary transformation $\mathbf{E} = e^{i\alpha\mathbf{n}}$, applied to Qbit 1, multiplies by the phase factor $e^{i\alpha}$ if and only if the

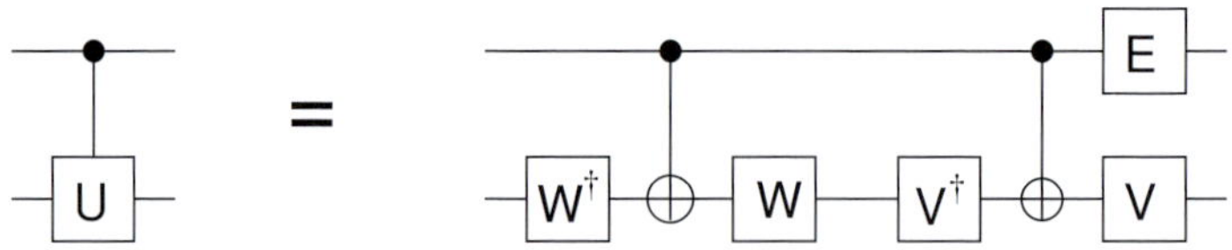

Fig 2.11 How to construct a controlled-U gate $\mathbf{C}^U$ from unitary 1-Qbit gates and two controlled-NOT gates. If the control Qbit is in the state $|0\rangle$, the operations on the target wire combine to give $(\mathbf{V}\mathbf{V}^\dagger)(\mathbf{W}\mathbf{W}^\dagger) = \mathbf{1}$. But if the control Qbit is in the state $|1\rangle$ then the operations combine to give $\mathbf{U} = (\mathbf{V}\sigma_x\mathbf{V}^\dagger)\,(\mathbf{W}\sigma_x\mathbf{W}^\dagger)$, where $\sigma_x = \vec{x}\cdot\vec{\sigma} = \mathbf{X}$. To within an overall numerical phase factor a general two-dimensional unitary transformation can always be put in this form for appropriate $\mathbf{V}$ and $\mathbf{W}$. The $\mathbf{E}$ on the control wire is the unitary transformation $\mathbf{E} = e^{i\alpha\mathbf{n}} = \begin{pmatrix} 1 & 0 \\ 0 & e^{i\alpha} \end{pmatrix}$, which supplies such a phase factor when the state of the control Qbit is $|1\rangle$. The two unitary gates between the cNOT gates on the lower wire, $\mathbf{W}$ and $\mathbf{V}^\dagger$, can be combined into the single unitary gate $\mathbf{V}^\dagger\mathbf{W}$, so in addition to the two cNOT gates the construction uses four 1-Qbit unitaries.

computational-basis state of Qbit 1 is $|1\rangle$. The resulting circuit for constructing $\mathbf{C}^U$ is shown in Figure 2.11.

(b) Given such a controlled-U gate $\mathbf{C}^U_{ij}$ with Qbit i the control and j the target, a doubly-controlled-U^2 gate, controlled by Qbits 2 and 1 and targeting Qbit 0, can be constructed out of three such controlled-U gates and two more cNOT gates:

$$\mathbf{C}^{U^2} = \mathbf{C}^U_{10}\mathbf{C}_{21}\mathbf{C}^{U^\dagger}_{10}\mathbf{C}_{21}\mathbf{C}^U_{20}. \tag{2.47}$$

The corresponding circuit diagram is shown in Figure 2.12. It is straightforward to establish that the sequence of operators on the right of (2.47) acts on the 3-Qbit computational-basis states as $\mathbf{1}$ unless Qbits 2 and 1 are both in the state $|1\rangle$, in which case it acts on Qbit 0 as $\mathbf{U}^2$.

(c) Finally, note that

$$\sqrt{\mathbf{Z}} = \begin{pmatrix} 1 & 0 \\ 0 & i \end{pmatrix}, \tag{2.48}$$

which is clearly unitary. Therefore, since $\mathbf{X} = \mathbf{HZH}$ and $\mathbf{H}^2 = 1$, we have

$$\sqrt{\mathbf{X}} = \mathbf{H}\sqrt{\mathbf{Z}}\mathbf{H}. \tag{2.49}$$

This plays the role of $\mathbf{U}$ in (b) to make the Toffoli gate.

The alternative construction of the Toffoli gate that uses only six cNOT gates has an action that is somewhat more transparent. It is illustrated in Figure 2.13. If $\mathbf{A}$ and $\mathbf{B}$ are any two unitaries with $\mathbf{A}^2 = \mathbf{B}^2 = \mathbf{1}$ then the 3-Qbit gate

$$\mathbf{C}^B_{10}\mathbf{C}^A_{20}\mathbf{C}^B_{10}\mathbf{C}^A_{20} \tag{2.50}$$

clearly acts as the identity on the computational basis, unless the states of Qbits 1 and 2 are both $|1\rangle$, in which case it acts as $(\mathbf{BA})^2$ on Qbit 0, so it is a doubly-controlled-$(BA)^2$ gate. Take $\mathbf{A} = \vec{a}\cdot\vec{\sigma}$ and $\mathbf{B} = \vec{b}\cdot\vec{\sigma}$ for unit vectors $\vec{a}$ and $\vec{b}$. Since $\vec{n}\cdot\vec{\sigma}$ can be expressed as $\mathbf{V}^\dagger(\vec{x}\cdot\vec{\sigma})\mathbf{V} = \mathbf{V}^\dagger\mathbf{X}\mathbf{V}$ for appropriate unitary $\mathbf{V}$, each controlled-A and controlled-B gate can be constructed with a single controlled-NOT gate and 1-Qbit unitaries. The product $\mathbf{BA}$ is the unitary $(\vec{b}\cdot\vec{\sigma})(\vec{a}\cdot\vec{\sigma}) = (\vec{b}\cdot\vec{a})\mathbf{1} + i(\vec{b}\times\vec{a})\cdot\vec{\sigma}$. Pick the unit vectors $\vec{b}$ and $\vec{a}$ with the angle between them $\pi/4$, lying in the plane perpendicular to x with their vector product directed along $\vec{x}$, so that $(\vec{b}\cdot\vec{\sigma})(\vec{a}\cdot\vec{\sigma}) = \cos(\pi/4)\mathbf{1} + i\sin(\pi/4)\vec{x}\cdot\vec{\sigma}$. (For

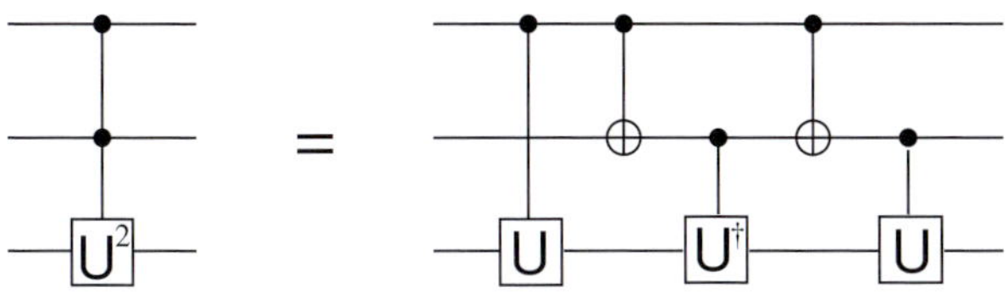

Fig 2.12 How to construct a 3-Qbit controlled-controlled-U^2 gate from 2-Qbit controlled-NOT, controlled-U, and controlled-$U^\dagger$ gates. If Qbits 2 and 1 (top and middle wires) are both in the state $|1\rangle$ then $\mathbf{U}$ acts twice on Qbit 0 (bottom wire) but $\mathbf{U}^\dagger$ does not. If Qbits 2 and 1 are both in the state $|0\rangle$ nothing acts on Qbit 0. If Qbits 2 and 1 are in the states $|1\rangle$ and $|0\rangle$ then only the $\mathbf{U}$ on the left and the $\mathbf{U}^\dagger$ act on Qbit 0 (and their product is $\mathbf{1}$), and if Qbits 2 and 1 are in the states $|0\rangle$ and $|1\rangle$ only the $\mathbf{U}$ on the right and the $\mathbf{U}^\dagger$ act on Qbit 0 (and their product is again $\mathbf{1}$.)

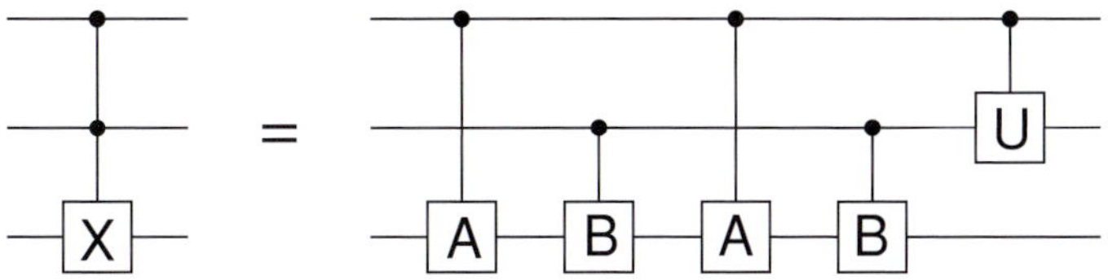

Fig 2.13 How to make a doubly-controlled-NOT (Toffoli) gate using six cNOT gates and 1-Qbit unitaries. The unitary operators $\mathbf{A}$ and $\mathbf{B}$ are given by $\mathbf{A} = \vec{a} \cdot \vec{\sigma}$ and $\mathbf{B} = \vec{b} \cdot \vec{\sigma}$ for appropriately chosen real unit vectors $\vec{a}$ and $\vec{b}$. Because $\vec{n} \cdot \vec{\sigma} = \mathbf{V}^\dagger(\vec{x} \cdot \vec{\sigma})\mathbf{V}$ for appropriate unitary $\mathbf{V}$, each controlled-A and controlled-B gate can be constructed with a single controlled-NOT gate and 1-Qbit unitaries. Because $\mathbf{A}^2 = \mathbf{B}^2 = \mathbf{1}$, the controlled-$A$ and controlled-B gates act together as a doubly-controlled-$(\mathbf{BA})^2$ gate. One can pick the directions $\vec{a}$ and $\vec{b}$ so that $(BA)^2 = i\mathbf{X}$. The controlled-U gate on the right corrects for this unwanted factor of i. Here $\mathbf{U}$ is the 1-Qbit unitary $e^{-i(\pi/2)\mathbf{n}}$. Since any controlled-U gate can be constructed with two cNOT gates and 1-Qbit unitaries, this adds two more cNOT gates to the construction, making a total of six.

example take $\vec{b} = \vec{z}$ and $\vec{a} = (1/\sqrt{2})(\vec{z} - \vec{y})$.) Then $(\mathbf{BA})^2 = \cos(\pi/2)\mathbf{1} + i\sin(\pi/2)\vec{x} \cdot \vec{\sigma} = i\,\vec{x} \cdot \vec{\sigma} = i\mathbf{X}$.

Thus (2.50) produces a doubly-controlled-NOT gate except for an extra factor of i accompanying the NOT. We can correct for this by applying an additional $\mathbf{C}_{21}^U$ gate, where $\mathbf{U}$ is the 1-Qbit unitary $e^{-i(\pi/2)\mathbf{n}}$. This controlled-U gate acts as the identity on the computational basis unless the states of Qbits 2 and 1 are both $|1\rangle$, in which case it multiplies the state by $e^{-i\pi/2} = -i$, thereby getting rid of the unwanted factor of i. Since we have just established that any controlled-U gate can be constructed with two cNOT gates and 1-Qbit unitaries, correcting the phase adds two more cNOT gates to the construction, making a total of six.

Alternatively, one can view this as a way to construct a Toffoli gate from four cNOT gates and a single controlled-phase gate of precisely the kind that plays a central role in the quantum Fourier transform described in Chapter 3. If quantum computation ever becomes a working technology, it might well be easier to construct controlled-phase gates as fundamental gates in their own right – pieces of 2-Qbit hardware as basic as cNOT gates.

As this and subsequent examples reveal, the cNOT gate is of fundamental importance in quantum computation. Appendix H gives some examples of how such gates might actually be realized. That appendix is addressed primarily to physicists, but readers with other backgrounds might find it an interesting illustration of the rather different questions that arise when one starts thinking about how actually to produce some of the basic quantum-computational hardware.

Chapter 3

Breaking RSA encryption

3.1 Period finding, factoring, and cryptography

Simon's problem (Section 2.5) starts with a subroutine that calculates a function $f(x)$, which satisfies $f(x) = f(y)$ for distinct x and y if and only if $y = x \oplus a$, where $\oplus$ denotes the bitwise modulo-2 sum of the n-bit integers a and x. The number of times a classical computer must invoke the subroutine to determine a grows exponentially with n, but with a quantum computer it grows only linearly.

This is a rather artificial example, of interest primarily because it gives a simple demonstration of the remarkable computational power a quantum computer can possess. It amounts to finding the unknown period a of a function on n-bit integers that is "periodic" under bitwise modulo-2 addition. A more difficult, but much more natural problem is to find the period r of a function f on the integers that is periodic under *ordinary* addition, satisfying $f(x) = f(y)$ for distinct x and y if and only if x and y differ by an integral multiple of r. Finding the period of such a periodic function turns out to be the key to factoring products of large prime numbers, a mathematically natural problem with quite practical applications.

One might think that finding the period of such a periodic function ought to be easy, but that is only because when one thinks of periodic functions one tends to picture slowly varying continuous functions (like the sine function) whose values at a small sample of points within a period can give powerful clues about what that period might be. But the kind of periodic function to keep in mind here is a function on the integers whose values within a period r are virtually random from one integer to the next, and therefore give no hint of the value of r.

The best known classical algorithms for finding the period r of such a function take a time that grows faster than any power of the number n of bits of r (exponentially with $n^{1/3}$). But in 1994 Peter Shor discovered that one can exploit the power of a quantum computer to learn the period r, in a time that scales only a little faster than n^3.

Because the ability to find periods efficiently, combined with some number-theoretic tricks, enables one to factor efficiently the product of two large prime numbers, Shor's discovery of super-efficient quantum period finding is of considerable practical interest. The very great computational effort required by all known classical factorization techniques underlies the security of the widely used RSA[1] method of encryption. Any computer that can efficiently find periods would be an enormous threat to the security of both military and commercial communications. This is why research into the feasibility of quantum computers is a matter of considerable interest in the worlds of war and business.

Although the elementary number-theoretic tricks that underlie the RSA method of encryption have nothing directly to do with how a quantum computer finds periods, they motivate the problem that Shor's quantum-computational algorithm so effectively solves. Furthermore, examining the number-theoretic basis of RSA encryption reveals that Shor's period-finding algorithm can be used to defeat it directly, without any detour into factoring. We therefore defer the number-theoretic connection between period finding and factoring to Section 3.10. If you are interested only in applying Shor's period-finding algorithm to decoding RSA encryption, Section 3.10 can be skipped. If you are not interested in the application of period finding to commerce and espionage, you can also skip the number theory in Sections 3.2 and 3.3 and go directly to the quantum-computational part of the problem – super-efficient period finding – in Section 3.4.

3.2 Number-theoretic preliminaries

The basic algebraic entities behind RSA encryption are finite groups, where the group operation is multiplication modulo some fixed integer N. In modulo-N arithmetic all integers that differ by multiples of N are identified, so there are only N distinct quantities, which can be represented by $0, 1, \ldots, N-1$. For example $5 \times 6 \equiv 2 \pmod 7$ since $5 \times 6 = 30 = 4 \times 7 + 2$. One writes $\equiv \pmod N$ to emphasize that the equality is only up to a multiple of N, reserving $=$ for strict equality. One can develop the results that follow using arithmetic rather than group theory, but the group-theoretic approach is simpler and uses properties of groups so elementary that they can be derived from the basic definitions in hardly more than a page. This is done in Appendix I, which readers unacquainted with elementary group theory should now read.

Let G_N be the set of all positive integers less than N (including 1) that have no factors in common with N. Since factoring into primes is unique, the product of two numbers in G_N (either the ordinary or the

1 Named after the people who invented it in 1977, Ronald Rivest, Adi Shamir, and Leonard Adleman. RSA encryption was independently invented by Clifford Cocks four years earlier, but his discovery was classified top secret by British Intelligence and he was not allowed to reveal his priority until 1997. For this and other fascinating tales about cryptography, see Simon Singh, *The Code Book*, New York, Doubleday (1999).

modulo-N product) also has no factors in common with N, so G_N is closed under multiplication modulo N. If a, b, and c are in G_N with $ab \equiv ac \pmod{N}$, then $a(b-c)$ is a multiple of N, and since a has no factors in common with N, it must be that $b-c$ is a multiple of N, so $b \equiv c \pmod{N}$. Thus the operation of multiplication modulo N by a fixed member a of G_N takes distinct members of G_N into distinct members, so the operation simply permutes the members of the finite set G_N. Since 1 is a member of G_N, there must be some d in G_N satisfying $ad = 1$ – i.e. a must have a multiplicative inverse in G_N. Thus G_N satisfies the conditions, listed in Appendix I, for it to be a group under modulo-N multiplication.

Every member a of a finite group G is characterized by its *order* k, the smallest integer for which (in the case of G_N)

$$a^k \equiv 1 \pmod{N}. \tag{3.1}$$

As shown in Appendix I, the order of every member of G is a divisor of the number of members of G (the *order* of G). If p is a prime number, then the group G_p contains $p-1$ numbers, since no positive integer less than p has factors in common with p. Since $p-1$ is then a multiple of the order k of any a in G_p, it follows from (3.1) that any integer a less than p satisfies

$$a^{p-1} \equiv 1 \pmod{p}. \tag{3.2}$$

This relation, known as *Fermat's little theorem,* extends to arbitrary integers a not divisible by p, since any such a is of the form $a = mp + a'$ with m an integer and a' less than p.

RSA encryption exploits an extension of Fermat's little theorem to a case characterized by *two* distinct primes, p and q. If an integer a is divisible neither by p nor by q, then no power of a is divisible by either p or q. Since, in particular, a^{q-1} is not divisible by p, we conclude from (3.2) that

$$[a^{q-1}]^{p-1} \equiv 1 \pmod{p}. \tag{3.3}$$

For the same reason

$$[a^{p-1}]^{q-1} \equiv 1 \pmod{q}. \tag{3.4}$$

The relations (3.3) and (3.4) state that $a^{(q-1)(p-1)} - 1$ is a multiple both of p and of q. Since p and q are distinct primes, it must therefore be a multiple of pq, and therefore

$$a^{(q-1)(p-1)} \equiv 1 \pmod{pq}. \tag{3.5}$$

(You are urged to check relations like (3.5) for yourself in special cases. If, for example $p = 3$ and $q = 5$ then (3.5) requires $2^8 - 1$ to be divisible by 15, and indeed, $255 = 17 \times 15$.)

As an alternative derivation of (3.5), note that since a is divisible neither by p nor by q, it has no factors in common with pq and is therefore in G_{pq}. The number of elements of G_{pq} is $pq - 1 - (p - 1) - (q - 1) = (p - 1)(q - 1)$, since there are $pq - 1$ integers less than pq, among which are $p - 1$ multiples of q and another distinct $q - 1$ multiples of p. Equation (3.5) follows because the order $(p - 1)(q - 1)$ of G_{pq} must be a multiple of the order of a.

We get the version of (3.5) that is the basis for RSA encryption by taking any integral power s of (3.5) and multiplying both sides by a:

$$a^{1+s(q-1)(p-1)} \equiv a \pmod{pq}. \tag{3.6}$$

(The relation (3.6) holds even for integers a that are divisible by p or q. It holds trivially when a is a multiple of pq. And if a is divisible by just one of p and q, let $a = kq$. Since a is not divisible by p neither is any power of a, and therefore Fermat's little theorem tells us that $[a^{s(q-1)}]^{p-1} = 1 + np$ for some integer n. On multiplying both sides by a we have $a^{1+s(q-1)(p-1)} \equiv a + nap \equiv a + nkqp$, so (3.6) continues to hold.)

Note finally that if c is an integer having no factor in common with $(p - 1)(q - 1)$ then c is in $G_{(p-1)(q-1)}$ and therefore has an inverse in $G_{(p-1)(q-1)}$; i.e. there is a d in $G_{(p-1)(q-1)}$ satisfying

$$cd \equiv 1 \,\big(\mathrm{mod}\,(p-1)(q-1)\big). \tag{3.7}$$

So for some integer s,

$$cd = 1 + s(p - 1)(q - 1). \tag{3.8}$$

In view of (3.8) and (3.6), any integer a must satisfy

$$a^{cd} \equiv a \pmod{pq}. \tag{3.9}$$

So if

$$b \equiv a^c \pmod{pq}, \tag{3.10}$$

then

$$b^d \equiv a \pmod{pq}. \tag{3.11}$$

The elementary arithmetical facts summarized in this single paragraph constitute the entire basis for RSA encryption.

3.3 RSA encryption

Bob wants to receive a message from Alice encoded so that he alone can read it. To do this he picks two large (say 200-digit) prime numbers p and q. He gives Alice, through a public channel, their product $N = pq$ and a large encoding number c that he has picked to have no factors in common with[2] $(p-1)(q-1)$. He does not, however, reveal the separate values of p and q and, given the practical impossibility of factoring a 400-digit number with currently available computers, he is quite confident that neither Alice nor any eavesdropper Eve will be able to calculate p and q knowing only their product N. Bob, however, because he does know p and q, and therefore $(p-1)(q-1)$, can find the multiplicative inverse d of c mod $(p-1)(q-1)$, which satisfies (3.7).[3] He keeps d strictly to himself for use in decoding.

Alice encodes a message by representing it as a string of fewer than 400 digits using, for example, some version of ASCII coding. If her message requires more than 400 digits she chops it up into smaller pieces. She interprets each such string as a number a less than N. Using the coding number c and the value of $N = pq$ she received from Bob, she then calculates $b \equiv a^c \pmod{pq}$, and sends it on to Bob through a public channel. With c typically a 200-digit number, you might think that this would itself be a huge computational task, but it is not, as noted in Section 3.8. When he receives b, Bob exploits his private knowledge of d to calculate $b^d \pmod{pq}$, which (3.11) assures him is Alice's original message a.

Were the eavesdropper Eve able to find the factors p and q of N, she could calculate $(p-1)(q-1)$ and find the decoding integer d from the publicly available coding integer c, the same way Bob did. But factoring a number as large as N is far beyond her classical computational powers. Efficient period finding is of interest in this cryptographic setting not only because it leads directly to efficient factoring (as described in Section 3.10), but also because it can lead Eve directly to an alternative way to decode Alice's message b without her knowing or having to compute the factors p and q of N. Here is how it works:

Eve uses her efficient period-finding machine to calculate the order r of Alice's publicly available encoded message $b = a^c$ in[4] G_{pq}. Now the order r of Alice's encoded message $b = a^c$ in G_{pq} is the same

2 As shown in Appendix J, the probability that two large random numbers have no common factor is greater than $\frac{1}{2}$, so such c are easily found. Whether two numbers do have any factors in common (and what their greatest common factor is) can be determined by a simple algorithm known to Euclid and easily executed by Bob on a classical computer. The Euclidean algorithm is described in Appendix J.

3 This can easily be done classically as a straightforward embellishment of the Euclidean algorithm. See Appendix J.

4 I assume that Alice's unencoded message a, and hence her coded message b, is in G_{pq} – i.e. that a is not a multiple of p or q. Since p and q are huge prime numbers, the odds against a being such a multiple are astronomical. But if Eve wants to be insanely careful she can find the greatest common factor of b and N, using the Euclidean algorithm. In the grossly improbable case that it turns out not to be 1, Eve will have factored N and can decode Alice's message the same way Bob does.

as the order of a. This is because the subgroup of G_{pq} generated by a contains $a^c = b$, and hence it contains the subgroup generated by b; but the subgroup generated by b contains $b^d = a$, and hence the subgroup generated by a. Since each subgroup contains the other, they must be identical. Since the order of a or b is the number of elements in the subgroup it generates, their orders are the same. So if Eve can find the order r of Alice's code message b, then she has also learned the order of Alice's original text a.

Since Bob has picked c to have no factors in common with $(p-1)(q-1)$, and since r divides the order $(p-1)(q-1)$ of G_{pq}, the coding integer c can have no factors in common with r. So c is congruent modulo r to a member c' of G_r, which has an inverse d' in G_r, and d' is also a modulo-r inverse of c:

$$cd' \equiv 1 \pmod{r}. \tag{3.12}$$

Therefore, given c (which Bob has publicly announced) and r (which Eve can get with her period-finding program from Alice's encoded message b and the publicly announced value of $N = pq$), it is easy for Eve to calculate d' with a classical computer, using, modulo r, the same extension of the Euclidean algorithm as Bob used to find d, modulo $(p-1)(q-1)$. It then follows that for some integer m

$$b^{d'} \equiv a^{cd'} = a^{1+mr} = a\left(a^r\right)^m \equiv a \pmod{pq}. \tag{3.13}$$

Eve has thus used her ability to find periods to decode Alice's encoded message $b = a^c$ to reveal Alice's original message a.

This use of period finding to defeat RSA encryption is summarized in Table 3.1.

3.4 Quantum period finding: preliminary remarks

So we can crack the RSA code if we have a fast way to find the period r of the known periodic function

$$f(x) = b^x \pmod{N}. \tag{3.14}$$

This might appear to be a simple task, especially since periodic functions of the special form (3.14) have the simplifying feature that $f(x+s) = f(x)$ *only* if s is a multiple of the period r. But $b^x \pmod{N}$ is precisely the kind of function whose values within a period hop about so irregularly as to offer no obvious clues about the period. One could try evaluating $f(x)$ for random x until one found two different values of x for which f agreed. Those values would differ by a multiple of the period, which would provide some important information about the value of the period itself. But this is an inefficient way to proceed, even classically.

Table 3.1. A summary of RSA encryption and how to break it with a fast period-finding routine on a quantum computer. Bob has chosen the encoding number c to have an inverse d modulo $(p-1)(q-1)$ so c can have no factors in common with $(p-1)(q-1)$. Since Alice's encoded message b is in G_{pq}, its order r is a factor of the order $(p-1)(q-1)$ of G_{pq}. So c can have no factors in common with r, and therefore has an inverse d' modulo r. Because b is a power of a and vice versa, each has the same order r in G_{pq}. Therefore $b^{d'} \equiv a^{cd'} \equiv a^{1+mr} \equiv a$ modulo N.

Bob knows	Alice knows	Public knows
p and q (primes); c and d satisfying $cd \equiv 1 \pmod{(p-1)(q-1)}$; b (encoded message).	a (her message); only c (not d) and only $N = pq$; $b \equiv a^c \pmod N$ (encoded message).	b (encoded message); only c (not d); only $N = pq$.
Decoding: $a \equiv b^d \pmod N$.		**Quantum decoding:** Quantum computer finds r: $b^r \equiv 1 \pmod N$; classical computer finds d': $cd' \equiv 1 \pmod r$; $a \equiv b^{d'} \pmod N$.

Let n_0 be the number of bits in $N = pq$, so that 2^{n_0} is the smallest power of 2 that exceeds N. If N is a 500-digit number – a typical size for cryptographic applications – n_0 will be around 1700. This also sets the scale for the typical number of bits in the other relevant numbers a, b, and their modulo-N period r. To have an appreciable probability of finding r by random searching requires a number of evaluations of f that is exponential in n_0 (just as in the classical approach to Simon's problem, described in Chapter 2). There are classical ways to improve on random searching, using, for example, Fourier analysis, but no classical approach is known that does not require a time that grows faster than any power of n_0. With a quantum computer, however, quantum parallelism gets us tantalizingly close (but, as in Simon's problem, not close enough) to solving the problem with a single application of $\mathbf{U}_f$, and enables us to solve it completely with probability arbitrarily close to unity in a time that grows only as a low-order polynomial in n_0.

To deal with values of x and $f(x) = b^x \pmod N$ between 0 and N, both the input and output registers must contain at least n_0 Qbits. For reasons that will emerge in Section 3.7, however, to find the period r efficiently the input register must actually have $n = 2n_0$ Qbits. Doubling the number of Qbits in the input register ensures that the range of values of x for which $f(x)$ is calculated contains at least N full periods of f. This redundancy turns out to be essential for a successful determination of the period by Shor's method. (We shall see in Section 3.7 that if p and q both happen to be primes of the form $2^j + 1$ then – and only then – the method works without doubling the size of the input register. Thus $N = 15 = (2+1)(2^2+1)$ does not provide a realistic test case for laboratory attempts to demonstrate Shor's algorithm for small p and q with real Qbits.)

We begin the quantum period-finding algorithm by using our quantum computer in the familiar way to construct the state

$$\frac{1}{2^{n/2}} \sum_{x=0}^{2^n-1} |x\rangle_n |f(x)\rangle_{n_0} \tag{3.15}$$

with a single application of $\mathbf{U}_f$. In Section 3.8 we take a closer look at how this might efficiently be done in the case of interest, $f(x) = b^x$ (mod N). Once the state of the registers has become (3.15), we can measure the n-Qbit output register.[5] If the measurement yields the value f_0, then the generalized Born rule tells us that the state of the n-Qbit input register can be taken to be

$$|\Psi\rangle_n = \frac{1}{\sqrt{m}} \sum_{k=0}^{m-1} |x_0 + kr\rangle_n. \tag{3.16}$$

Here x_0 is the smallest value of x $(0 \leq x_0 < r)$ for which $f(x_0) = f_0$, and m is the smallest integer for which $mr + x_0 \geq 2^n$, so

$$m = \left[\frac{2^n}{r}\right] \quad \text{or} \quad m = \left[\frac{2^n}{r}\right] + 1, \tag{3.17}$$

depending on the value of x_0 (where $[x]$ is the integral part of x – the largest integer less than or equal to x). As in the examples of Chapter 2, if we could produce a small number of identical copies of the state (3.16) the job would be done, for a measurement in the computational basis would yield a random one of the values $x_0 + kr$, and the difference between the results of pairs of measurements on such identical copies would give us a collection of random multiples of r from which r itself could straightforwardly be extracted. But this possibility is ruled out by the no-cloning theorem. All we can extract is a single value of $x_0 + kr$ for unknown random x_0, which is useless for determining r. And, of course, if we ran the whole algorithm again, we would end up with a state of the form (3.16) for another random value of x_0, which would permit no useful comparison with what we had learned from the first run.

But, as with Simon's problem, we can do something more clever to the state (3.16) before making our final measurement. The problem is the displacement by the unknown random x_0, which prevents any information about r from being extracted in a single measurement. We need a unitary transformation that transforms the x_0 dependence into

5 It is not, in fact, necessary to measure the output register. One can continue to work with the full state (3.15) in which one breaks down the sum on x into a sum over all the different values of f and a sum over all the values of x associated with each value of f. The only purpose of the measurement is to clarify the analysis by eliminating a lot of uninteresting additional structure, coming from the sum on the values of f, that plays no role beyond making many of the subsequent expressions somewhat lengthier.

a harmless overall phase factor. This is accomplished with the *quantum Fourier transform*.

3.5 The quantum Fourier transform

The heart of Shor's algorithm is a superfast quantum Fourier transform, which can be carried out by a spectacularly efficient quantum circuit built entirely out of 1-Qbit and 2-Qbit gates. The n-Qbit quantum Fourier transform is defined to be that unitary transformation $\mathbf{U}_{\mathrm{FT}}$ whose action on the computational basis is given by

$$\mathbf{U}_{\mathrm{FT}}|x\rangle_n = \frac{1}{2^{n/2}} \sum_{y=0}^{2^n-1} e^{2\pi i xy/2^n} |y\rangle_n. \tag{3.18}$$

The product xy is here ordinary multiplication.[6] One easily verifies that $\mathbf{U}_{\mathrm{FT}}|x\rangle$ is normalized to unity and that $\mathbf{U}_{\mathrm{FT}}|x\rangle$ is orthogonal to $\mathbf{U}_{\mathrm{FT}}|x'\rangle$ unless $x = x'$, so $\mathbf{U}_{\mathrm{FT}}$ is unitary. Unitarity also emerges directly from the analysis that follows, which explicitly constructs $\mathbf{U}_{\mathrm{FT}}$ out of 1- and 2-Qbit unitary gates. The unitary $\mathbf{U}_{\mathrm{FT}}$ is useful because, as one also easily verifies, applied to a superposition of states $|x\rangle$ with complex amplitudes $\gamma(x)$, it produces another superposition with amplitudes that are related to $\gamma(x)$ by the appropriate discrete Fourier transform:

$$\mathbf{U}_{\mathrm{FT}}\left(\sum_{x=0}^{2^n-1} \gamma(x)|x\rangle\right) = \sum_{x=0}^{2^n-1} \tilde{\gamma}(x)|x\rangle, \tag{3.19}$$

where

$$\tilde{\gamma}(x) = \frac{1}{2^{n/2}} \sum_{y=0}^{2^n-1} e^{2\pi i xy/2^n} \gamma(y). \tag{3.20}$$

The celebrated classical fast Fourier transform is an algorithm requiring a time that grows with the number of bits as $n2^n$ (rather than $(2^n)^2$ as the obvious direct approach would require) to evaluate $\tilde{\gamma}$. But there is a quantum algorithm for executing the unitary transformation $\mathbf{U}_{\mathrm{FT}}$ exponentially faster than fast, in a time that grows only as n^2. The catch, as usual, is that one does not end up knowing the complete

6 A warning to physicists (which others can ignore). This looks deceptively like a (discretized) transformation from a position to a momentum representation, and one's first reaction might be that it is (perhaps disappointingly) familiar. But it has, in fact, an entirely different character. The number x is the integer represented by the state $|x\rangle$; it is not the position of anything. Changing x to $x+1$ induces an arithmetically natural but physically quite unnatural transformation on the computational-basis states, determined by the laws of binary addition, including carrying. It bears no resemblance to anything that could be associated with a spatial translation in the physical space of Qbits. So your eyes should not glaze over, and you should regard $\mathbf{U}_{\mathrm{FT}}$ as a new and unfamiliar physical transformation of Qbits.

set of Fourier coefficients, as one does after applying the classical fast Fourier transform. One just has n Qbits described by the state given by the right side of (3.19), and as we have repeatedly noted, having a collection of Qbits in a given state does not enable one to learn what that state actually is. There is no way to extract all the Fourier coefficients $\tilde{\gamma}$, given an n-Qbit register in the state (3.19). But if γ is a periodic function with a period that is no bigger than $2^{n/2}$, then a register in the state (3.19) can give powerful clues about the precise value of the period r, even though r can be hundreds of digits long.

Notice the resemblance of the quantum Fourier transform (3.18) to the n-fold Hadamard transformation. Since $-1 = e^{\pi i}$, the n-fold Hadamard (2.30) assumes the form

$$\mathbf{H}^{\otimes n}|x\rangle_n = \frac{1}{2^{n/2}} \sum_{y=0}^{2^n-1} e^{\pi i x \cdot y}|y\rangle_n. \tag{3.21}$$

Aside from the different powers of 2 appearing in the quantum Fourier transform (3.18) – so the factors of modulus 1 in the superposition are not just 1 and -1 – the only other difference between the two transforms is that xy is ordinary multiplication in the quantum Fourier transform, whereas $x \cdot y$ is the bitwise inner product in the n-fold Hadamard. Because the arithmetic product xy is a more elaborate function of x and y than $x \cdot y$, the quantum Fourier transformation cannot be built entirely out of 1-Qbit unitary gates as the n-fold Hadamard is. But, remarkably, it can be constructed entirely out of 1- and 2-Qbit gates. Even more remarkably, when the procedure is used for period finding all of the 2-Qbit gates can be replaced by 1-Qbit measurement gates followed by additional 1-Qbit unitary gates whose application is contingent on the measurement outcomes.

To construct a circuit to execute the quantum Fourier transform $\mathbf{U}_{\mathrm{FT}}$, it is convenient to introduce an n-Qbit unitary operator $\mathcal{Z}$, diagonal in the computational basis:

$$\mathcal{Z}|y\rangle_n = e^{2\pi i y/2^n}|y\rangle_n. \tag{3.22}$$

This can be viewed as a generalization to n Qbits of the 1-Qbit operator $\mathbf{Z}$, to which it reduces when $n = 1$. Using the familiar relation

$$\mathbf{H}^{\otimes n}|0\rangle_n = \frac{1}{2^{n/2}} \sum_{y=0}^{2^n-1} |y\rangle_n, \tag{3.23}$$

we can reexpress the definition (3.18) as

$$\mathbf{U}_{\mathrm{FT}}|x\rangle_n = \mathcal{Z}^x \mathbf{H}^{\otimes n}|0\rangle_n. \tag{3.24}$$

This gives $\mathbf{U}_{\mathrm{FT}}|x\rangle_n$ as an x-dependent operator acting on the state $|0\rangle$.

We next reexpress the right side of (3.24) as an x-independent linear operator acting on the state $|x\rangle_n$. Since the computational-basis states

$|x\rangle_n$ are a basis, this will give us an alternative expression for $\mathbf{U}_{\mathrm{FT}}$ itself. The construction of this alternative form for (3.24) is made more transparent by specializing to the case of four Qbits. The structure that emerges in the case $n = 4$ has an obvious extension to general n. Dealing with the case of general n from the start only obscures things.

When $n = 4$ we want to find an appropriate form for

$$\mathbf{U}_{\mathrm{FT}}|x_3\rangle|x_2\rangle|x_1\rangle|x_0\rangle = \mathcal{Z}^x \mathbf{H}_3\mathbf{H}_2\mathbf{H}_1\mathbf{H}_0|0\rangle|0\rangle|0\rangle|0\rangle. \tag{3.25}$$

As usual, we number the Qbits by the power of 2 with which they are associated, with the least significant on the right, so that, reading from right to left, the Qbits are labeled 0, 1, 2, and 3; $\mathbf{H}_i$ acts on the Qbit labeled i (and as the identity on all other Qbits). If $|y\rangle_4 = |y_3\rangle|y_2\rangle|y_1\rangle|y_0\rangle$ in the definition (3.22) of $\mathcal{Z}$, so that $y = 8y_3 + 4y_2 + 2y_1 + y_0$, then the operator $\mathcal{Z}$ can be constructed out of single-Qbit number operators:

$$\mathcal{Z} = \exp\left(\frac{i\pi}{8}(8\mathbf{n}_3 + 4\mathbf{n}_2 + 2\mathbf{n}_1 + \mathbf{n}_0)\right). \tag{3.26}$$

The operator $\mathcal{Z}^x$ appearing in (3.25) then becomes

$$\mathcal{Z}^x = \exp\left(\frac{i\pi}{8}(8x_3 + 4x_2 + 2x_1 + x_0)(8\mathbf{n}_3 + 4\mathbf{n}_2 + 2\mathbf{n}_1 + \mathbf{n}_0)\right). \tag{3.27}$$

Because the 1-Qbit operator $\exp(2\pi i\mathbf{n})$ acts as the identity on either of the 1-Qbit states $|0\rangle$ and $|1\rangle$, and because any 1-Qbit state is a superposition of these two, $\mathbf{n}$ obeys the operator identity

$$\exp(2\pi i\mathbf{n}) = \mathbf{1}. \tag{3.28}$$

Therefore, in multiplying out the two terms

$$(8x_3 + 4x_2 + 2x_1 + x_0)(8\mathbf{n}_3 + 4\mathbf{n}_2 + 2\mathbf{n}_1 + \mathbf{n}_0) \tag{3.29}$$

appearing in the exponential (3.27), we can drop all products $x_i\mathbf{n}_j$ whose coefficients are a power of 2 greater than 8, getting

$$\begin{aligned}\mathcal{Z}^x = \exp\big[i\pi\big(&x_0\mathbf{n}_3 + (x_1 + \tfrac{1}{2}x_0)\mathbf{n}_2 + (x_2 + \tfrac{1}{2}x_1 + \tfrac{1}{4}x_0)\mathbf{n}_1 \\ &+ (x_3 + \tfrac{1}{2}x_2 + \tfrac{1}{4}x_1 + \tfrac{1}{8}x_0)\mathbf{n}_0\big)\big].\end{aligned} \tag{3.30}$$

Note next that the number and Hadamard operators for any single Qbit obey the relation

$$\exp(i\pi x\mathbf{n})\mathbf{H}|0\rangle = \mathbf{H}|x\rangle. \tag{3.31}$$

This is trivial when $x = 0$, and when $x = 1$ it reduces to the correct statement

$$(-1)^{\mathbf{n}}\tfrac{1}{\sqrt{2}}(|0\rangle + |1\rangle) = \tfrac{1}{\sqrt{2}}(|0\rangle - |1\rangle). \tag{3.32}$$

(Alternatively, note that $\exp(i\pi\mathbf{n}) = \mathbf{Z}$ and $\mathbf{ZH} = \mathbf{HX}$.) The effect on (3.25) of the four terms in (3.30) that do not contain factors of $\frac{1}{2}$, $\frac{1}{4}$, or $\frac{1}{8}$ is to produce the generalization of (3.31) to several Qbits:

$$\begin{aligned}
&\exp\bigl[i\pi\bigl(x_0\mathbf{n}_3 + x_1\mathbf{n}_2 + x_2\mathbf{n}_1 + x_3\mathbf{n}_0\bigr)\bigr]\mathbf{H}_3\mathbf{H}_2\mathbf{H}_1\mathbf{H}_0|0\rangle|0\rangle|0\rangle|0\rangle \\
&\quad = \bigl[\exp\bigl(i\pi x_0\mathbf{n}_3\bigr)\mathbf{H}_3\bigr]\bigl[\exp\bigl(i\pi x_1\mathbf{n}_2\bigr)\mathbf{H}_2\bigr]\bigl[\exp\bigl(i\pi x_2\mathbf{n}_1\bigr)\mathbf{H}_1\bigr] \\
&\qquad \times\bigl[\exp\bigl(i\pi x_3\mathbf{n}_0\bigr)\mathbf{H}_0\bigr]|0\rangle|0\rangle|0\rangle|0\rangle \\
&\quad = \mathbf{H}_3\mathbf{H}_2\mathbf{H}_1\mathbf{H}_0|x_0\rangle|x_1\rangle|x_2\rangle|x_3\rangle. \qquad (3.33)
\end{aligned}$$

We have used the fact that number operators associated with different Qbits commute with one another. Note also that because the number operator $\mathbf{n}_i$ is multiplied by x_{3-i} on the left side of (3.33), the state of the Qbit labeled i on the right is $|x_{3-i}\rangle$.

The remaining six terms in (3.30) (containing fractional coefficients) further convert (3.25) to the form

$$\begin{aligned}
\mathbf{U}_{\mathrm{FT}}|x_3\rangle|x_2\rangle|x_1\rangle|x_0\rangle &= \exp\bigl[i\pi\bigl(\tfrac{1}{2}x_0\mathbf{n}_2 + (\tfrac{1}{2}x_1 + \tfrac{1}{4}x_0)\mathbf{n}_1 \\
&\quad + (\tfrac{1}{2}x_2 + \tfrac{1}{4}x_1 + \tfrac{1}{8}x_0)\mathbf{n}_0\bigr)\bigr]\mathbf{H}_3\mathbf{H}_2\mathbf{H}_1\mathbf{H}_0|x_0\rangle|x_1\rangle|x_2\rangle|x_3\rangle. \qquad (3.34)
\end{aligned}$$

Since the Hadamard transformation $\mathbf{H}_i$ commutes with the number operator $\mathbf{n}_j$ when $i \neq j$, we can regroup the terms in (3.34) so that each number operator $\mathbf{n}_i$ appears immediately to the left of its corresponding Hadamard operator $\mathbf{H}_i$:

$$\begin{aligned}
\mathbf{U}_{\mathrm{FT}}|x_3\rangle|x_2\rangle|x_1\rangle|x_0\rangle = \mathbf{H}_3 &\exp\bigl[i\pi\mathbf{n}_2\tfrac{1}{2}x_0\bigr]\mathbf{H}_2\exp\bigl[i\pi\mathbf{n}_1(\tfrac{1}{2}x_1 + \tfrac{1}{4}x_0)\bigr]\mathbf{H}_1 \\
&\times \exp\bigl[i\pi\mathbf{n}_0(\tfrac{1}{2}x_2 + \tfrac{1}{4}x_1 + \tfrac{1}{8}x_0)\bigr]\mathbf{H}_0 \\
&\times |x_0\rangle|x_1\rangle|x_2\rangle|x_3\rangle. \qquad (3.35)
\end{aligned}$$

The state $|x_0\rangle|x_1\rangle|x_2\rangle|x_3\rangle$ is an eigenstate of the number operators $\mathbf{n}_3, \mathbf{n}_2, \mathbf{n}_1, \mathbf{n}_0$ with respective eigenvalues x_0, x_1, x_2, x_3. If we did not have to worry about Hadamard operators interposing themselves between number operators and their eigenstates, we could replace each x_i in (3.35) by the number operator $\mathbf{n}_{3-i}$ of which it is the eigenvalue to get

$$\begin{aligned}
\mathbf{U}_{\mathrm{FT}}|x_3\rangle|x_2\rangle|x_1\rangle|x_0\rangle = \mathbf{H}_3 &\exp\bigl[i\pi\tfrac{1}{2}\mathbf{n}_2\mathbf{n}_3\bigr]\mathbf{H}_2\exp\bigl[i\pi\mathbf{n}_1(\tfrac{1}{2}\mathbf{n}_2 + \tfrac{1}{4}\mathbf{n}_3)\bigr] \\
&\times \mathbf{H}_1\exp\bigl[i\pi\mathbf{n}_0(\tfrac{1}{2}\mathbf{n}_1 + \tfrac{1}{4}\mathbf{n}_2 + \tfrac{1}{8}\mathbf{n}_3)\bigr]\mathbf{H}_0 \\
&\times |x_0\rangle|x_1\rangle|x_2\rangle|x_3\rangle. \qquad (3.36)
\end{aligned}$$

But as (3.36) makes clear, we do indeed not have to worry, because every $\mathbf{H}_i$ appears safely to the *left* of every $\mathbf{n}_i$ that has replaced an x_{3-i}.

If we define 2-Qbit unitary operators by

$$\mathbf{V}_{ij} = \exp\bigl(i\pi\mathbf{n}_i\mathbf{n}_j/2^{|i-j|}\bigr), \qquad (3.37)$$

then (3.36) assumes the more readable form

$$\mathbf{U}_{\mathrm{FT}}|x_3\rangle|x_2\rangle|x_1\rangle|x_0\rangle = \mathbf{H}_3\big(\mathbf{V}_{32}\mathbf{H}_2\big)\big(\mathbf{V}_{31}\mathbf{V}_{21}\mathbf{H}_1\big)\big(\mathbf{V}_{30}\mathbf{V}_{20}\mathbf{V}_{10}\mathbf{H}_0\big)|x_0\rangle|x_1\rangle|x_2\rangle|x_3\rangle. \quad (3.38)$$

I have put in unnecessary parentheses to guide the eye to the simple structure, whose generalization to more than four Qbits is, as promised, obvious.

If we define the unitary operator $\mathbf{P}$ to bring about the permutation of computational basis states

$$\mathbf{P}|x_3\rangle|x_2\rangle|x_1\rangle|x_0\rangle = |x_0\rangle|x_1\rangle|x_2\rangle|x_3\rangle, \quad (3.39)$$

then (3.38) becomes

$$\mathbf{U}_{\mathrm{FT}}|x_3\rangle|x_2\rangle|x_1\rangle|x_0\rangle = \mathbf{H}_3\big(\mathbf{V}_{32}\mathbf{H}_2\big)\big(\mathbf{V}_{31}\mathbf{V}_{21}\mathbf{H}_1\big)\big(\mathbf{V}_{30}\mathbf{V}_{20}\mathbf{V}_{10}\mathbf{H}_0\big)\mathbf{P}|x_3\rangle|x_2\rangle|x_1\rangle|x_0\rangle. \quad (3.40)$$

Since (3.40) holds for all computational-basis states it holds for arbitrary states and is therefore equivalent to the operator identity

$$\mathbf{U}_{\mathrm{FT}} = \mathbf{H}_3(\mathbf{V}_{32}\mathbf{H}_2)(\mathbf{V}_{31}\mathbf{V}_{21}\mathbf{H}_1)(\mathbf{V}_{30}\mathbf{V}_{20}\mathbf{V}_{10}\mathbf{H}_0)\mathbf{P}. \quad (3.41)$$

The form (3.41) expresses $\mathbf{U}_{\mathrm{FT}}$ as a product of unitary operators, thereby independently establishing what we have already noted directly from its definition, that $\mathbf{U}_{\mathrm{FT}}$ is unitary. More importantly it gives an explicit construction of $\mathbf{U}_{\mathrm{FT}}$ entirely out of *one-* and *two-*Qbit unitary gates, whose number grows only quadratically with the number n of Qbits. (The permutation $\mathbf{P}$ can be constructed out of cNOT gates and one additional Qbit, initially in the state $|0\rangle$ – an instructive exercise to think about – but in the application that follows it is much easier to build directly into the circuitry the rearranging of Qbits accomplished by $\mathbf{P}$.)

The permutation operator $\mathbf{P}$ plays an important role in establishing that the circuit (3.41) that produces the quantum Fourier transform $\mathbf{U}_{\mathrm{FT}}$ has an inverse $\mathbf{U}_{\mathrm{FT}}^\dagger$ possessing the structure one expects for an inverse Fourier transform. Since the adjoint of a product is the product of the adjoints in the opposite order, and since Hadamards and $\mathbf{P}$ are self-adjoint, we have from (3.41)

$$\mathbf{U}_{\mathrm{FT}}^\dagger = \mathbf{P}(\mathbf{H}_0\mathbf{V}_{10}^\dagger\mathbf{V}_{20}^\dagger\mathbf{V}_{30}^\dagger)(\mathbf{H}_1\mathbf{V}_{21}^\dagger\mathbf{V}_{31}^\dagger)(\mathbf{H}_2\mathbf{V}_{32}^\dagger)\mathbf{H}_3. \quad (3.42)$$

One can insert $\mathbf{1} = \mathbf{PP}$ on the extreme right of (3.42) and then note that the effect of sandwiching all the Hadamards and 1-Qbit unitaries between two $\mathbf{P}$s is simply to alter all their indices by the permutation taking $0123 \to 3210$. Therefore

$$\mathbf{U}_{\mathrm{FT}}^\dagger = (\mathbf{H}_3\mathbf{V}_{23}^\dagger\mathbf{V}_{13}^\dagger\mathbf{V}_{03}^\dagger)(\mathbf{H}_2\mathbf{V}_{12}^\dagger\mathbf{V}_{02}^\dagger)(\mathbf{H}_1\mathbf{V}_{01}^\dagger)\mathbf{H}_0\mathbf{P}. \quad (3.43)$$

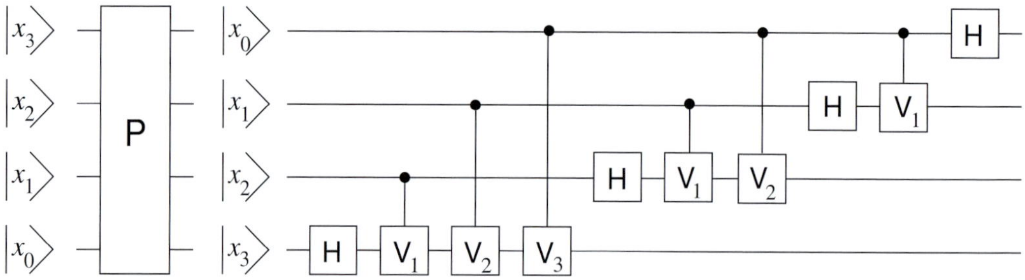

Fig 3.1 A diagram of a circuit that illustrates, for four Qbits, the construction of the quantum Fourier transform $\mathbf{U}_{\mathrm{FT}}$ defined in (3.18) as the product of 1- and 2-Qbit gates given in (3.40).

If we now move every $\mathbf{V}^\dagger$ to the right past as many Hadamards as we can, keeping in mind that each $\mathbf{V}$ commutes with all Hadamards except those sharing either of its indices, then we have

$$\mathbf{U}_{\mathrm{FT}}^\dagger = (\mathbf{H}_3\mathbf{V}_{23}^\dagger)(\mathbf{H}_2\mathbf{V}_{13}^\dagger\mathbf{V}_{12}^\dagger)(\mathbf{H}_1\mathbf{V}_{03}^\dagger\mathbf{V}_{02}^\dagger\mathbf{V}_{01}^\dagger)\mathbf{H}_0\mathbf{P}. \tag{3.44}$$

Finally, if we note from (3.37) that each $\mathbf{V}$ is symmetric in its indices, and rearrange the parentheses in (3.44) to make easier the comparison with the form (3.41) of $\mathbf{U}_{\mathrm{FT}}$, we have

$$\mathbf{U}_{\mathrm{FT}}^\dagger = \mathbf{H}_3(\mathbf{V}_{32}^\dagger\mathbf{H}_2)(\mathbf{V}_{31}^\dagger\mathbf{V}_{21}^\dagger\mathbf{H}_1)(\mathbf{V}_{30}^\dagger\mathbf{V}_{20}^\dagger\mathbf{V}_{10}^\dagger\mathbf{H}_0)\mathbf{P}. \tag{3.45}$$

This is precisely the form (3.41) of $\mathbf{U}_{\mathrm{FT}}$ itself, except that each $\mathbf{V}$ is replaced by its adjoint, which (3.37) shows amounts to replacing each i by $-i$ in the arguments of all the phase factors. This is exactly what one does to invert the ordinary functional Fourier transform.

3.6 Eliminating the 2-Qbit gates

A circuit diagram that compactly expresses the content of (3.40) is shown in Figure 3.1. As is always the case in such diagrams, the order in which the gates act is from left to right, although in the equation (3.40) that the diagram represents, the order in which the gates act is from right to left. The diagram introduces an artificial asymmetry into the 2-Qbit unitary gate $\mathbf{V}_{ij}$, treating one Qbit as a control bit, which determines whether or not the unitary operator $e^{i\pi\mathbf{n}/2^{|i-j|}}$ acts on the other Qbit, taken to be the target. Although this is the most common way of representing the circuit for the quantum Fourier transform, the figure could equally well have been drawn with the opposite convention, as in Figure 3.2.

Both Figure 3.1 and Figure 3.2 follow the usual convention, in which Qbits representing more significant bits are represented by lines higher in the figure. Acting on the computational basis, however, the first gate

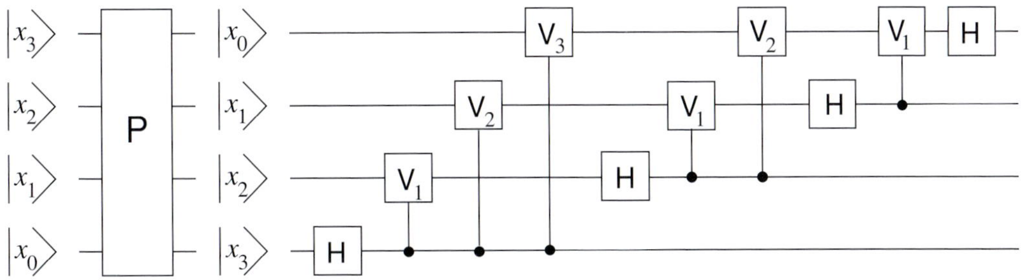

Fig 3.2 Since the action (3.37) of the controlled-V gates is symmetric in i and j, Figure 3.1 can be redrawn with control and target Qbits interchanged.

on the left, **P**, permutes the states of the Qbits, exchanging the states of the most and least significant Qbits, the states of the next most significant and next least significant Qbits, etc. Rather than introducing such a permutation gate, it makes more sense simply to reverse the convention for the input state, associating Qbits that represent more significant bits with lower lines in the figure. The gate **P** is then omitted, and the conventional ordering of significant bits is reversed for the input. The complete figure thus reduces to the portion to the right of the permutation gate **P**. For the output, of course, the conventional ordering remains in effect: Qbits on higher lines represent more significant bits.

If the input on the left of the complete Figure 3.1 or 3.2 (with the gate **P**) is the computational-basis state $|x\rangle_4 = |x_3\rangle|x_2\rangle|x_1\rangle|x_0\rangle$ the output on the right will be $\mathbf{U}_{\mathrm{FT}}|x\rangle_4$, the superposition (3.18) of computational-basis states $|y\rangle_4 = |y_3\rangle|y_2\rangle|y_1\rangle|y_0\rangle$, defined in (3.18).

There is no need for the figures to have subscripts on the Hadamard gates appearing in (3.40), since each is explicitly attached to the line associated with the Qbit on which it acts. For the same reason each 2-Qbit controlled-V gate requires only a single subscript, which specifies the unitary operator $\mathbf{V}_k$ that acts on the target Qbit when the computational-basis state of the control Qbit is $|1\rangle$; the subscript k is the number of "wires" the target Qbit is away from the control Qbit. The explicit form of $\mathbf{V}_k$ is $e^{i\pi\mathbf{n}/2^k}$, where $\mathbf{n}$ is the number operator for the target Qbit.

Figure 3.2 reveals a further simplification of great practical interest, if all the Qbits are measured as soon as the action of the quantum Fourier transformation is completed. This simplification, pointed out by Griffiths and Niu, allows the 2-Qbit controlled-V gates to be replaced by 1-Qbit gates that act or not, depending on the outcome of a prior measurement of the control Qbit, as shown in Figure 3.3. The simplification is made possible by the following general fact.

If a controlled operation $\mathbf{C}^U$, or a series of consecutive controlled operations all with the same control Qbit, is immediately followed by

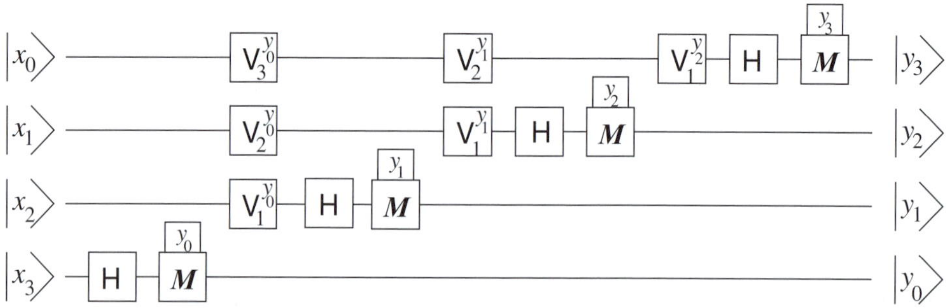

Fig 3.3 If the Qbits are all measured immediately after all the gates of the quantum Fourier transform have acted, then the 1-Qbit measurement gates can be applied to each Qbit immediately after the action of the Hadamard gate on that Qbit, and the controlled-V gates that follow the action of the Hadamards in Figure 3.2 can be replaced by 1-Qbit gates that act or not depending on whether the outcome y of the 1-Qbit measurement is 1 or 0.

a measurement of the control Qbit, then the possible final states of all the Qbits and the probabilities of those states are exactly the same as they would be if the measurement of the control Qbit took place before the application of the controlled operation, and then the target Qbit(s) were acted upon or not by $\mathbf{U}$, depending on whether the result of the prior measurement was 0 or 1. To confirm this, write an n-Qbit state as

$$|\Psi\rangle_n = \alpha_0|0\rangle_1|\Phi_0\rangle_{n-1} + \alpha_1|1\rangle_1|\Phi_1\rangle_{n-1}, \tag{3.46}$$

where the state of the control Qbit is on the left, the states $|\Phi_i\rangle$ are unit vectors, and the unitary operation $\mathbf{U}$ acts on some or all of the remaining $n-1$ Qbits. Applying the controlled-U operation $\mathbf{C}^U$ to Ψ_n gives

$$\mathbf{C}^U|\Psi\rangle_n = \alpha_0|0\rangle_1|\Phi_0\rangle_{n-1} + \alpha_1|1\rangle_1\mathbf{U}|\Phi_1\rangle_{n-1}. \tag{3.47}$$

If this is immediately followed by a measurement of the control Qbit, the post-measurement states and associated probabilities are

$$|0\rangle|\Phi_0\rangle,\ p = |\alpha_0|^2; \qquad |1\rangle\mathbf{U}|\Phi_1\rangle,\quad p = |\alpha_1|^2, \tag{3.48}$$

according to the generalized Born rule. On the other hand if we measure the control Qbit *before* applying the controlled-U, the resulting states and associated probabilities are

$$|0\rangle|\Phi_0\rangle,\ p = |\alpha_0|^2; \qquad |1\rangle|\Phi_1\rangle,\ p = |\alpha_1|^2, \tag{3.49}$$

so if we then apply $\mathbf{U}$ to the remaining $n-1$ Qbits if and only if the result of the earlier measurement was 1, we end up with exactly the same states and probabilities as in (3.48).

We shall see that if one's aim is to find the period of the function f, one can indeed measure each Qbit immediately after applying the quantum Fourier transform. So this replacement of controlled unitary gates by 1-Qbit unitary gates, which act or not depending on the outcome of the measurement, is of great importance from the technological point of view, 1-Qbit unitaries being far easier to implement than 2-Qbit controlled gates.

To see how the general procedure works in this particular case, consider first the bottom wire in Figure 3.2. Once $\mathbf{H}$ and the three controlled-V gates have acted on it, nothing further happens to that Qbit until its final measurement. If the result of that measurement is 1, the state of all four Qbits reduces to that component of the full superposition in which $\mathbf{V}_1$, $\mathbf{V}_2$, and $\mathbf{V}_3$ have acted on the three wires above the bottom wire; if the result of the measurement is 0, the 4-Qbit state reduces to the component in which they have not acted. We can produce exactly the same effect if we measure the least significant output Qbit immediately after $\mathbf{H}$ has acted on the bottom wire, before any of the other gates have acted, and then apply or do not apply the three unitary transformations to the other three Qbits, depending on whether the outcome of the measurement is 1 or 0. Next, we apply the Hadamard transformation to the second wire from the bottom. We then immediately measure that Qbit and, depending on the outcome, apply or do not apply the appropriate 1-Qbit unitary transformations to each of the remaining two Qbits. Continuing in this way, we end up producing exactly the same statistical distribution of measurement results as we would have produced had we used the 2-Qbit controlled-V gates, measuring none of the Qbits until the full unitary transformation $\mathbf{U}_{\mathrm{FT}}$ had been produced. Thus Figure 3.2, followed by measurements of all four Qbits on the right yielding the values y_3, y_2, y_1, and y_0, is equivalent to Figure 3.3.

The most attractive (but least common) way of representing the quantum Fourier transform with a circuit diagram is shown in Figure 3.4.[7] In this form the inversion in order from most to least significant Qbits between the input and the output is shown by bending the Qbit lines, rather than by inverting the order in the state symbols. The 2-Qbit gates $\mathbf{V}$ are also displayed in a symmetric way that does not suggest an artificial distinction between control and target Qbits.

3.7 Finding the period

The period r of f appears in the state (3.16) of the input-register Qbits produced from a single application of $\mathbf{U}_f$. To get valuable information

7 The figure is based on one drawn by Robert B. Griffiths and Chi-Sheng Niu in their paper setting forth the Griffiths–Niu trick, "Semiclassical Fourier transform for quantum computation," *Physical Review Letters* **76**, 3228–3231 (1996) (`http://arxiv.org/abs/quant-ph/9511007`).

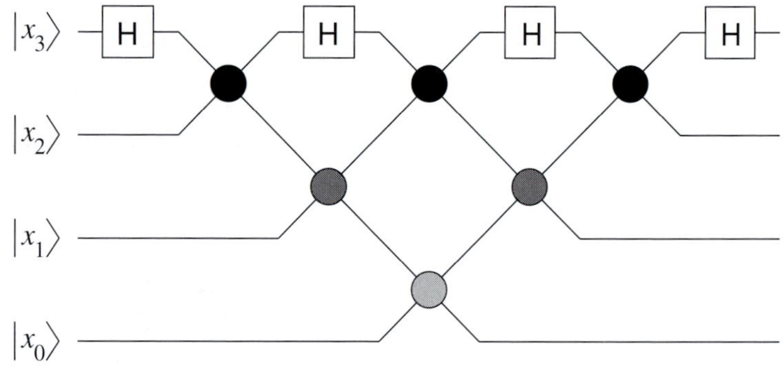

Fig 3.4 A more symmetric way of drawing Figure 3.1 or 3.2, due to Griffiths and Niu. Although it is superior to the conventional diagram, it does not seem to have caught on. The permutation **P** that in effect permutes the Qbits in the input register is now built into the diagram by using lines that no longer connect input-register Qbits to output-register Qbits at the same horizontal level. Because the lines now cross one another, the unitary operators **V** can be represented by the circles at the intersections of the lines associated with the Qbits that they couple, eliminating the artificial distinction between control and target Qbits used in Figures 3.1 and 3.2. The form of each such operator is $\mathbf{V} = \exp(i\pi \mathbf{n}\mathbf{n}'/2^k)$, where $\mathbf{n}$ and $\mathbf{n}'$ are the Qbit number operators associated with the two lines that cross at the dot, and $k = 1, 2$, or 3 depending on whether the dot lies in the first, second, or third horizontal row below the top row of Hadamard transformations. The larger the phase produced by **V**, the blacker the circle.

about r we apply the quantum Fourier transformation (3.18) to the input register:

$$\begin{aligned}\mathbf{U}_{\mathrm{FT}} \frac{1}{\sqrt{m}} \sum_{k=0}^{m-1} |x_0 + kr\rangle &= \frac{1}{2^{n/2}} \sum_{y=0}^{2^n-1} \frac{1}{\sqrt{m}} \sum_{k=0}^{m-1} e^{2\pi i (x_0+kr)y/2^n} |y\rangle \\ &= \sum_{y=0}^{2^n-1} e^{2\pi i x_0 y/2^n} \frac{1}{\sqrt{2^n m}} \left(\sum_{k=0}^{m-1} e^{2\pi i k r y/2^n} \right) |y\rangle .\end{aligned} \tag{3.50}$$

If we now make a measurement, the probability $p(y)$ of getting the result y is just the squared magnitude of the coefficient of $|y\rangle$ in (3.50). The factor $e^{2\pi i x_0 y/2^n}$, in which the formerly troublesome x_0 explicitly occurs, drops out of this probability[8] and we are left with

$$p(y) = \frac{1}{2^n m} \left| \sum_{k=0}^{m-1} e^{2\pi i k r y/2^n} \right|^2 . \tag{3.51}$$

This completes the quantum-computational part of the process, except that, as noted below, we may have to repeat the procedure a small number of times (of order ten or so) to achieve a high probability of learning the period r. To see why the form (3.51) of $p(y)$ makes this possible, we require some further purely mathematical analysis, that at a certain point will exploit yet another branch of elementary number theory.

The probability (3.51) is a simple explicit function of the integer y, whose magnitude has maxima when y is close[9] to integral multiples of

8 The random value of $x_0 < r$ also determines whether m is given by rounding the enormous value of $2^n/r$ up or down to the nearest integer – see Equation (3.17) and the surrounding text – but this makes a negligible difference in what follows.

9 Such sums of phase factors are familiar to physicists (to whom this cautionary footnote is addressed), particularly in the context of time-dependent perturbation theory, where one approximates them in terms of Dirac delta-functions concentrated in the maximum values. The analysis

$2^n/r$. In fact we now show that the probability is at least 0.4 that the measured value of y will be as close as possible to – i.e. within $\frac{1}{2}$ of – an integral multiple of $2^n/r$. To see this we calculate a lower bound for $p(y)$ when

$$y = y_j = j\,2^n/r + \delta_j, \tag{3.52}$$

with $|\delta_j| \leq \frac{1}{2}$. Only the term in δ_j contributes to the exponentials in (3.51). The summation is a geometric series, which can be explicitly summed to give

$$p(y_j) = \frac{1}{2^n m}\frac{\sin^2(\pi\delta_j mr/2^n)}{\sin^2(\pi\delta_j r/2^n)}. \tag{3.53}$$

Since (3.17) tells us that m is within an integer of $2^n/r$, and since $2^n/r \geq N^2/r > N$, we can with negligible error replace $mr/2^n$ by 1 in the numerator of (3.53), and replace the sine in the denominator by its (extremely small) argument. This gives

$$p(y_j) = \frac{1}{2^n m}\left(\frac{\sin(\pi\delta_j)}{\pi\delta_j r/2^n}\right)^2 = \frac{1}{r}\left(\frac{\sin(\pi\delta_j)}{\pi\delta_j}\right)^2. \tag{3.54}$$

When x is between 0 and $\pi/2$, the graph of $\sin x$ lies above the straight line connecting the origin to the maximum at $x = \pi/2$:

$$x\big/\left(\tfrac{1}{2}\pi\right) \leq \sin x, \quad 0 \leq x \leq \pi/2. \tag{3.55}$$

Since $\delta_j \leq \frac{1}{2}$ the probability (3.54) is bounded below by

$$p(y_j) \geq (4/\pi^2)/r. \tag{3.56}$$

Since there are at least $r - 1$ different values of j, and since r is a large number,[10] one has at least a 40% chance ($4/\pi^2 = 0.4053$) of getting one of the special values (3.52) for y – a value that is within $\frac{1}{2}$ of an integral multiple of $2^n/r$.

Note, in passing, that as $\delta_j \to 0$ in (3.54) the probability $p(y_j)$ becomes $1/r$, so that if all the δ_j are 0 – i.e. if the period r is exactly a power of 2 – then the probability of measuring an integral multiple of $2^n/r$ is essentially 1. Indeed, you can easily check that in this (highly unlikely) case the probability remains 1 even if we do not double the number of Qbits in the input register and take $n = n_0$. Thus the case $r = 2^j$ avoids some of the major complications of quantum period

required here is different in two important ways. Because we need to know the enormous integer r exactly we must pay much more careful attention to just how much of the probability is concentrated in those special values of y, and we must also solve the subtle problem of how to get from such maximum values to the precise period r itself.

10 One can easily test with a classical computer all values of r less than, say, 100, to see whether they are periods of f; one need resort to the quantum computation only if r itself is enormous.

finding. Since r divides $(p-1)(q-1)$, all periods modulo pq must be powers of 2 if p and q are both primes of the form 2^n+1. The smallest such primes are 3, 5, 17, and 257. Hence claims to have realized the Shor algorithm for factoring 15 are to be taken *cum grano salis*, as should possible future claims based on factoring 51, 85, and 771.

Note also that the derivation of (3.56) requires only that the argument of the sine in the denominator of (3.53) be small. This will be the case if 2^n is any large multiple of N – i.e. if the input register is large enough to contain many periods of b^x (mod N). The stronger requirement that 2^n should be as large as N^2 – that the input register should actually be able to accomodate at least N full periods – emerges when we examine whether it is possible to learn r itself, given an integral multiple of $2^n/r$.

Suppose that we have found a y that is within $\frac{1}{2}$ of $j2^n/r$ for some integer j. It follows that

$$\left|\frac{y}{2^n}-\frac{j}{r}\right| \le \frac{1}{2^{n+1}}. \tag{3.57}$$

Since y is the result of our measurement and we know n, the number of input-register Qbits, we have an estimate for the fraction j/r. It is here that our use of an n-Qbit input register with $2^n > N^2$ is crucial. By using twice as many Qbits as needed to represent all the integers up to N, we have ensured that our estimate (3.57) of j/r is off by no more than $1/(2N^2)$. But since $r < N$, and since any two distinct fractions with denominators less than N must differ[11] by at least $1/N^2$, the measured value of y and the fact that r is less than N is enough to determine a unique value of the rational number j/r.

That value of j/r can be efficiently extracted from the known value of $y/2^n$ by an application of the theory of continued fractions. This exploits the theorem that if x is an estimate for j/r that differs from it by less than $1/(2r^2)$, then j/r will appear as one of the partial sums in the continued-fraction expansion of x. The application of the theorem in this context is illustrated in Appendix K. The continued-fraction expansion of $y/2^n$ gives us not j and r separately, but the fraction j/r reduced to lowest terms – i.e. it gives us integers j_0 and r_0 with no common factors that satisfy $j_0/r_0 = j/r$. The r_0 we learn is thus a divisor of r.

Since r is r_0 times the factors j has in common with r, if we were lucky enough to get a j that is coprime to r, then $r_0 = r$. Since, as shown in Appendix J, two random numbers j and r have a better than even chance of having no common factors, we do not have to be terribly lucky.

11 For

$$\left|\frac{a}{b}-\frac{c}{d}\right| \ge \frac{1}{bd}$$

unless the two fractions are identical.

We can easily check to see whether r_0 itself is the period r by computing (with a classical computer) b^{r_0} (mod N) and seeing whether or not it is b. If it is not, we can try several low multiples, $2r_0, 3r_0, 4r_0, \ldots$, since it is unlikely that j will share a large factor with r.

If this fails, we can repeat the entire quantum computation from the beginning. We now get j'/r, where j' is another (random) integer, yielding another divisor r_0' of r, which is r divided by the factors it has in common with j'. If j and j' have no factors in common – which has a better than even chance of happening – then r will be the least common multiple[12] of its two divisors r_0 and r_0'. We can again test to see whether we have the right r by evaluating b^r (mod N) to see whether it is indeed equal to b. If it is not, we can again try some of the lower multiples of our candidate for r and, if necessary, go through the whole business one more time to get yet another random multiple of $1/r$.

Because we are not certain that our measurement gives us one of the y_j and thus a divisor of r, we may have to repeat the whole procedure several (but not a great many) times before succeeding, carrying out some not terribly taxing mathematical detective work, with the aid of a classical computer, to find the period r. The detective work is greatly simplified by the fact (established in Appendix L) that when N is the product of two primes, the period r is not only less than N, but also less than $\frac{1}{2}N$. As a result, a more extended analysis shows that the probability of learning a divisor of r from the measured value of y is bounded from below not just by 0.4, but by more than 0.9. Furthermore, by adding just a small number q of additional Qbits to the input register, so that n exceeds $2n_0 + q$, the probability of learning a divisor of r in a single run can be made quite close to 1. These refinements are described in Appendix L.

3.8 Calculating the periodic function

We have assumed the existence of an efficient subroutine that calculates b^x (mod N). You might think that calculating $f(x) = b^x$ (mod N) for arbitrary values of x less than, say, $2^n = 10^{800}$ would require astronomical numbers of multiplications, but it does not. We simply square b (mod N), square the result (mod N), square that, etc., calculating the comparatively small number of powers b^{2^j} (mod N) with $j < n$. The binary expansion of $x = x_{n-1}x_{n-2}\ldots x_1x_0$ tells us which of these must be multiplied together to get $b^x = \prod_j (b^{2^j})^{x_j}$.

So if we start with x in the input register, 1 (i.e. $000\ldots001$) in the output register, and b in an additional work register, then we can proceed as follows:

12 The least common multiple of two numbers is their product divided by their greatest common divisor; the greatest common divisor can be found with the Euclidean algorithm, as shown in Appendix J.

(a) multiply the ouput register by the work register if and only if $x_0 = 1$;
(b) replace the contents of the work register by its modulo-N square;
(a′) repeat (a) with the multiplication now conditional on $x_1 = 1$;
(b′) repeat (b);
(a″) repeat (a) with the multiplication now conditional on $x_2 = 1$; etc.

At the end of this process we will still have x in the input register (which serves only as a set of control bits for the n controlled multiplications), and we will have b^x (mod N) in the output register. The work register will contain b^{2^n} whatever the value of x in the input register, and it will therefore be unentangled with the input and output registers and can be ignored when we take our starting point to be a superposition of classical inputs.[13]

Note the striking difference between classical and quantum programming styles. One's classical computational instincts would direct one to make a look-up table of all n modulo-N multiple squares of b, since (a) Cbits are cheap and stable and (b) otherwise to get b^x (mod N) for all the needed values of x one would have to recalculate the successive squares so many times that this would become ridiculously inefficient. But the situation is quite the opposite with a quantum computer, since (a) Qbits are expensive and fragile and (b) "quantum parallelism" makes it possible to produce the state (3.15) with only a single execution of the procedure that does the successive squarings, thereby relieving us of any need to store all the modulo-N squares, at a substantial saving in Qbits.

As usual with quantum parallelism, there is the major catch that an immediate measurement of Qbits in the state (3.15) can reveal only the value of a single (random) one of the modulo-N powers of b. But by applying $\mathbf{U}_{\mathrm{FT}}$ to the input register of the state (3.15) and only then making the measurement, we can get important collective information about the modulo-N values of b^x – in this case a divisor of the crucial period r – at the (unimportant) price of losing all information about the individual values of b^x.

3.9 The unimportance of small phase errors

To execute the quantum Fourier transform one needs 2-Qbit gates $\mathbf{V}_{ij} = e^{i\pi \mathbf{n}_i \mathbf{n}_j / 2^{|i-j|}}$ or, if one exploits the Griffiths–Niu trick, 1-Qbit gates $\mathbf{V}_j = e^{i\pi \mathbf{n}_j / 2^j}$. Since we need to deal with numbers of many hundreds of digits, the 2^j appearing in these phase gates can be larger than 10^{100}. Producing such tiny phase shifts requires a degree of control over the gates that is impossible to achieve. Typically such phase-shift

13 As noted in Chapter 2, any additional registers used in the squaring and multiplication subroutines must also be restored to their initial states to insure that they are also disentangled from the input and output registers.

gates would allow two Qbits to interact in a carefully controlled way for an interval of time that was specified very precisely, but obviously not to hundreds of significant figures. It is therefore crucial that the effectiveness of the period-finding algorithm not be greatly affected by small errors in the phase shifts.

On the face of it this seems worrisome. Since we need to know the period r to hundreds of digits, don't we have to get the phase shifts right to a comparable precision? Here the fundamentally digital character of the actual output of a quantum computation saves the day. To learn r we require the outcomes of several hundreds of 1-Qbit measurements, each of which has just two possible outcomes (0 or 1). The action of the unitary gates that precede the measurements is like that of an analog computer, involving continuously variable phase shifts that cannot be controlled with perfect precision. But this analog evolution affects only the *probabilities* of the sharply defined digital outputs. Small alterations in the phases produce small alterations in the probabilities of getting that extremely precise digital information, but not the precision of the information itself, once it is acquired.[14]

Suppose that the phase of each term in the quantum Fourier transform (3.18) is incorrect by an amount $\varphi(x, y)$, and that each of these altered phases is bounded in magnitude by $\varphi \ll 1$. The probability $p(y)$ in (3.51) will be changed to

$$p_\varphi(y) = \frac{1}{2^n m}\left|\sum_{k=0}^{m-1} e^{2\pi i k r y/2^n} e^{i\varphi_k(y)}\right|^2, \tag{3.58}$$

where $\varphi_k(y) = \varphi(x_0 + kr, y)$. Since all the phases $\varphi_k(y)$ are small compared with unity,

$$e^{i\varphi_k(y)} \approx 1 + i\varphi_k(y), \tag{3.59}$$

and therefore

$$p_\varphi(y) \approx \frac{1}{2^n m}\left|\sum_{k=0}^{m-1} e^{2\pi i k r y/2^n}\big(1 + i\varphi_k(y)\big)\right|^2. \tag{3.60}$$

What effect does this have on the probability of learning from the measurement one of the special values y^j given in (3.52)?

We have

$$p_\varphi(y_j) \approx \frac{1}{2^n m}\left|\sum_{k=0}^{m-1} e^{2\pi i k r \delta_j/2^n}\big(1 + i\varphi_{jk}\big)\right|^2, \tag{3.61}$$

14 For a long time this crucial point seems to have been discussed only in an unpublished internal IBM report by D. Coppersmith. In 2002 that 1994 report finally appeared: D. Coppersmith, "An approximate Fourier transform useful in quantum factoring," `http://arxiv.org/abs/quant-ph/0201067`.

where $\varphi_{jk} = \varphi_k(y_j)$. If we expand to linear order in the small quantities φ_{jk}, we get

$$p_\varphi(y_j) \approx p(y_j) + \frac{2}{2^n m}\,\mathrm{Im}\left[\left(\sum_{k=0}^{m-1} e^{-2\pi i k r \delta_j/2^n}\varphi_{jk}\right)\left(\sum_{k'=0}^{m-1} e^{2\pi i k' r \delta_j/2^n}\right)\right]. \tag{3.62}$$

We can get an upper bound on the magnitude of the difference between the exact and approximate probabilities by replacing the imaginary part of the product of the two sums by the product of the absolute values of the sums, and then replacing each term in each sum by its absolute value. Since the absolute value of each φ_{jk} is bounded by φ, we can conclude that

$$|p(y_j) - p_\varphi(y_j)| \le \frac{2m}{2^n}\varphi = \frac{2}{r}\varphi. \tag{3.63}$$

Since there are r different y_j, the probability of getting one of the special values y_j is altered by less than 2φ. So if one is willing to settle for a probability of getting a special value that is at worst 1% less than the ideal value of about 0.4, then one can tolerate phase errors up to $\varphi = 0.4/200 = 1/500$. If one leaves out of the quantum Fourier transform circuit all controlled-phase gates $e^{\pi i \mathbf{n}_i \mathbf{n}_j/2^{|i-j|}}$ with $|i - j| > \ell$, the maximum phase error φ this can produce in any term is $\varphi = n\pi/2^\ell$, and therefore the probability will be within 1% of its ideal value if $1/2^\ell < 1/(500n\pi)$.

The number n of Qbits in the input register might be as large as 3000 for problems of interest (factoring a 500-digit N). Consequently for all practical purposes one can omit from the quantum Fourier transform all controlled-phase gates connecting Qbits that are more than about $\ell = 22$ wires apart in the circuit diagram. This has two major advantages. Of crucial importance, quantum engineers will not have to produce impossibly precise phase changes. Furthermore, the size of the circuit executing the quantum Fourier transform has to grow only linearly with large n rather than quadratically. Since n is likely to be of order 10^3 for practical code breaking, this too is a significant improvement.

3.10 Period finding and factoring

Since Shor's period-finding quantum algorithm is always described as a factoring algorithm, we conclude this chapter by noting how period finding leads to factoring. We consider only the case relevant to RSA encryption, where one wants to factor the product of two large primes, $N = pq$, although the connection between period finding and factoring is more general.

If we have a way to determine periods (such as Shor's algorithm) and want to find the large prime factors of $N = pq$, we pick a random number a coprime to N. The odds that a random a happens to be a multiple of p or of q are minuscule when p and q are enormous, but if you are the worrying kind you can check that it isn't, using the Euclidean algorithm. (In the overwhelmingly unlikely event that a is a multiple of p or q then the Euclidean algorithm applied to a and N will give you p or q directly, and you will have factored N.) Using our period-finding routine, we find the order of a in G_{pq}: the smallest r for which

$$a^r \equiv 1 \ (\mathrm{mod}\ pq). \tag{3.64}$$

We can use this information to factor N if our choice of a was lucky in two ways.

Suppose first that we are fortunate enough to get an r that is even. We can then calculate

$$x = a^{r/2} \ (\mathrm{mod}\ pq) \tag{3.65}$$

and note that

$$0 \equiv x^2 - 1 \equiv (x-1)(x+1) \quad (\mathrm{mod}\ pq). \tag{3.66}$$

Now $x - 1 = a^{r/2} - 1$ is not congruent to 0 modulo pq, since r is the *smallest* power of a congruent to 1. Suppose in addition – our second piece of good fortune – that

$$x + 1 = a^{r/2} + 1 \not\equiv 0 \ (\mathrm{mod}\ pq). \tag{3.67}$$

In that case neither $x - 1$ nor $x + 1$ is divisible by $N = pq$, but (3.66) tells us that their product is. Since p and q are prime this is possible only if one of them, say p, divides $x - 1$ and the other, q, divides $x + 1$. Because the only divisors of N are p and q, it follows that p is the greatest common divisor of N and $x - 1$, while q is the greatest common divisor of N and $x + 1$. We can therefore find p or q by a straightforward application of the Euclidean algorithm.

So it all comes down to the likelihood of our being lucky. We show in Appendix M that the probability is at least 0.5 that a random number a in G_{pq} has an order r that is even with $a^{r/2} \not\equiv -1 \ (\mathrm{mod}\ pq)$. So we do not have to repeat the procedure an enormous number of times to achieve a very high probability of success. If you're willing to accept the fact that you don't have to try out very many random numbers a in order to succeed, then this elementary argument is all you need to know about why period finding enables you to factor $N = pq$. But if you're curious about why the probability of good fortune is so high, then you must contend with Appendix M, where I have constructed an elementary but rather elaborate argument, by condensing a fairly large body of number-theoretic lore into the comparatively simple form it assumes when applied to the special case in which the number N is the product of two primes.

Chapter 4

Searching with a quantum computer

4.1 The nature of the search

Suppose you know that exactly one n-bit integer satisfies a certain condition, and suppose you have a black-boxed subroutine that acts on the $N = 2^n$ different n-bit integers, outputting 1 if the integer satisfies the condition and 0 otherwise. In the absence of any other information, to find the special integer you can do no better with a classical computer than to apply the subroutine repeatedly to different random numbers until you hit on the special one. If you apply it to M different integers the probability of your finding the special number is M/N. You must test $\frac{1}{2}N$ different integers to have a 50% chance of success.

If, however, you have a quantum computer with a subroutine that performs such a test, then you can find the special integer with a probability that is very close to 1 when N is large, using a method that calls the subroutine a number of times no greater than $(\pi/4)\sqrt{N}$.

This very general capability of quantum computers was discovered by Lov Grover, and goes under the name of *Grover's search algorithm*. Shor's period-finding algorithm and Grover's search algorithm, together with their various modifications and extensions, constitute the two masterpieces of quantum-computational software.

One can think of Grover's black-boxed subroutine in various ways. The subroutine might perform a mathematical calculation to determine whether the input integer is the special one. Here is a simple example. If an odd number p can be expressed as the sum of two squares, $m^2 + n^2$, then since one of m or n must be even and the other odd, p must be of the form $4k + 1$. It is a fairly elementary theorem of number theory that if p is a *prime* number of the form $4k + 1$ then it can always be expressed as the sum of two squares, and in exactly one way. (Thus $5 = 4 + 1$, $13 = 9 + 4$, $17 = 16 + 1$, $29 = 25 + 4$, $37 = 36 + 1$, $41 = 25 + 16$, $53 = 49 + 4$, $61 = 36 + 25$, etc.) Given any such prime p, the simple-minded way to find the two squares is to take randomly selected integers x with $1 \le x \le N$, with N the largest integer less than $\sqrt{p/2}$, until you find the one for which $\sqrt{p - x^2}$ is an integer a. If p is of the order of a trillion, then following the simple-minded procedure you would have to calculate $\sqrt{p - x^2}$ for nearly a million x to have a better than even chance of succeeding. But using Grover's procedure with an appropriately programmed quantum computer you could succeed with a probability of success extremely close to 1 by calling the quantum subroutine that evaluated $\sqrt{p - x^2}$ fewer than a thousand times.

Mathematically well-informed friends tell me that for this particular example there are ways to proceed with a classical computer that are much more efficient than random testing, but the quantum algorithm to be described below enables even mathematical ignoramuses, equipped with a quantum computer, to do better than random testing by a factor of $1/\sqrt{N}$. And Grover's algorithm will provide this speed-up on arbitrary problems.

Alternatively, the black box could contain Qbits that have been loaded with a body of data – for example alphabetically ordered names and phone numbers – and one might be looking for the name that went with a particular phone number. It is with this kind of application in mind that Grover's neat trick has been called searching a database. Using as precious a resource as Qbits, however, merely to store classical information would be insanely extravagant, given our current or even our currently foreseeable ability to manufacture Qbits. Finding a unique solution – or one of a small number of solutions, as described in Section 4.3 – to a tough mathematical puzzle seems a more promising application.

4.2 The Grover iteration

Grover's algorithm assumes that we have been given a quantum search subroutine that indicates, when presented with any n-bit integer x, whether or not x is the special a being sought, returning this information as the value of a function $f(x)$ satisfying

$$f(x) = 0, \quad x \neq a; \qquad f(x) = 1, \quad x = a. \tag{4.1}$$

Grover discovered a completely general way to do significantly better than the classical method of merely letting the subroutine operate on different numbers from the list of 2^n candidates until it produces the output 1. The quantum-computational speed-up relies on the usual implementation of the subroutine that calculates f, in the form of a unitary transformation $\mathbf{U}_f$ that acts on an n-Qbit input register that contains x and a 1-Qbit output register that is or is not flipped from 0 to 1, depending on whether x is or is not the special number a:

$$\mathbf{U}_f\big(|x\rangle_n|y\rangle_1\big) = |x\rangle_n|y \oplus f(x)\rangle_1. \tag{4.2}$$

An example of a simple circuit that has precisely this action is shown in Figure 4.1. The figure can be viewed as providing a minimalist version of Grover's algorithm, reminiscent of the Bernstein–Vazirani problem (Section 2.4), though not susceptible to the special trick that worked in that simpler case. In this minimalist example we are given a black box containing the circuit depicted in Figure 4.1, but are not told

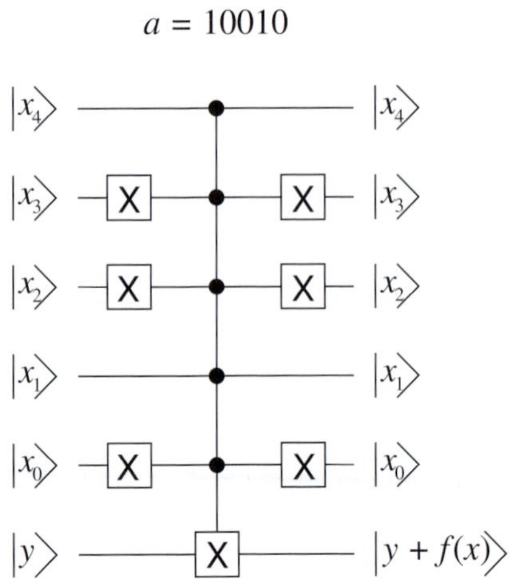

Fig 4.1 A possible realization of a black box that executes the unitary transformation $\mathbf{U}_f(|x\rangle_n|y\rangle_1) = |x\rangle_n|y \oplus f(x)\rangle_1$, where $f(x) = 0,\ x \neq a$; $f(x) = 1,\ x = a$. The input register has $n = 5$ Qbits and the special number a is 10010. The 6-Qbit gate in the center of the figure is a five-fold-controlled-NOT, which acts on the computational basis to flip the target bit if and only if every one of the five control bits is in the state $|1\rangle$. The construction of such a gate out of more elementary gates is shown in Figures 4.4–4.7.

which of the n control Qbits are acted on by NOT gates – information specified by the unknown n-bit integer a. If there were n Qbits in the input register and the computer were classical, we could do no better than to try each of the $N = 2^n$ possible inputs until we found the one for which the output register was flipped. But using Grover's algorithm we can determine this information with probability quite close to 1, by invoking the search subroutine no more than $\sqrt{N} = 2^{n/2}$ times – more precisely $(\pi/4)\sqrt{N}$ times – when N is large.

As in the Bernstein–Vazirani problem, it is useful to alter the flip of the state of the output register into an overall sign change, by transforming the 1-Qbit output register into the state

$$\mathbf{H}|1\rangle = \tfrac{1}{\sqrt{2}}(|0\rangle - |1\rangle) \tag{4.3}$$

prior to the application of $\mathbf{U}_f$. The action of $\mathbf{U}_f$ is then to multiply the $(n+1)$-Qbit state by -1 if and only if $x = a$:

$$\mathbf{U}_f(|x\rangle \otimes \mathbf{H}|1\rangle) = (-1)^{f(x)}|x\rangle \otimes \mathbf{H}|1\rangle. \tag{4.4}$$

In this form, the effect of $\mathbf{U}_f$ on the states $|x\rangle \otimes \mathbf{H}|1\rangle$ is exactly the same as doing nothing at all to the 1-Qbit output register, while acting on the n-Qbit input register with an n-Qbit unitary transformation $\mathbf{V}$ that acts on the computational basis as follows:

$$\mathbf{V}|x\rangle = (-1)^{f(x)}|x\rangle = \begin{cases} |x\rangle, & x \neq a, \\ -|a\rangle, & x = a. \end{cases} \tag{4.5}$$

Since $\mathbf{U}_f$ is linear, so is $\mathbf{V}$. Acting on a general superposition $|\Psi\rangle = \sum_x |x\rangle\langle x|\Psi\rangle$ of computational basis states, $\mathbf{V}$ changes the sign of the component of the state along $|a\rangle$, while leaving unchanged the component orthogonal to $|a\rangle$:

$$\mathbf{V}|\Psi\rangle = |\Psi\rangle - 2|a\rangle\langle a|\Psi\rangle. \tag{4.6}$$

So we can write $\mathbf{V}$ as

$$\mathbf{V} = \mathbf{1} - 2|a\rangle\langle a|, \tag{4.7}$$

where $|a\rangle\langle a|$ is the projection operator[1] on the state $|a\rangle$.

As we shall see, $\mathbf{U}_f$ is the only unitary transformation appearing in Grover's algorithm that acts as anything other than the identity on the output register. Because the output register starts in the state $\mathbf{H}|1\rangle$, unentangled with the input register, and because $\mathbf{U}_f$ maintains the output register in this particular state, the output register remains unentangled with the input register and in the state $\mathbf{H}|1\rangle$ throughout Grover's algorithm. We could continue to describe things in terms of $\mathbf{U}_f$ and retain the 1-Qbit output register, expanding (4.6), for example, to the form

$$\mathbf{U}_f(|\Psi\rangle \otimes \mathbf{H}|1\rangle) = [|\Psi\rangle - 2|a\rangle\langle a|\Psi\rangle] \otimes \mathbf{H}|1\rangle. \tag{4.8}$$

But it is simpler to suppress all explicit reference to the unaltered output register, which is always unentangled with the input register and always in the state $\mathbf{H}|1\rangle$. We simply replace the $(n+1)$-Qbit unitary $\mathbf{U}_f$ with the n-Qbit unitary $\mathbf{V}$ that acts on the n-Qbit input register, and define all other operators that appear in the algorithm only by their action on the input register, with the implicit understanding that they act as the identity on the output register.

To execute Grover's algorithm, we once again initially transform the n-Qbit input register into the uniform superposition of all possible inputs,

$$|\phi\rangle = \mathbf{H}^{\otimes n}|0\rangle_n = \frac{1}{2^{n/2}}\sum_{x=0}^{2^n-1}|x\rangle_n. \tag{4.9}$$

In addition to $\mathbf{V}$, Grover's algorithm requires a second n-Qbit unitary $\mathbf{W}$ that acts on the input register in a manner similar to $\mathbf{V}$, but with a fixed form that does not depend on a. The unitary transformation $\mathbf{W}$ preserves the component of any state along the standard state $|\phi\rangle$ given in (4.9), while changing the sign of its component orthogonal to $|\phi\rangle$:

$$\mathbf{W} = 2|\phi\rangle\langle\phi| - \mathbf{1}, \tag{4.10}$$

where $|\phi\rangle\langle\phi|$ is the projection operator on the state $|\phi\rangle$. We defer to Section 4.3 the not entirely obvious question of how to build $\mathbf{W}$ out of 1- and 2-Qbit unitary gates.

Given implementations of $\mathbf{V}$ and $\mathbf{W}$, Grover's algorithm is quite straightforward. It consists of simply applying many times the product $\mathbf{WV}$ to the input register, taken initially to be in the state $|\phi\rangle$. Each such application requires one invocation of the search subroutine.

1 This notation for projection operators is developed in Appendix A.

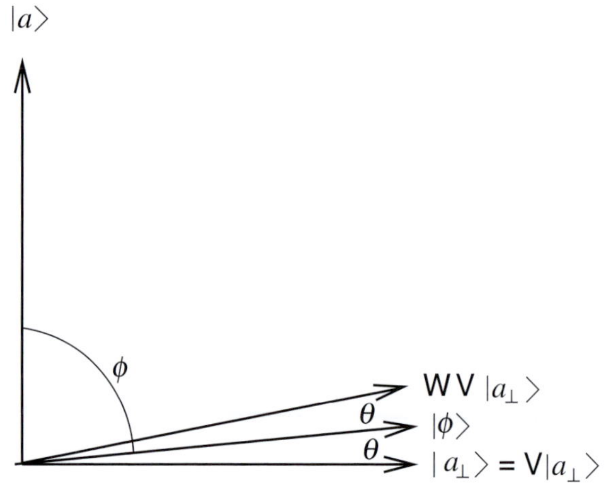

Fig 4.2 Real linear combinations of the special state $|a\rangle$, and the uniform superposition $|\phi\rangle = 2^{-n/2}\sum|x\rangle$, define a plane in which these two states are very nearly orthogonal. The state $|a_\perp\rangle$ in that plane is orthogonal to $|a\rangle$, and therefore makes a small angle θ with $|\phi\rangle$. The unitary transformation **V** takes any vector in the plane into its reflection in the line through the origin along $|a_\perp\rangle$, so it leaves $|a_\perp\rangle$ invariant. The unitary transformation **W** takes any vector in the plane into its reflection in the line through the origin along $|\phi\rangle$, so it rotates $|a_\perp\rangle$ counterclockwise through the angle 2θ. Therefore the combined operation **WV** rotates $|a_\perp\rangle$ counterclockwise through 2θ, and since **WV** is a rotation it does the same to any vector in the plane.

To see what is accomplished by repeatedly applying **WV** to the initial state $|\phi\rangle$, note that both **V** and **W** acting on either $|\phi\rangle$ or $|a\rangle$ give linear combinations of these two states. Since $\langle a|\phi\rangle = \langle\phi|a\rangle = 1/2^{n/2}$, whatever the value of a, the linear combinations have real coefficients and are given by

$$\begin{aligned} \mathbf{V}|a\rangle &= -|a\rangle, & \mathbf{V}|\phi\rangle &= |\phi\rangle - \frac{2}{2^{n/2}}|a\rangle; \\ \mathbf{W}|\phi\rangle &= |\phi\rangle, & \mathbf{W}|a\rangle &= \frac{2}{2^{n/2}}|\phi\rangle - |\rangle a. \end{aligned} \tag{4.11}$$

So if we start with the state $|\phi\rangle$ and let any sequence of these two operators act successively, the states that result will always remain in the two-dimensional plane spanned by real linear combinations of $|\phi\rangle$ and $|a\rangle$. Finding the result of repeated applications of **WV** to the initial state $|\phi\rangle$ reduces to an exercise in plane geometry.

It follows from the form (4.9) of $|\phi\rangle$ that $|\phi\rangle$ and $|a\rangle$, considered as vectors in the plane of their real linear combinations, are very nearly perpendicular, since the cosine of the angle γ between them is given by

$$\cos\gamma = \langle a|\phi\rangle = 2^{-n/2} = 1/\sqrt{N}, \tag{4.12}$$

which is small when N is large. It is convenient to define $|a_\perp\rangle$ to be the normalized real linear combination of $|\phi\rangle$ and $|a\rangle$ that is strictly orthogonal to $|a\rangle$ and makes the small angle $\theta = \pi/2 - \gamma$ with $|\phi\rangle$, as illustrated in Figures 4.2 and 4.3. Since

$$\sin\theta = \cos\gamma = 2^{-n/2} = 1/\sqrt{N}, \tag{4.13}$$

θ is very accurately given by

$$\theta \approx 2^{-n/2} \tag{4.14}$$

when $\sqrt{N}$ is large.

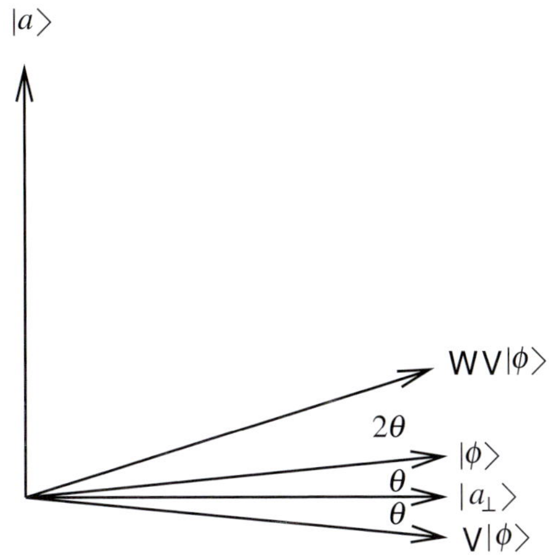

Fig 4.3 Since the rotation **WV** rotates any vector in the plane of real linear combinations of $|a\rangle$ and $|\phi\rangle$ clockwise through an angle 2θ, it takes $|\phi\rangle$ into a vector $\mathbf{WV}|\phi\rangle$ that makes an angle 3θ with $|a_\perp\rangle$. This can also be seen directly from the separate behaviors of **V** and **W**: **V** takes $|\phi\rangle$ into its mirror image in $|a_\perp\rangle$, and **W** then takes $\mathbf{V}|\phi\rangle$ into its mirror image in $|\phi\rangle$.

Since **W** leaves $|\phi\rangle$ invariant and reverses the direction of any vector orthogonal to $|\phi\rangle$, its geometrical action on any vector in the two-dimensional plane containing $|\phi\rangle$, $|a\rangle$, and $|a_\perp\rangle$ is simply to replace the vector by its reflection in the mirror line through the origin along $|\phi\rangle$. On the other hand **V** reverses the direction of $|a\rangle$ while leaving any vector orthogonal to $|a\rangle$ invariant, so it acts on a general vector in the two-dimensional plane by replacing it with its reflection in the mirror line through the origin along $|a_\perp\rangle$. The product **WV**, being a product of two two-dimensional reflections, is a two-dimensional rotation.[2] The angle of that rotation is most easily seen by considering the effect of **WV** on $|a_\perp\rangle$ (see Figure 4.2). The application of **V** leaves $|a_\perp\rangle$ invariant, and the subsequent action of **W** on $|a_\perp\rangle$ reflects it in the line through the origin along the direction of $|\phi\rangle$. So the net effect of the rotation **WV** on $|a_\perp\rangle$ is to rotate $|a_\perp\rangle$ past $|\phi\rangle$ through a total angle that is twice the angle θ between $|a_\perp\rangle$ and $|\phi\rangle$.

Because **WV** is a rotation, the result of applying it to any other vector in the plane is also to rotate that vector through the angle 2θ in the direction from $|a_\perp\rangle$ to $|\phi\rangle$. So applying **WV** to the initial state $|\phi\rangle$ gives a vector rotated away from $|a_\perp\rangle$ by 3θ, since $|\phi\rangle$ is already rotated away from $|a_\perp\rangle$ by θ (Figure 4.3). Applying **WV** a second time results in a vector rotated away from $|a_\perp\rangle$ by 5θ, and each subsequent application of **WV** increases the angle between the final state and $|a_\perp\rangle$

2 A two-dimensional reflection can be achieved by adding a third dimension perpendicular to the plane and performing a 180° rotation with the mirror line as axis. This reverses the irrelevant direction orthogonal to the plane. The product of two such three-dimensional rotations is also a rotation, takes the plane into itself, and does not reverse the third orthogonal direction, so it is a two-dimensional rotation in the plane.

by another 2θ. Since θ is very close to $2^{-n/2}$, after an integral number of applications as close as possible to

$$(\pi/4)2^{n/2}, \tag{4.15}$$

the resulting state will be very nearly orthogonal to $|a_\perp\rangle$ in the plane spanned by $|\phi\rangle$ and $|a\rangle$ – i.e. it will be very nearly equal to $|a\rangle$ itself.

Consequently a measurement of the input register in the computational basis will yield a with a probability very close to 1. We can check to see whether we have been successful by "querying the oracle." If $f(a)$ is 1, as it will be with very high probability, this confirms that we have found the desired a. If we were unlucky we might have to repeat the whole procedure a few more times before achieving success.

4.3 How to construct W

It remains to specify how to construct **W** out of 1- and 2-Qbit unitary gates. Now $-\mathbf{W}$ works just as well as **W** for purposes of the search algorithm, since it leads to a final state that differs, if at all, only by a harmless overall minus sign. It follows from (4.9) and (4.10) and the fact that $\mathbf{H}^{\otimes n}$ is its own inverse that

$$-\mathbf{W} = \mathbf{1} - 2|\phi\rangle\langle\phi| = \mathbf{H}^{\otimes n}(\mathbf{1} - 2|00\ldots00\rangle\langle 00\ldots00|)\mathbf{H}^{\otimes n}, \tag{4.16}$$

so we need a gate that acts as the identity on every computational-basis state except $|00\ldots00\rangle$, which it multiplies by -1. This is just the action of an $(n-1)$-fold-controlled-Z gate, with the roles of the 1-Qbit states $|0\rangle$ and $|1\rangle$ interchanged. This interchange is accomplished by sandwiching the $(n-1)$-fold-controlled-Z between $\mathbf{X}^{\otimes n}$ gates, and we therefore have

$$-\mathbf{W} = \mathbf{H}^{\otimes n}\mathbf{X}^{\otimes n}(\mathbf{c}^{n-1}\mathbf{Z})\mathbf{X}^{\otimes n}\mathbf{H}^{\otimes n}. \tag{4.17}$$

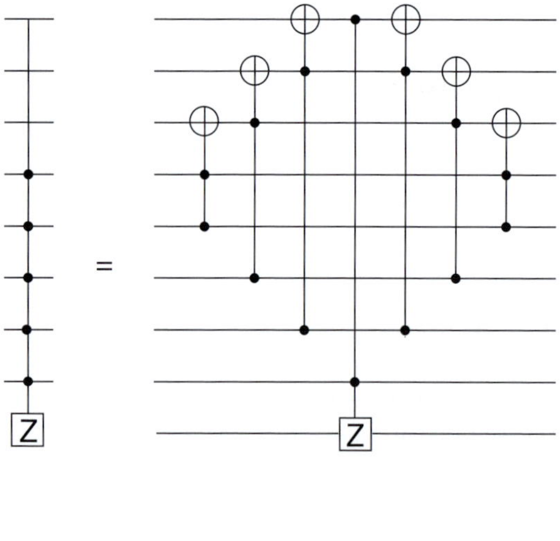

Fig 4.4 The n-fold-controlled-Z transformation, $\mathbf{c}^n\mathbf{Z}$, acts as the identity on states of the computational basis unless all n control Qbits are in the state $|1\rangle$, when it acts on the target Qbit as **Z**. Here it is constructed out of doubly controlled gates, using an additional $n-2$ ancilliary Qbits, all initially in the state $|0\rangle$. One uses $2(n-2)$ $\mathbf{c}^2\mathbf{X}$ (Toffoli) gates and a $\mathbf{c}^2\mathbf{Z}$ gate. The construction is illustrated for the case $n = 5$. The top three wires are the three ancillary Qbits. The next five wires from the top are the five control Qbits, and the bottom wire is the target Qbit. One easily verifies (by applying the circuit to computational-basis states, with each of the ancillary Qbits in the state $|0\rangle$) that **Z** acts on the target Qbit if and only if every one of the five control Qbits is in the state $|1\rangle$. The Toffoli gates are symmetrically disposed on both sides of the diagram to ensure that at the end of the process each of the three ancillary Qbits is set back to its initial state $|0\rangle$. This is essential if the ancillary Qbits are not to become entangled with the Qbits on which the Grover iteration acts, represented by the bottom six wires.

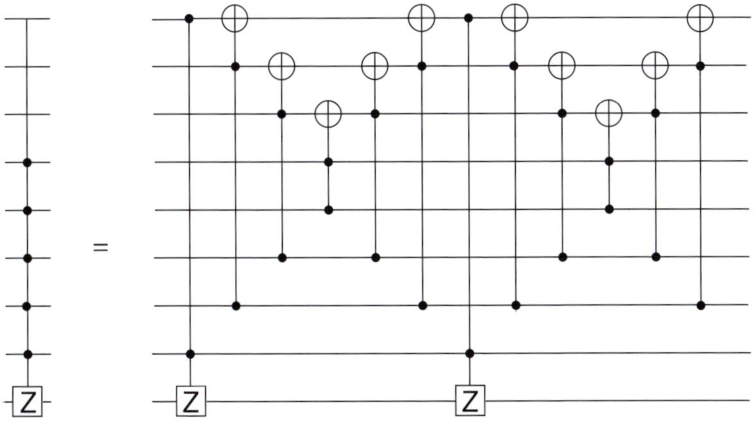

Fig 4.5 An improved version of Figure 4.4, with twice as many gates. Gates have been added on the left and right to ensure that the circuit works for *arbitrary* initial computational-basis states of the three ancillary Qbits at the top, restoring them to their initial states at the end of the computation. To see this note that because Toffoli gates or $\mathbf{c}^2\mathbf{Z}$ gates are their own inverses, the circuit acts as the identity on those computational-basis states of all nine Qbits in which any one of the five control Qbits (second through sixth wires from the bottom) is in the state $|0\rangle$, regardless of the computational-basis states of the other Qbits. This is because, as an examination of the figure reveals, replacing the gate governed by any one of the five control Qbits by the identity always results in a pairwise cancellation of all the remaining gates. It remains only to confirm that when all five control Qbits are in the state $|1\rangle$, the circuit acts as $\mathbf{Z}$ on the target Qbit at the bottom, and the state of the three ancillary Qbits at the top is unchanged. This is established in Figure 4.6, which shows the operation of the gates in Figure 4.5 when the five control Qbits are all in the state $|1\rangle$. Because $\mathbf{X} = \mathbf{HZH}$ one can also use this circuit to produce a multiply-controlled-NOT gate, by applying Hadamard gates to the bottom wire on the far right and left.

We can construct **W** by constructing $\mathbf{c}^{n-1}\mathbf{Z}$, the $(n-1)$-fold-controlled-Z.

Figure 4.4 shows a straightforward but not terribly efficient way to make a $\mathbf{c}^{n-1}\mathbf{Z}$ gate for the case $n = 6$. We use $n - 3$ ancillary Qbits, all initially in the state $|0\rangle$, $2(n-3)$ $\mathbf{c}^2\mathbf{X}$ (Toffoli) gates, and one $\mathbf{c}^2\mathbf{Z}$ gate. As explained in Section 2.6, these can all be built out of 1- and 2-Qbit gates. It is essential for the success of the algorithm that each ancillary Qbit be restored to its initial state $|0\rangle$, since our analysis of the Grover algorithm assumes that the input and output registers have states of their own, unentangled with any other Qbits, after each application of **W** and **V**.

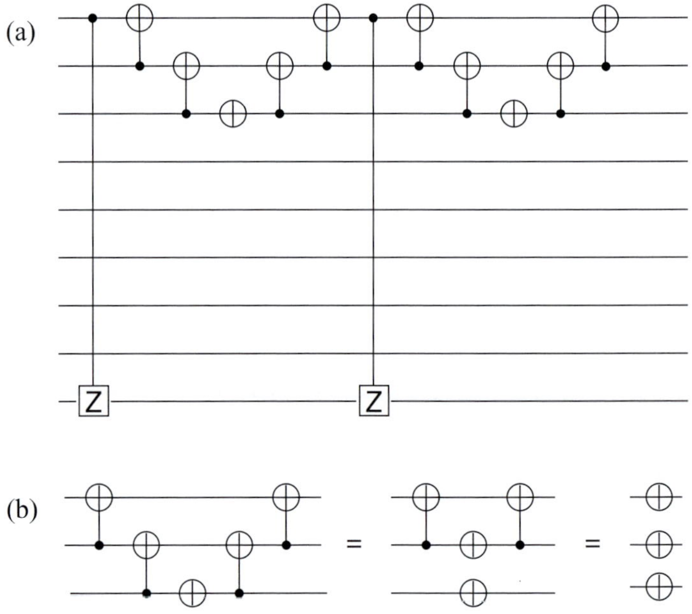

Fig 4.6 Part (a) reproduces what remains of Figure 4.5 when all five control Qbits are in the state $|1\rangle$. One easily verifies that two identical cNOT gates, separated by a NOT acting on their control Qbit, have exactly the same action on the computational basis as NOT gates acting on both the control and target Qbits. As a result each of the two identical sets of five adjacent gates acting on the three ancillary Qbits at the top of part (a) reduces simply to three NOT gates, as shown in part (b). Making this further simplification in part (a), note that because each of the three ancillary Qbits is acted on by two NOT gates, its state is unaltered. The two NOT gates acting on the upper wire also ensure that precisely one of the two **cZ** gates applies **Z** to the bottom Qbit, whatever the state of the upper wire.

The construction of Figure 4.4 is rather expensive in Qbits, requiring $n-3$ ancillas to apply the algorithm to an n-bit set of possibilities for the special number a. At a cost of four times as many Toffoli gates, one can reduce the number of ancillas to a single one, regardless of the size of n. The way to do this is developed in Figures 4.5–4.7. Figures 4.5 and 4.6 show how nearly doubling the number of gates makes it possible for the construction of Figure 4.4 to work for *arbitrary* initial states of the ancillas. Figure 4.7 then shows how, by an additional doubling, one can, with the aid of a single ancilla, divide an n-fold-controlled-Z into two multiply-controlled-NOT gates and two multiply-controlled-Z gates, each acting on about $\frac{1}{2}n$ Qbits. (Since $\mathbf{X}=\mathbf{HZH}$, one can convert a multiply-controlled-Z gate into a multiply-controlled-NOT gate by applying Hadamard gates to the target Qbit at the beginning and end of the circuit.) The multiply-controlled-Z gates in Figure 4.7 are able nondisruptively to use the control Qbits of the multiply-controlled-NOT gates as their ancillary Qbits in the construction of Figure 4.5. And the multiply-controlled-NOT gates in Figure 4.7 can make similar use of the control Qbits of the multiply-controlled-Z gates.

4.4 Generalization to several special numbers

If there are several special numbers, essentially the same algorithm can be used to find one of them, if we know how many there are. The function f in (4.1) now becomes

$$f(x)=0,\quad x\neq a_1,\ldots,a_m;\qquad f(x)=1,\quad x=a_1,\ldots,a_m. \tag{4.18}$$

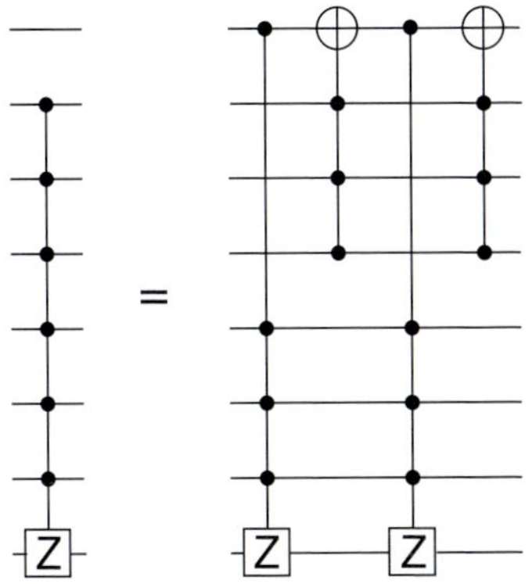

Fig 4.7 The identity illustrated by the circuit is easily confirmed. There is only one ancilla, whose state is left unchanged. By introducing circuits of the form in Figure 4.5 into this circuit one can produce $\mathbf{c}^n\mathbf{Z}$ or $\mathbf{c}^n\mathbf{X}$ gates with the aid of just a single ancilla. (Since $\mathbf{X} = \mathbf{HZH}$ Figure 4.5 works for either type.) In constructing each of the multiply-controlled-NOT gates in Figure 4.7 out of Toffoli gates, one can borrow the control Qbits of the multiply-controlled-Z gates to use as ancillary Qbits in the expansions of Figure 4.5, since those expansions work whatever the state of their ancillary Qbits, and restore that state to its original form. For the same reasons one can also borrow the control Qbits of the multiply-controlled-NOT gates to construct the multiply-controlled-Z gates.

The n-Qbit unitary transformation $\mathbf{V}$ extracted from (4.4) becomes one whose action on computational-basis states in the input register is given by

$$\mathbf{V}|x\rangle = |x\rangle, \quad x \neq a_1, \ldots, a_m; \qquad \mathbf{V}|x\rangle = -|x\rangle, \quad x = a_1, \ldots, a_m. \tag{4.19}$$

If we replace the state $|a\rangle$ by

$$|\psi\rangle = \frac{1}{\sqrt{m}} \sum_{i=1}^{m} |a_i\rangle, \tag{4.20}$$

then starting with $|\phi\rangle$, which continues to have the form (4.9), the transformations $\mathbf{V}$ and $\mathbf{W}$ now keep the state of the input register in the two-dimensional plane spanned by the real linear combinations of $|\psi\rangle$ and $|\phi\rangle$. The unitary transformation $\mathbf{V}$ changes the sign of $|\psi\rangle$ but preserves the linear combination of $|\phi\rangle$ and $|\psi\rangle$ orthogonal to $|\psi\rangle$, so $\mathbf{V}$ is now a reflection in the line through the origin along the vector $|\psi_\perp\rangle$ perpendicular to $|\psi\rangle$ in the plane. Everything else is just as in the case of a single special number except that now the angle Θ between $|\psi_\perp\rangle$ and $|\phi\rangle$ satisfies

$$\sin\Theta = \cos(\pi/2 - \Theta) = \langle\psi|\phi\rangle = \sqrt{m/2^n}. \tag{4.21}$$

When $m/2^n \ll 1$, we can arrive at a state very close to $|\psi\rangle$ with

$$(\pi/4)2^{n/2}/\sqrt{m} \tag{4.22}$$

applications of **WV**. A measurement then gives us, with a probability very close to 1, a random one of the special values a_i. Note that the mean number of invocations of the subroutine decreases only as $1/\sqrt{m}$ with the number m of marked items, in contrast to a classical search, for which doubling the number of acceptable solutions would halve the time of the search. When $m/2^n$ is not small we have to reexamine the expression (4.22) for the optimal number of iterations, but at that point the quantum search offers little significant advantage over a classical one.

We must know how many special numbers there are for the procedure to work, since we have to know how many times to do the Grover iteration before making our measurement. By exploiting the fact that the Grover iteration is periodic, restoring the initial state after about $\pi 2^{n/2}/\sqrt{m}$ iterations, it is possible to combine Grover iterations with a clever application of the quantum Fourier transform to learn the value of m with enough accuracy to enable one then to apply the Grover iteration the right number of times to ensure a high probability of success, even when m is not known at the start.

4.5 Searching for one out of four items

The simplest nontrivial application of Grover's algorithm is to the case $n = 2$, or $N = 4$. (When $n = 1$ a single invocation of the subroutine suffices to identify a even with a classical computer.) When $n = 2$, (4.13) tells us that $\sin\theta = \frac{1}{2}$, so $\theta = 30°$. Consequently $3\theta = 90°$, and the probability of identifying a with a single invocation of the subroutine is *exactly* 1.

This is a significant improvement on the classical computer, with which one can do no better than trying each of the four possibilities for a in random order. It is equally likely that the marked item will be the first, second, third, or fourth on the list. Since the probability is $\frac{1}{4}$ that the marked item is first on the list, $\frac{1}{4}$ that it is second, and $\frac{1}{4} + \frac{1}{4} = \frac{1}{2}$ that it is third or fourth, the mean number of attempts is $\frac{1}{4} \times 1 + \frac{1}{4} \times 2 + \frac{1}{2} \times 3 = 2\frac{1}{4}$. (It is not necessary to make a fourth attempt, since if the first three attempts fail to produce a, then one knows that a is the one remaining untested number.)

The case $n = 2$ is also special in that one does not have to resort to the elaborate procedure specified in Figures 4.4–4.7 to produce the n-fold-controlled-Z gate. A single Toffoli gate sandwiched between Hadamards on the target Qbit does the job.

Chapter 5

Quantum error correction

5.1 The miracle of quantum error correction

Correcting errors might sound like a dreary practical problem, of little aesthetic or conceptual interest. But aside from being of crucial importance for the feasibility of quantum computation, it is also one of the most beautiful and surprising parts of the subject. The surprise is that error correction is possible at all, since the only way to detect errors is to make measurements, but measurement gates disruptively alter the states of the measured Qbits, apparently making things even worse. "Quantum error correction" would seem to be an oxymoron. The beauty lies in the ingenious ways that people have found to get around this apparently insuperable obstacle.

The discovery in 1995 of quantum error correction by Peter Shor and, independently, Andrew Steane had an enormous impact on the prospects for actual quantum computation. It changed the dream of building a quantum computer capable of useful computation from a clearly unattainable vision to a program that poses an enormous but not necessarily insuperable technological challenge.

Error correction is not a major issue for classical computation. In a classical computer the physical systems that embody individual bits – the Cbits – are immense on the atomic scale. The two states of a Cbit representing 0 and 1 are so grossly different that the probability is infinitesimal for flipping from one to the other as a result of thermal fluctuations, mechanical vibrations, or other irrelevant extraneous interactions.

Error correction does become important, even classically, in the transmission of information over large distances, because the farther the signal travels, the more it attenuates. One can deal with this in a variety of straightforward or ingenious ways. One of the crudest is to encode each logical bit in three actual bits, replacing $|0\rangle$ and $|1\rangle$ by the codewords

$$|\bar{0}\rangle = |0\rangle|0\rangle|0\rangle = |000\rangle, \qquad |\bar{1}\rangle = |1\rangle|1\rangle|1\rangle = |111\rangle. \tag{5.1}$$

One can then monitor each codeword, checking for flips in any of the individual Cbits and restoring them by applying the principle of majority rule whenever a flip is detected. Monitoring has to take place often enough to make negligible the probability that more than a single bit has flipped in a single codeword between inspections.

Quantum error correction also uses multi-Qbit codewords and also requires monitoring at a rate that renders certain kinds of compound errors highly improbable. But there are several ways in which error correction in a quantum computer is quite different.

(a) A quantum computer, unlike a classical computer, requires error correction. The physical Qbits are individual atomic-scale physical systems such as atoms, photons, trapped ions, or nuclear magnetic moments. Any coupling to anything not under the explicit control of the computer and its program can substantially disrupt the state associated with those Qbits, entangling them with computationally irrelevant features of the computer or the world outside the computer, thereby destroying the computation. For a quantum computer to work without error correction, each Qbit would have to be impossibly well isolated from irrelevant interactions with other parts of the computer and anything else in its environment.

(b) In contrast to classical error correction, checking for errors in a quantum computer is problematic. The obvious way to monitor a Qbit is to measure it. But the result of measuring a Qbit is to alter its state, if it has one of its own, and, more generally, to destroy its quantum correlations with other Qbits with which it might be entangled. Such disruptions are stochastic – i.e. unpredictable – and introduce major errors of their own. One must turn to less obvious forms of monitoring.

(c) Bit flips are not the only errors. There are entirely nonclassical sources of trouble. For example phase errors, such as the alteration of $|0\rangle + |1\rangle$ to $|0\rangle - |1\rangle$, can be just as damaging.

(d) Unlike the discrete all-or-nothing bit-flip errors suffered by Cbits, errors in the state of Qbits grow continuously out of their uncorrupted state.

We begin our discussion of error correction by examining in Section 5.2 a simple model of quantum error correction that works when the possible errors are artificially limited to a few specific kinds of disruption. Although this is clearly unrealistic, the error-correction procedure is easy to follow. It also introduces in a simple setting most of the tricks that continue to work in the more realistic case.

5.2 A simplified example

Much of the flavor of quantum error correction is conveyed by an artificially simple model in which the only errors a collection of Qbits is allowed to experience are the classically meaningful errors: random

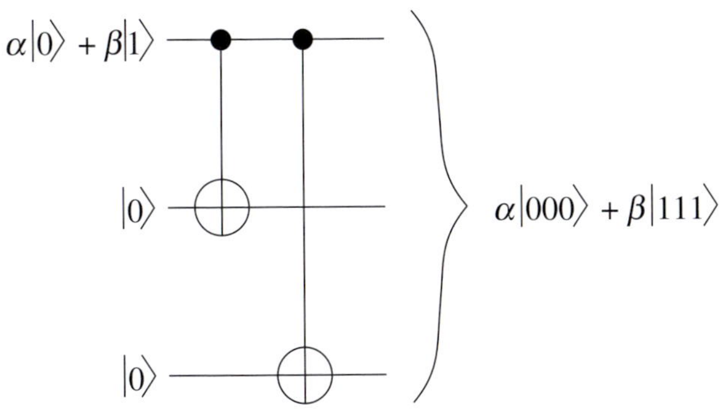

Fig 5.1 A unitary circuit that encodes the 1-Qbit state $\alpha|0\rangle + \beta|1\rangle$ into the 3-Qbit code state $\alpha|000\rangle + \beta|111\rangle$, using two cNOT gates and two other Qbits each initially in the state $|0\rangle$. The circuit clearly works for the computational-basis states $|0\rangle$ and $|1\rangle$, and therefore, by linearity, it works for arbitrary superpositions.

flips of individual Qbits. We shall examine the more general possibilities for quantum errors in Section 5.3 below.

Bit-flip errors in a computation can be modeled by a circuit that differs from the ideal error-free circuit only in the occasional presence of extraneous 1-Qbit NOT gates. If such randomly occurring error-producing NOT gates are sufficiently rare, then since the only allowed errors are bit-flip errors, one might hope to be able to correct the corruptions they introduce by tripling the number of Qbits and using precisely the 3-Qbit code (5.1) that corrects for bit-flip errors in the classical case. Because of the disruptive effect of measurement gates in diagnosing errors, it is not at all clear that such a 3-Qbit code can be effective for bit-flip errors in the quantum case. It can nevertheless be made to work, though the way in which one does the encoding and performs the error correction is much subtler for Qbits than it is for Cbits.

To begin with, there is the question of encoding. Classically one merely replaces each of the two computational-basis states $|x\rangle$ by the codeword states $|\overline{x}\rangle = |x\rangle|x\rangle|x\rangle$, for $x = 0$ or 1. Qbits, however, can also be in superpositions $\alpha|0\rangle + \beta|1\rangle$, and one requires a circuit that automatically encodes this into $\alpha|\overline{0}\rangle + \beta|\overline{1}\rangle = \alpha|0\rangle|0\rangle|0\rangle + \beta|1\rangle|1\rangle|1\rangle$ for arbitrary α and β, in the absence of any knowledge of what the values of α and β might be. This can be done with two cNOT gates that target two additional Qbits initially both in the state $|0\rangle$, as illustrated in Figure 5.1:

$$\alpha|\overline{0}\rangle + \beta|\overline{1}\rangle = \alpha|0\rangle|0\rangle|0\rangle + \beta|1\rangle|1\rangle|1\rangle = \mathbf{C}_{21}\mathbf{C}_{20}\big(\alpha|0\rangle + \beta|1\rangle\big)|0\rangle|0\rangle. \tag{5.2}$$

Having produced such a 3-Qbit codeword state, we must then guard against its corruption by the possible action of an extraneous NOT gate that acts on at most one of the three Qbits, as illustrated in Figure 5.2. This is easily done for Cbits, for which there are only two possible uncorrupted initial states, $|000\rangle$ and $|111\rangle$, and examining them is unproblematic. To see whether either initial state has been corrupted by the action of a single NOT gate, one nondisruptively reads

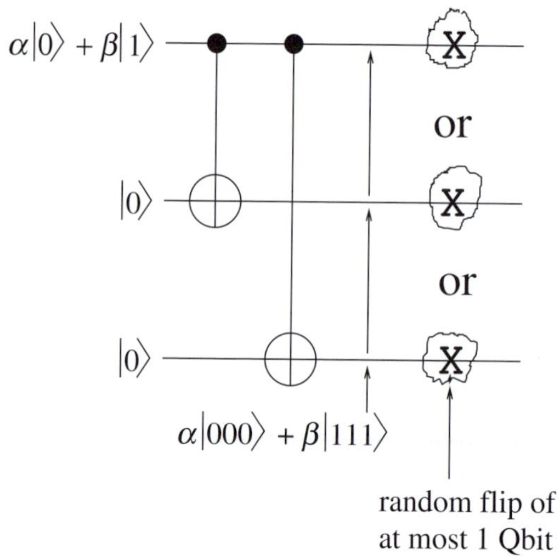

Fig 5.2 The encoded state of Figure 5.1 may or may not be corrupted by the action of a single extraneous NOT gate. The error-inducing gates are depicted in a lighter font – X instead of **X** – and inside a noisy-looking corrupted box.

the three Cbits. If this reveals all three Cbits to be in the same state, there is no corruption to correct. If one of them is found to be in a different state from the other two, that particular Cbit is the one that was acted upon by the extraneous NOT gate, and applying a second NOT gate to that Cbit restores the initial state.

One cannot, however, nondisruptively "read" the state of a collection of Qbits. The only way to extract information is by the action of measurement gates. But measuring any of the three Qbits immediately destroys the uncorrupted superposition

$$|\Psi\rangle = \alpha|000\rangle + \beta|111\rangle, \tag{5.3}$$

converting it either to $|000\rangle$ (with probability $|\alpha|^2$) or to $|111\rangle$ (with probability $|\beta|^2$). There is a similar coherence-destroying effect on each of the three possible corrupted states,

$$\begin{aligned}|\Psi_0\rangle &= \mathbf{X}_0|\Psi\rangle = \alpha|001\rangle + \beta|110\rangle,\\ |\Psi_1\rangle &= \mathbf{X}_1|\Psi\rangle = \alpha|010\rangle + \beta|101\rangle,\\ |\Psi_2\rangle &= \mathbf{X}_2|\Psi\rangle = \alpha|100\rangle + \beta|011\rangle,\end{aligned} \tag{5.4}$$

obliterating any dependence of the post-measurement state on the complex amplitudes α and β. This might appear (and for some time was thought) to be the end of the story: quantum error correction is impossible because of the disruptive effect of the measurement needed to diagnose the error.

But there are subtler ways to extract the information needed to diagnose and correct possible errors. Although there continues to be a disruption in these refined procedures, the damaging effects are shifted from the codeword Qbits to certain ancillary Qbits. By coupling the codeword Qbits to these ancillary Qbits with appropriate 2-Qbit unitary gates, and then applying measurement gates only to the ancillas, one can extract information about certain *relations* prevailing among the codeword Qbits. This more limited information turns out to be enough to diagnose and correct certain errors in a coherence-preserving manner, without revealing anything about the original uncorrupted codeword state. Acquiring no information about the uncorrupted state is a necessary restriction on any error-correction procedure capable of perfectly restoring the uncorrupted state. If one could get even partial information about the structure of a state without disrupting it, one could continue collecting additional information nondisruptively until one was well on the way to violating the no-cloning theorem.

Note that all possible forms for the uncorrupted 3-Qbit codeword (5.3) – given by assigning all possible values to the amplitudes α and β – lie in a two-dimensional subspace of the full eight-dimensional space containing all possible 3-Qbit states. Furthermore, each of the three allowed corruptions (5.4) also lies in a two-dimensional subspace of the full 3-Qbit space, and the three subspaces containing the three allowed corruptions are each orthogonal to the subspace containing the uncorrupted codeword, and orthogonal to each other. This turns out to be crucial to the success of the enterprise.

More generally, if we wanted to use an n-Qbit codeword in a model in which the only allowed errors were flips of a single Qbit, then we would require $2(1+n)$ dimensions to accommodate the $n+1$ mutually orthogonal two-dimensional subspaces associated with a general uncorrupted state and its n different 1-Qbit corruptions. Since all possible states of n Qbits span a 2^n-dimensional space, a necessary condition for an n-Qbit bit-flip-error-correcting code to be possible is

$$2^{n-1} \geq 1 + n. \tag{5.5}$$

The smallest n satisfying (5.5) is $n = 3$, for which it holds as an equality. This shows that the 3-Qbit code is, in this sense, perfect for the purpose of correcting errors limited to flips of a single Qbit.

Figure 5.3 shows that 3-Qbit codewords, as well as meeting this necessary condition for the correction of quantum bit-flip errors, actually do permit it to be carried out. The error detection and correction requires two additional ancillary Qbits (the upper two wires), initially both in the state $|0\rangle$. Both ancillas are targeted by pairs of cNOT gates controlled by subsets of the three codeword Qbits. Note first that if the 3-Qbit codeword has not been corrupted, so its state remains (5.3), then both the ancillary Qbits remain in the state $|0\rangle$ after the action of the cNOT gates, since the term $|000\rangle$ in the codeword results in none of the target Qbits being flipped, while the term $|111\rangle$ results in both of the target Qbits being flipped twice, which is equivalent to no flip.

In a similar way each of the three corruptions (5.4) results in a different unique final state for the ancillary Qbits. The first of those

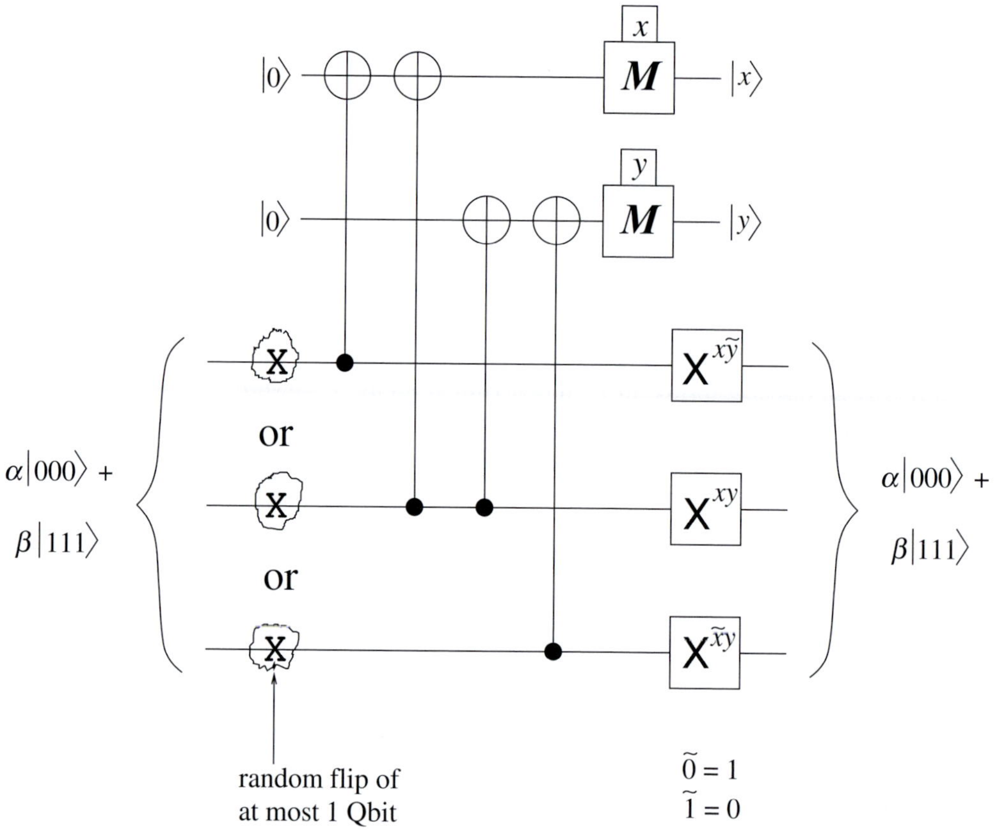

Fig 5.3 How to detect and correct the three possible single-bit-flip errors shown in Figure 5.2. One requires two ancillary Qbits (the upper two wires), each initially in the state $|0\rangle$, coupled to the codeword Qbits by cNOT gates. After the cNOT gates have acted each ancilla is measured. If both measurements give 0, then none of the erroneous NOT gates on the left have acted and none of the error-correcting NOT gates on the right need to be applied. If the upper measurement gate shows $x = 1$ and the lower one shows $y = 0$, then the uppermost of the three erroneous NOT gates has acted on the left. Its action is undone by applying the uppermost of the three NOT gates on the right. The other two possible 1-Qbit errors are similarly corrected.

corruptions results in $|0\rangle$ for the upper ancilla and $|1\rangle$ for the lower, since either term in the superposition $\alpha|001\rangle + \beta|110\rangle$ results in zero or two flips for the upper ancilla, and a single flip for the lower ancilla. The next form in (5.4) produces a single flip for both ancillas, resulting in $|1\rangle$ for both. The third results in $|1\rangle$ for the upper and $|0\rangle$ for the lower ancilla.

So if the two ancillary Qbits are measured after the cNOT gates have acted, the four possible readings, 00, 01, 10, and 11, of the two measurement gates reveal whether or not a random one of the codeword Qbits has been flipped and, in the event of a flip, which of the three has suffered it. On the basis of this information one can either accept the codeword as uncorrupted or apply a NOT gate to the Qbit that has been identified as the flipped one, thereby restoring the initial uncorrupted state. One easily confirms that this is precisely what is accomplished by the NOT gates on the extreme right of Figure 5.3.

This accomplishes what any valid quantum error-correction procedure must do: it restores the original uncorrupted state without revealing any information whatever about what the form of that state – the particular values of the amplitudes α and β – might actually be. The procedure succeeds in preserving the superposition by extracting information only about correlations among the Qbits making up the codeword, without ever extracting information about individual Qbits. Working only with correlations makes it possible to apply a single linear test that works equally well for diagnosing 1-Qbit errors in either $|000\rangle$ or $|111\rangle$, and therefore also works for any superposition of those states.

This simple example of quantum error correction requires the use of measurement gates to diagnose the error. The outputs of the measurement gates are noted, and then used to determine which, if any, of a collection of error-correcting NOT gates should be applied. The procedure can be automated into a bigger quantum circuit that eliminates (or almost eliminates) the need to use measurement gates combined with unitary gates, which are or are not applied depending on the readings of the measurement gates. This can be achieved by a combination of cNOT and Toffoli gates, controlled by the ancillary Qbits, as shown in Figure 5.4.

Replacing measurement gates by additional cNOT gates does not entirely eliminate the need for measurement, because at the end of the process the state of the ancillary Qbits will depend on the character of the error and will in general no longer be the state $|0\rangle|0\rangle$ with which the error-correction procedure starts. If one wishes to reuse these ancillary Qbits for further error correction, it is necessary to reset them to their initial state $|0\rangle|0\rangle$. This can efficiently be done by *measuring* them and applying the appropriate NOT gates if either is found to be in the state $|1\rangle$. Thus measurement gates followed by NOT gates that depend on the measurement outcome are still needed to prepare the circuit for a possible future error correction.

This procedure (automated or not) will continue to work even when the codeword Qbits are entangled with many other codeword Qbits, as they will be in the course of a nontrivial computation. In such a case the codeword Qbits have no state of their own, the state of all the many codeword Qbits being of the form

$$\alpha|000\rangle|\Psi\rangle + \beta|111\rangle|\Phi\rangle, \tag{5.6}$$

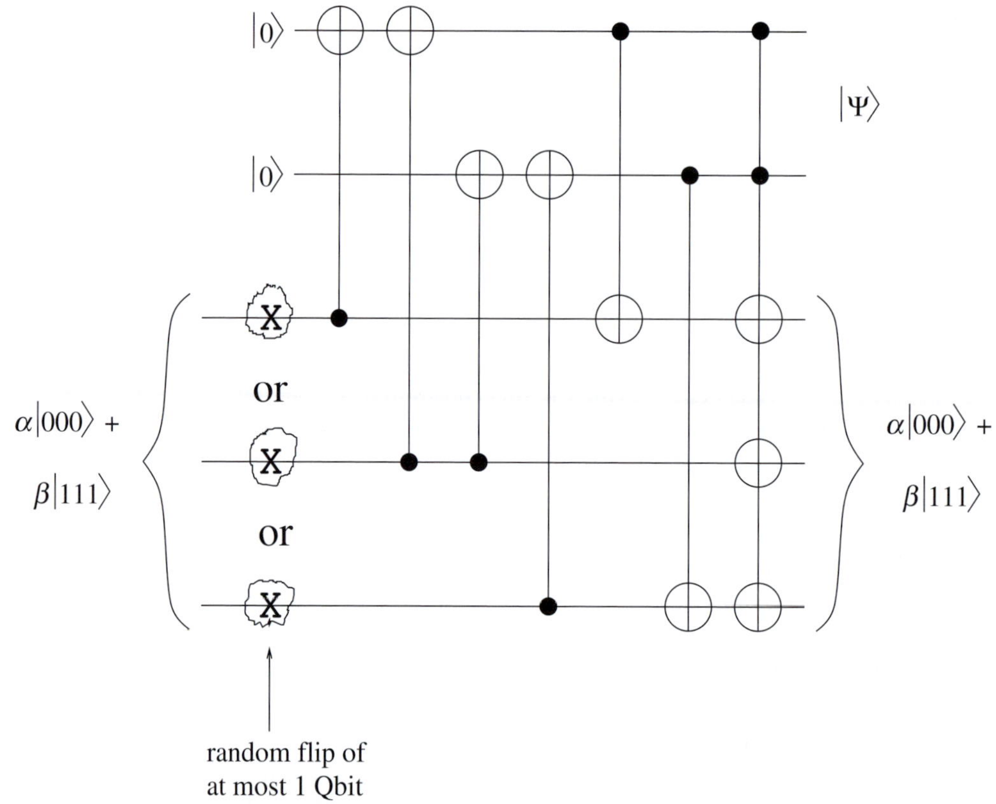

Fig 5.4 Automation of the error-correction process of Figure 5.3. The three controlled gates on the right – one of them a doubly controlled Toffoli gate with multiple targets – have precisely the same error-correcting effect on the three codeword Qbits as does the application of NOT gates contingent on measurement outcomes in Figure 5.3. The final state $|\Psi\rangle$ of the ancillas (which is also the state that determines the action of the three controlled gates on the right) is $|00\rangle$ if none of the erroneous NOT gates on the left has acted. It is $|10\rangle$ if only the upper erroneous NOT gate has acted, $|11\rangle$ if only the middle one has acted, and $|01\rangle$ if only the lower one has acted.

with the error correction applied to the three Qbits on the left. One easily confirms that the added complication of entanglement with other Qbits has no effect on the validity of the error-correction procedure.

There is an alternative way of representing the use of cNOT gates in Figure 5.3 to diagnose the error, which is useful in correcting quantum errors in more realistic cases. The alternative point of view is based on the easily confirmed fact that the uncorrupted state (5.3) is left unaltered by either of the operators $\mathbf{Z}_2\mathbf{Z}_1$ and $\mathbf{Z}_1\mathbf{Z}_0$, while the three corruptions (5.4) are each eigenstates of $\mathbf{Z}_2\mathbf{Z}_1$ and $\mathbf{Z}_1\mathbf{Z}_0$ with distinct

Table 5.1. Two operators that diagnose the error syndrome for the 3-Qbit code that protects against bit-flip errors. The four entries in each of the two rows indicate whether the operator for that row commutes (+) or anticommutes (−) with the operators at the top of the four columns

	$\mathbf{X}_2$	$\mathbf{X}_1$	$\mathbf{X}_0$	$\mathbf{1}$
$\mathbf{Z}_2\mathbf{Z}_1$	−	−	+	+
$\mathbf{Z}_1\mathbf{Z}_0$	+	−	−	+

sets of eigenvalues: 1 and −1 in the case of $|\Psi_0\rangle$; −1 and −1 in the case of $|\Psi_1\rangle$; and −1 and 1 in the case of $|\Psi_2\rangle$.

While these last three facts can be confirmed directly from the explicit forms of $|\Psi_0\rangle$, $|\Psi_1\rangle$, and $|\Psi_2\rangle$ on the right of (5.4), it is worth noting, for purposes of comparison with some of the more complex cases that follow, that they also follow from the facts that $\mathbf{Z}_2\mathbf{Z}_1$ and $\mathbf{Z}_1\mathbf{Z}_0$ act as the identity on the uncorrupted state $|\Psi\rangle$, that the corrupted states are of the form $|\Psi_j\rangle = \mathbf{X}_j|\Psi\rangle$, and that $\mathbf{X}_j$ *commutes* with $\mathbf{Z}_i$ when $i \neq j$, while $\mathbf{X}_j$ *anticommutes* with $\mathbf{Z}_j$: $\mathbf{Z}_j\mathbf{X}_j = -\mathbf{X}_j\mathbf{Z}_j$. The resulting pattern of commutations (+) or anticommutations (−) is summarized in Table 5.1.

Thus the joint eigenvalues of the commuting operators $\mathbf{Z}_2\mathbf{Z}_1$ and $\mathbf{Z}_1\mathbf{Z}_0$ distinguish among the uncorrupted state and each of the three possible corruptions. A procedure that takes advantage of this by sandwiching controlled $\mathbf{Z}_2\mathbf{Z}_1$ and controlled $\mathbf{Z}_1\mathbf{Z}_0$ gates between Hadamards acting on the control Qbits is shown in Figure 5.5. Although it takes a little thought to confirm directly from the figure that Figure 5.5 does indeed accomplish error correction – we shall work this out in Section 5.4 as a special case of a much more general procedure – one can confirm that it does by simply noting that Figure 5.5 is mathematically equivalent to Figure 5.3. This equivalence follows from the facts that $\mathbf{X} = \mathbf{HZH}$, that $\mathbf{H}^2 = \mathbf{1}$, and that the action of controlled-Z is unaltered by exchanging the target and control Qbits.

This oversimplified example, in which only bit-flip errors are allowed, illustrates most of the features of quantum error correction that one encounters in more realistic cases. The more general procedure is complicated by the fact that, as noted above and made precise in Section 5.3 below, the general error a Qbit can experience is more complicated than just a single bit flip. As a result, one needs codewords containing more than three Qbits to correct general single-Qbit errors, and one requires more complicated diagnostic and corrective procedures than those of Figures 5.3–5.5, involving more than just a pair of ancillary Qbits. But although the codewords and error-correcting

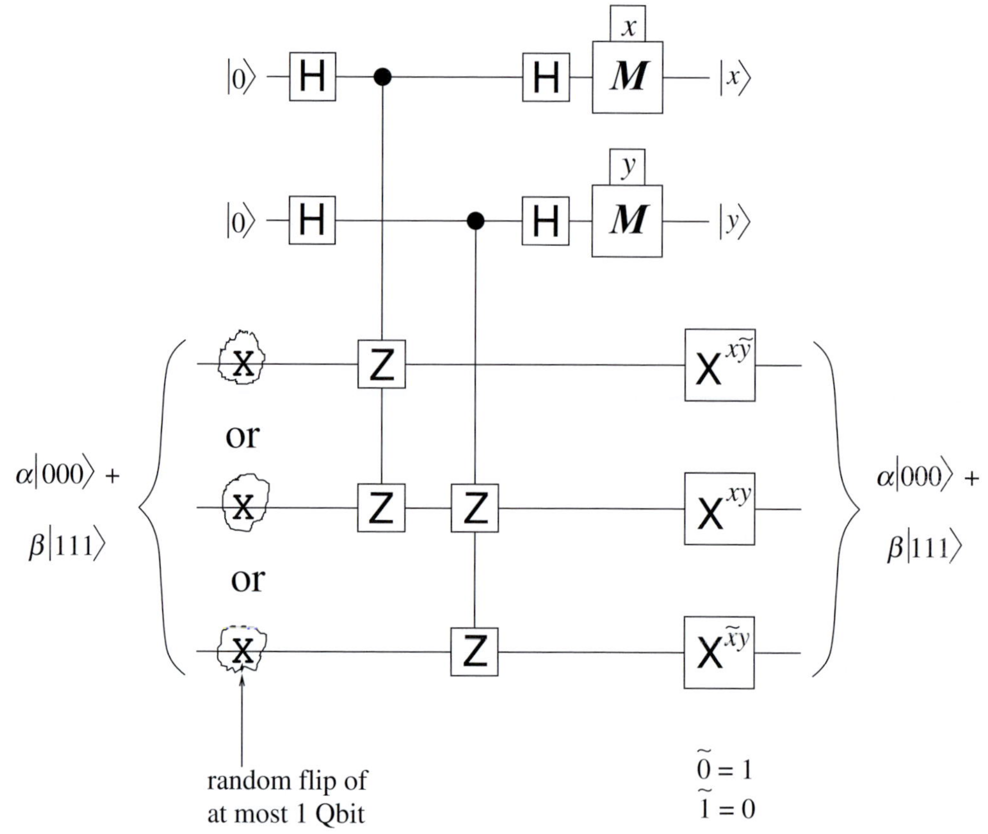

Fig 5.5 An apparently unnecessary complication of the error-correcting circuit in Figure 5.3, which transforms it into the more general form described in Section 5.4. The circuit is equivalent to that in Figure 5.3: (1) the cNOT gates in Figure 5.3 can be replaced by controlled-Z gates if Hadamard gates act on each ancilla before and after the controlled gates act; (2) each of the four controlled-Z gates acts in the same way if its control and target Qbits are interchanged; and (3) pairs of controlled gates with the same control Qbit and two different targets can be combined into a single controlled gate with that control Qbit and a 2-Qbit target operation that is just the product of the two 1-Qbit target operations. The part of the circuit between and including the pairs of Hadamards on the right and left is a simple example of the more complex error-diagnosing circuits that appear in Figures 5.8, 5.9, and N.2 (in Appendix N).

circuits are more elaborate, once we have identified the more general form of quantum errors there are no further conceptual complications in understanding the kinds of procedures that can correct them.

The more general form Qbit errors can assume is discussed in Section 5.3. Somewhat surprisingly, it turns out that the general 1-Qbit error can be viewed as a simple extension of what we have just described: in addition to the possibility of an extraneous $\mathbf{X}$ gate acting on the Qbit, there might also be an extraneous $\mathbf{Z}$ gate or an extraneous $\mathbf{Y} = \mathbf{ZX}$ gate. If we can diagnose and correct for each of these three possible corruptions, then we can correct for arbitrary 1-Qbit errors.

Section 5.4 describes a generalization of the diagnostic scheme we have just exploited for extracting relational information about the Qbits that make up a codeword, by coupling groups of them to ancillary Qbits, which are then measured. It turns out that the operators needed to diagnose the error – generalizations of the operators $\mathbf{Z}_2\mathbf{Z}_1$ and $\mathbf{Z}_1\mathbf{Z}_0$ for the 3-Qbit code – are also useful for defining the more general codewords.

In Section 5.5 we examine two of the most important n-Qbit codes with $n > 3$ that are able to correct general single-Qbit errors: the 5-Qbit and 7-Qbit codes. The 5-Qbit code is the ideal code for general 1-Qbit errors in the same way that the 3-Qbit code is ideal for bit-flip errors. The 7-Qbit code is more likely to be of practical interest, for reasons we shall describe. The earliest quantum error-correcting code – the 9-Qbit code discovered by Shor – is now of only historical interest, and is relegated to Appendix N.

5.3 The physics of error generation

Errors are not, of course, produced by extra gates accidentally appearing in a circuit, as in the oversimplified example of Section 5.2. They are produced by extraneous interactions with the world external to the computer or with computationally irrelevant degrees of freedom of the computer itself. Although one would like the state of the Qbits to evolve only under the action of the unitary transformations imposed by the gates of the computer, inevitably Qbits will interact, even if only weakly, with other physical systems or degrees of freedom having nothing to do with the computation in which the Qbits are participating. In a well-designed computer such spurious interactions will be kept to a minimum, but their disruptive effects on the quantum state of the Qbits can grow continuously from zero, in contrast to disruptive effects on Cbits, which have to exceed a large threshold before a Cbit can change its state. In a quantum computer such spurious changes of state will eventually accumulate to the point where the calculation falls apart, unless ongoing efforts are made to eliminate them.

To characterize the most general way in which a Qbit can be deflected from its computational task, we must finally acknowledge that Qbits are not the only things in the world that are described by quantum states. The quantum theory provides the most fundamental description we have of everything in the world, and it describes everything in the world – not just Qbits – by means of quantum states.

This spectacular expansion of the scope of quantum states might not come as a complete surprise to the nonphysicist reader. I have stressed all along that the quantum state of a Qbit or a collection of Qbits is not a property carried by those Qbits, but a way of concisely summarizing everything we know that has happened to them, to enable us to make statistical predictions about the information we might then be able to extract from them. If quantum states are not properties inherent in the system they describe, but states of the knowledge we have managed to acquire about the prior history of the system – if they somehow incorporate fundamental aspects of how we exchange information with the world outside of us – then they might indeed have an applicability going beyond the particular kinds of systems we have applied them to up until now.

Indeed, nowhere in this exposition of quantum computation has it been necessary to refer to the individual character of the Qbits. Whether they are spinning electrons, polarized photons, atoms in cavities, or any number of other things, the quantum-mechanical description of their computational behavior has been exactly the same. So insofar as the assignment of quantum states to physical systems is a general feature of how we come to grips with the external world, it might not be unreasonable to assign a quantum state $|e\rangle$ to whatever part of the world comes into interactive contact with the Qbit or Qbits – their *environment*. We will not make any specific assumptions about the character of that environment or of the quantum state $|e\rangle$ associated with it, beyond noting that, unlike the state of a single Qbit, the state of the environment is likely to be a state in a space of enormously many dimensions if there is any complexity to the environment that couples, however weakly, to the Qbit.

If, in spite of this recommended point of view, you still feel uncomfortable applying quantum states to noncomputational degrees of freedom, then I invite you to regard $|e\rangle$ as the state of some enormous collection of extra Qbits, from which one would like the computation to be completely decoupled, but which, for reasons beyond our control, somehow manage to interact weakly with the Qbits we are actually interested in. I offer this invitation as a conceptual aid to computer scientists uncomfortable with my claim that quantum states apply to the description of arbitrary physical systems. But I also note that in recent years a few physicists have suggested that the entire world should indeed be viewed as an enormous collection of Qbits – a position that has not attracted many adherents to date.

Returning from grand world views to the practical reality of errors in a quantum computation, we shall regard a single Qbit, initially in the state $|x\rangle$ ($x = 0$ or 1), as being part of a larger system consisting of the Qbit plus its environment, initially in the state $|e\rangle|x\rangle$. In the ideal case, as the Qbit evolves under 1-Qbit unitary gates or interacts with other Qbits under 2-Qbit unitary gates, it stays unentangled with its environment. The environmental component of the state is then irrelevant to the computational process and can be ignored, as we have been doing up to now.

Unfortunately, however, interactions with the environment will in general transform and entangle the states of the Qbit and its environment. The most general way in which this can come about can be expressed in the form

$$\begin{aligned} |e\rangle|0\rangle &\to |e_0\rangle|0\rangle + |e_1\rangle|1\rangle, \\ |e\rangle|1\rangle &\to |e_2\rangle|0\rangle + |e_3\rangle|1\rangle, \end{aligned} \tag{5.7}$$

where $|e\rangle$ is the initially uncorrelated state of the environment and $|e_0\rangle, \ldots, |e_3\rangle$ are possible final environmental states. The environmental final states are not necessarily orthogonal or normalized, and are constrained only by the requirement that the two states on the right side of (5.7) should be orthogonal, since the Qbit–environment interaction is required, like any other physical interaction, to lead to a unitary development in time. This corruption of a computation by the entanglement of the state of Qbits with the state of their environment is called *decoherence*. It is the primary enemy of quantum computation.

Included in (5.7) are cases like the oversimplified one we examined in Section 5.2, in which the Qbit remains isolated from the environment ($|e_i\rangle = a_i|e\rangle,\ i = 0, \ldots, 3$) but still suffers in that isolation an unintended unitary evolution. But (5.7) also includes the case of major practical interest. This is the case in which the interaction with the environment has a small but otherwise quite general entangling effect on the Qbit:

$$|e_0\rangle \approx |e_3\rangle \approx |e\rangle; \qquad \langle e_1|e_1\rangle,\ \langle e_2|e_2\rangle \ll 1. \tag{5.8}$$

In dealing with such entangling interactions with the environment, it is useful to introduce projection operators

$$\mathbf{P}_x = \frac{1 + (-1)^x \mathbf{Z}}{2}, \tag{5.9}$$

which project onto the 1-Qbit states $|x\rangle,\ x = 0, 1$. Using these projection operators, we can combine the two time evolutions in (5.7) into a single form:

$$|e\rangle|x\rangle \to \big(\big[|e_0\rangle\mathbf{1} + |e_1\rangle\mathbf{X}\big]\mathbf{P}_0\big)|x\rangle + \big(\big[|e_2\rangle\mathbf{X} + |e_3\rangle\mathbf{1}\big]\mathbf{P}_1\big)|x\rangle. \tag{5.10}$$

In (5.10) I have introduced the convenient notation $|e\rangle\mathbf{U}$ to describe the linear operator from a 1-Qbit to a many-Qbit space that takes the 1-Qbit state $|\psi\rangle$ into the many-Qbit state $|e\rangle \otimes \mathbf{U}|\psi\rangle$; like most embellishments of Dirac notation it is defined so that the appropriate form of the associative law holds:

$$\big(|e\rangle\mathbf{U}\big)|\psi\rangle = |e\rangle \otimes \mathbf{U}|\psi\rangle. \tag{5.11}$$

Using the explicit form (5.9) of the two projection operators, defining[1]

$$\mathbf{Y} = \mathbf{ZX}, \tag{5.12}$$

and continuing to use the notational convention (5.11), we can rewrite (5.10) as

$$|e\rangle|x\rangle \to \bigg(\frac{|e_0\rangle + |e_3\rangle}{2}\mathbf{1} + \frac{|e_0\rangle - |e_3\rangle}{2}\mathbf{Z} + \frac{|e_2\rangle + |e_1\rangle}{2}\mathbf{X} + \frac{|e_2\rangle - |e_1\rangle}{2}\mathbf{Y}\bigg)|x\rangle. \tag{5.13}$$

There is nothing special about the particular environmental states appearing in (5.13), so we can rewrite it more compactly in terms of four other (in general neither normalized nor orthogonal) states $|a\rangle$, $|b\rangle$, $|c\rangle$, and $|d\rangle$ of the environment as

$$|e\rangle|x\rangle \to \big(|d\rangle\mathbf{1} + |a\rangle\mathbf{X} + |b\rangle\mathbf{Y} + |c\rangle\mathbf{Z}\big)|x\rangle. \tag{5.14}$$

The time development represented by the arrow in (5.14) is unitary and therefore linear, so the combination of environmental states and unitary operators on the right side of (5.14) acts linearly on $|x\rangle$. Therefore (5.14) holds not only for $|e\rangle|0\rangle$ and $|e\rangle|1\rangle$ but also for any superposition $\alpha|e\rangle|0\rangle + \beta|e\rangle|1\rangle = |e\rangle\big(\alpha|0\rangle + \beta|1\rangle\big) = |e\rangle|\psi\rangle$, in the form

$$|e\rangle|\psi\rangle \to \big(|d\rangle\mathbf{1} + |a\rangle\mathbf{X} + |b\rangle\mathbf{Y} + |c\rangle\mathbf{Z}\big)|\psi\rangle. \tag{5.15}$$

The actions of $\mathbf{X}$, $\mathbf{Z}$, and $\mathbf{Y}$ on the uncorrupted state $|\psi\rangle$ are sometimes described as subjecting the Qbit to a bit-flip error, a phase error, and a combined bit-flip and phase error. Using this terminology, a general environmental degradation of the state of a Qbit, which can always be put in the form (5.15), can be viewed as a superposition of no error ($\mathbf{1}$), a bit-flip error ($\mathbf{X}$), a combined bit-flip and phase error ($\mathbf{Y}$), and a phase error ($\mathbf{Z}$). The oversimplified example of Section 5.2 ignored the possibility of phase errors ($\mathbf{Z}$) and combined errors ($\mathbf{Y}$).

If we were to extend this analysis to the corruption of an n-Qbit codeword $|\Psi\rangle_n$, we would end up with a combined state of the codeword and the environment of the form

$$|e\rangle|\Psi\rangle \to \sum_{\mu_1=0}^{3}\cdots\sum_{\mu_n=0}^{3} |e_{\mu_1\cdots\mu_n}\rangle \mathbf{X}^{(\mu_1)} \otimes \cdots \otimes \mathbf{X}^{(\mu_n)}|\Psi\rangle, \tag{5.16}$$

where

$$\mathbf{X}^{(0)} = \mathbf{1}, \qquad \mathbf{X}^{(1)} = \mathbf{X}, \qquad \mathbf{X}^{(2)} = \mathbf{Y}, \qquad \mathbf{X}^{(3)} = \mathbf{Z}. \tag{5.17}$$

The construction of error-correcting codewords requires a physical assumption analogous to the assumption in Section 5.2 that at most a

1 This $\mathbf{Y}$ differs by a factor of i from the $\mathbf{Y}$ briefly used in Section 1.4.

single Qbit in a codeword suffers a flip error. If $|\Psi\rangle$ is a state of a small number n of Qbits that make up an n-Qbit codeword, then the probability of corruption of the codeword is so small that the terms in (5.16) differing from the term $\mathbf{1} \otimes \cdots \otimes \mathbf{1}$ that reproduces the uncorrupted state are dominated by those in which only a single one of the $\mathbf{X}^{(\mu_i)}$ differs from $\mathbf{1}$. If this condition is met, then the general form (5.16) of a corrupted n-Qbit codeword is a superposition of terms in which each individual Qbit making up the word has suffered a degradation of the form (5.15):

$$|e\rangle|\Psi\rangle \to \left(|d\rangle\mathbf{1} + \sum_{i=0}^{n-1} |a_i\rangle\mathbf{X}_i + |b_i\rangle\mathbf{Y}_i + |c_i\rangle\mathbf{Z}_i\right)|\Psi\rangle. \tag{5.18}$$

One can allow for the more general possibility of two or more Qbits in a codeword being corrupted together if one is willing to use longer codewords to correct such errors. The examples of error correction given below are all at the level of single-Qbit errors of the form (5.18) in the codeword. The extent to which the dominant sources of error will actually be of this form may well depend on the kind of physical system used to realize the Qbits. Eventually the theory of quantum error correction will have to face this issue. Meanwhile this possible future source of difficulty should not distract you from appreciating how remarkable it is that an error-correction procedure exists at all, even in the restricted setting of single-Qbit errors.

To correct 1-Qbit errors we require a procedure that restores a corrupted state of the form

$$|d\rangle|\Psi\rangle + \sum_{i=0}^{n-1} \big(|a_i\rangle\mathbf{X}_i|\Psi\rangle + |b_i\rangle\mathbf{Y}_i|\Psi\rangle + |c_i\rangle\mathbf{Z}_i|\Psi\rangle\big) \tag{5.19}$$

to the uncorrupted form

$$|e\rangle|\Psi\rangle, \tag{5.20}$$

where $|e\rangle$ is the environmental state accompanying whichever of the $3n + 1$ terms in (5.19) our error-correction procedure has projected the corrupted state onto. If the term in $\mathbf{X}_i$ were the only one present in (5.19), we could use a 3-Qbit codeword and achieve this projection by applying precisely the error-correction technique described in Section 5.2. But to deal with the additional possibilities associated with the terms in $\mathbf{Y}_i$ and $\mathbf{Z}_i$ we require longer codewords and more elaborate diagnostic methods.

5.4 Diagnosing error syndromes

Before turning to specific quantum error-correcting codes, it is useful to anticipate the general structure of the gates we will be using to identify and project onto a particular term in the general 1-Qbit corruption (5.19) of a codeword. As noted earlier, these will be generalizations of the controlled $\mathbf{Z}_2\mathbf{Z}_1$ and $\mathbf{Z}_1\mathbf{Z}_0$ gates used to diagnose errors in the artificial case in which only bit-flip errors are allowed.

Let $\mathbf{A}$ be *any* n-Qbit Hermitian operator whose square is the unit operator:

$$\mathbf{A}^2 = \mathbf{1}. \tag{5.21}$$

It follows from (5.21) that $\mathbf{A}$ is unitary, since $\mathbf{A}^\dagger = \mathbf{A}$. The eigenvalues of $\mathbf{A}$ can only be 1 or -1, since $\mathbf{A}$ acting twice on an eigenstate must act as the identity $\mathbf{1}$. The projection operators onto the subspaces of states with eigenvalue $+1$ and -1 are, respectively,

$$\mathbf{P}_0^A = \frac{\mathbf{1}+\mathbf{A}}{2} \quad \text{and} \quad \mathbf{P}_1^A = \frac{\mathbf{1}-\mathbf{A}}{2}. \tag{5.22}$$

Since $\mathbf{P}_0 + \mathbf{P}_1 = \mathbf{1}$, any state $|\psi\rangle$ can be expressed as a superposition of its projections onto these two subspaces: $|\psi\rangle = \mathbf{P}_0|\psi\rangle + \mathbf{P}_1|\psi\rangle$.

The operators $\mathbf{Z}_2\mathbf{Z}_1$ and $\mathbf{Z}_1\mathbf{Z}_0$ encountered in the 3-Qbit code for correcting bit-flip errors are examples of such $\mathbf{A}$. In the more general cases we shall be examining, the operators $\mathbf{A}$ will be more general products of both $\mathbf{Z}$ and $\mathbf{X}$ operators associated with different Qbits in the codeword; for example $\mathbf{A} = \mathbf{Z}_4\mathbf{X}_3\mathbf{Z}_2\mathbf{X}_1\mathbf{X}_0$.

In addition to the n Qbits on which $\mathbf{A}$ acts, we introduce an ancillary Qbit and consider the controlled operator $\mathbf{C}^A$, which we write here in the alternative form $\mathbf{cA}$ to avoid having subscripts on superscripts, which acts as $\mathbf{A}$ on the n Qbits when the state of the ancilla is $|1\rangle$ and as the identity when the state of the ancilla is $|0\rangle$. If the state of the ancilla is a superposition of $|0\rangle$ and $|1\rangle$, the action of $\mathbf{cA}$ is defined by linearity. When $\mathbf{A}$ is a product of 1-Qbit operators, the operator $\mathbf{cA}$ can be taken to be a product of ordinary 2-Qbit controlled operators. If $\mathbf{A} = \mathbf{Z}_4\mathbf{X}_3\mathbf{Z}_2\mathbf{X}_1\mathbf{X}_0$, then $\mathbf{cA}$ would be $\mathbf{cZ}_4\mathbf{cX}_3\mathbf{cZ}_2\mathbf{cX}_1\mathbf{cX}_0$, where each of the five terms has a different target Qbit, but all are controlled by one and the same ancilla.

If the ancilla is initially in the state $|0\rangle$ and one applies a Hadamard transform $\mathbf{H}$ to the ancilla both before and after applying $\mathbf{cA}$ to the $n+1$ Qbits and the initial state of the n Qbits is $|\Psi\rangle$, then the n Qbits will end up entangled with the ancilla in the state

$$\begin{aligned}
(\mathbf{H}\otimes\mathbf{1})\mathbf{cA}(\mathbf{H}\otimes\mathbf{1})|0\rangle|\Psi\rangle &= (\mathbf{H}\otimes\mathbf{1})\mathbf{cA}\tfrac{1}{\sqrt{2}}\big(|0\rangle+|1\rangle\big)|\Psi\rangle \\
&= (\mathbf{H}\otimes\mathbf{1})\tfrac{1}{\sqrt{2}}\big(|0\rangle|\Psi\rangle+|1\rangle\mathbf{A}|\Psi\rangle\big) \\
&= \tfrac{1}{2}\big(|0\rangle+|1\rangle\big)|\Psi\rangle+\tfrac{1}{2}\big(|0\rangle-|1\rangle\big)\mathbf{A}|\Psi\rangle \\
&= |0\rangle\tfrac{1}{2}(\mathbf{1}+\mathbf{A})|\Psi\rangle+|1\rangle\tfrac{1}{2}(\mathbf{1}-\mathbf{A})|\Psi\rangle \\
&= |0\rangle\mathbf{P}_0^A|\Psi\rangle+|1\rangle\mathbf{P}_1^A|\Psi\rangle.
\end{aligned} \tag{5.23}$$

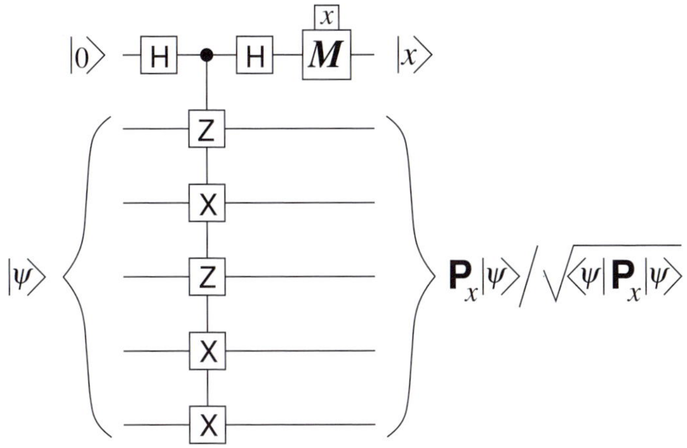

Fig 5.6 The way in which measurement gates are employed in quantum error correction. The ancilla (upper wire) is initially in the state zero. The remaining five Qbits are initially in the state $|\psi\rangle$. If the measurement gate acting on the ancilla gives the result x (0 or 1) then the final state of the five Qbits will be the (renormalized) projection $\mathbf{P}_x|\psi\rangle$ of the initial state into the subspace spanned by the eigenstates of $\mathbf{Z}_4\mathbf{X}_3\mathbf{Z}_2\mathbf{X}_1\mathbf{X}_0$ with eigenvalue $(-1)^x$.

If we now measure the ancilla, then according to the generalized Born rule, if the measurement gate indicates 0 or 1, then the state of the n Qbits becomes the (renormalized) projection of $|\Psi\rangle$ into the subspace of positive (eigenvalue $+1$) or negative (eigenvalue -1) eigenstates of $\mathbf{A}$. This is illustrated for the case $\mathbf{A} = \mathbf{Z}_4\mathbf{X}_3\mathbf{Z}_2\mathbf{X}_1\mathbf{X}_0$ in Figure 5.6.

This procedure is called *measuring* $\mathbf{A}$ or a *measurement* of $\mathbf{A}$. The terminology reflects the fact that it is a generalization of the ordinary process of measuring a single Qbit, to which it reduces when $n = 1$ and $\mathbf{A} = \mathbf{Z}$. In that case the subspaces spanned by the positive and negative eigenstates of $\mathbf{Z}$ are just the one-dimensional subspaces spanned by $|0\rangle$ and $|1\rangle$, and the probabilities of the two outcomes, as one can easily check, are indeed given by the Born rule.

In error correction one needs several such Hermitian operators, each squaring to unity, all acting on the same n Qbits. For concreteness consider the case of three such operators, $\mathbf{A}$, $\mathbf{B}$, and $\mathbf{C}$. Introduce an ancillary Qbit for each operator, labeling the ancillas 0, 1, and 2, and introduce controlled operators $\mathbf{cA}$, $\mathbf{cB}$, and $\mathbf{cC}$, each controlled by the corresponding ancilla. Now apply Hadamards to each of the ancillas (each initially taken to be in the state $|0\rangle$), both before and after the product of all the controlled operators acts. The result (see Figure 5.7) is the obvious generalization of (5.23), taking $|0\rangle|0\rangle|0\rangle|\Psi\rangle$ into

$$
\begin{aligned}
&\big(\mathbf{H}_2\mathbf{H}_1\mathbf{H}_0\big)\big(\mathbf{cCcBcA}\big)\big(\mathbf{H}_2\mathbf{H}_1\mathbf{H}_0\big)|0\rangle|0\rangle|0\rangle|\Psi\rangle \\
&= \sum_{x_2=0}^{1}\sum_{x_1=0}^{1}\sum_{x_0=0}^{1}|x_2\rangle|x_1\rangle|x_0\rangle\left(\frac{1+(-1)^{x_2}\mathbf{C}}{2}\right)\left(\frac{1+(-1)^{x_1}\mathbf{B}}{2}\right) \\
&\qquad\qquad \times\left(\frac{1+(-1)^{x_0}\mathbf{A}}{2}\right)|\Psi\rangle \\
&= \sum_{x_2=0}^{1}\sum_{x_1=0}^{1}\sum_{x_0=0}^{1}|x_2\rangle|x_1\rangle|x_0\rangle\mathbf{P}_{x_2}^{C}\mathbf{P}_{x_1}^{B}\mathbf{P}_{x_0}^{A}|\Psi\rangle.
\end{aligned}
\tag{5.24}
$$

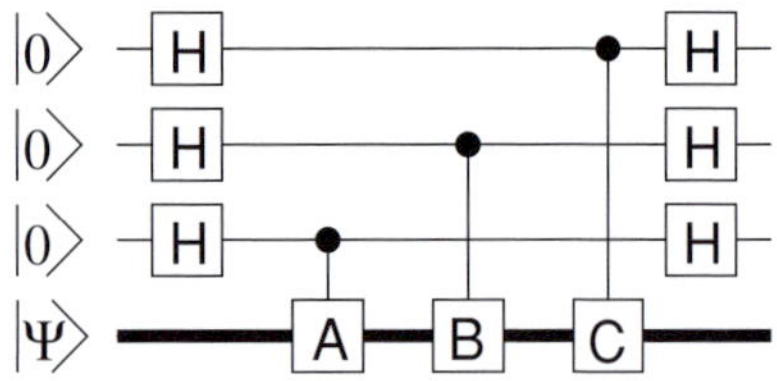

Fig 5.7 **A**, **B**, and **C** are commuting operators satisfying $\mathbf{A}^2 = \mathbf{B}^2 = \mathbf{C}^2 = \mathbf{1}$. They act on the n-Qbit state $|\Psi\rangle$ associated with the thick lower wire. The effect of measuring the three ancillas (top three wires) is to project the state of the n Qbits associated with the lower wire into its component in one of the eight eigenspaces of **A**, **B**, and **C**. If the results of measuring the control bits associated with **A**, **B**, and **C** are x_0, x_1, and x_2 then the projection is into the eigenspace with eigenvalues $(-1)^{x_0}$, $(-1)^{x_1}$, and $(-1)^{x_2}$. Such a process is called "measuring **A**, **B**, and **C**." When $n = 3$ and **A**, **B**, and **C** are three different 1-Qbit **Z** operators, the process is equivalent to an ordinary measurement of the three Qbits on which the three **Z** operators act.

If **A**, **B**, and **C** all *commute* – which is always the case in the examples relevant to error correction – then the state

$$\mathbf{P}^C_{x_2}\mathbf{P}^B_{x_1}\mathbf{P}^A_{x_0}|\Psi\rangle = \left(\frac{1+(-1)^{x_2}\mathbf{C}}{2}\right)\left(\frac{1+(-1)^{x_1}\mathbf{B}}{2}\right)\left(\frac{1+(-1)^{x_0}\mathbf{A}}{2}\right) \tag{5.25}$$

is an eigenstate of all the operators **C**, **B**, and **A**, with respective eigenvalues

$$(-1)^{x_2}, (-1)^{x_1}, \text{ and } (-1)^{x_0}. \tag{5.26}$$

This follows directly from the fact that if $\mathbf{V}^2 = 1$ then

$$\mathbf{V}\left(\frac{1+(-1)^x\mathbf{V}}{2}\right) = (-1)^x\left(\frac{1+(-1)^x\mathbf{V}}{2}\right). \tag{5.27}$$

So measurement of the three ancillas projects the n Qbits into one of the eight simultaneous eigenspaces of the three commuting operators **C**, **B**, and **A**, and the outcome $x_2x_1x_0$ of the measurement determines which eigenspace it is. This process is described as a joint measurement of **C**, **B**, and **A**.

Note that if **A**, **B**, and **C** are 1-Qbit operators $\mathbf{Z}_i$, $\mathbf{Z}_j$, and $\mathbf{Z}_k$ that act on the ith, jth, and kth of the n Qbits, then this process reduces to the ordinary measurement of those three Qbits, since $\frac{1}{2}(1+(-1)^x\mathbf{Z})$ projects onto the 1-Qbit state $|x\rangle$. The two equivalent error-correction circuits in Figures 5.3 and 5.5 are measurements, in this generalized sense, of the two commuting operators $\mathbf{A} = \mathbf{Z}_2\mathbf{Z}_1$ and $\mathbf{B} = \mathbf{Z}_1\mathbf{Z}_0$.

The form (5.18) of a general 1-Qbit error on an n-Qbit codeword reveals that to correct errors it is necessary to make a measurement, in this more general sense of the term, that projects a possibly corrupted codeword into an identifiable one of $1 + 3n$ orthogonal two-dimensional spaces: one two-dimensional subspace for the uncorrupted codeword $|\Psi\rangle$, and $3n$ additional two-dimensional subspaces for each of the 1-Qbit error terms $\mathbf{X}_i|\Psi\rangle$, $\mathbf{Y}_i|\Psi\rangle$, and $\mathbf{Z}_i|\Psi\rangle$, $i = 0, \ldots, n-1$, in (5.18). Thus the 2^n-dimensional space spanned by all the states of the n Qbits must be large enough to contain $1 + 3n$ orthogonal two-dimensional subspaces, giving us the condition

$$2^{n-1} \geq 3n + 1 \tag{5.28}$$

on an n-Qbit code capable of correcting a general 1-Qbit error. The lowest n satisfying this condition is $n = 5$, for which it holds as an equality. Remarkably, there is indeed a 5-Qbit code for which this can be done. This is reminiscent of the situation in Section 5.2, where it was necessary only to discriminate between the uncorrupted codeword $|\Psi\rangle$ and the n NOT-corruptions $\mathbf{X}_i|\Psi\rangle$. There the number of Qbits had to satisfy (5.5), which is first satisfied (again as an equality) when $n = 3$.

The 5-Qbit code is the most compact and elegant of the quantum error-correcting codes, but it suffers from the fact that it is difficult to construct the appropriate generalizations of 1- and 2-Qbit gates between codewords. I therefore go on to describe a second, 7-Qbit code, which overcomes this problem. The first quantum error-correcting code, discovered by Peter Shor, which uses a 9-Qbit generalization of the 3-Qbit code of Section 5.2, is now of solely historical interest. It is described in Appendix N.

5.5 The 5-Qbit error-correcting code

The two 5-Qbit code words $|\overline{0}\rangle$ and $|\overline{1}\rangle$ are most conveniently defined in terms of the very operators, described in general terms in Section 5.4, that will be used to diagnose the error syndrome. So we begin by specifying those operators.

To distinguish $1 + (3 \times 5) = 16$ mutually orthogonal two-dimensional subspaces we require four such mutually commuting Hermitian operators that square to unity, since each can independently have two eigenvalues (± 1) and $2^4 = 16$. These operators are defined as follows:

$$\begin{aligned}\mathbf{M}_0 &= \mathbf{Z}_1\mathbf{X}_2\mathbf{X}_3\mathbf{Z}_4,\\ \mathbf{M}_1 &= \mathbf{Z}_2\mathbf{X}_3\mathbf{X}_4\mathbf{Z}_0,\\ \mathbf{M}_2 &= \mathbf{Z}_3\mathbf{X}_4\mathbf{X}_0\mathbf{Z}_1,\\ \mathbf{M}_3 &= \mathbf{Z}_4\mathbf{X}_0\mathbf{X}_1\mathbf{Z}_2.\end{aligned} \tag{5.29}$$

Each of the $\mathbf{M}_i$ squares to unity because each is a product of commuting operators that square to unity. To check that the $\mathbf{M}_i$ are mutually commuting, note that all the individual $\mathbf{X}_i$ and $\mathbf{Z}_j$ operators commute with one another except for an $\mathbf{X}_i$ and $\mathbf{Z}_i$ with the same index, which anticommute: $\mathbf{X}_i\mathbf{Z}_i = -\mathbf{Z}_i\mathbf{X}_i$. But in converting the product of any two different $\mathbf{M}_i$ to the product in the reverse order by reversing the orders of the individual $\mathbf{X}_i$ and $\mathbf{Z}_i$ operators that make them up, one always encounters exactly two interchanges that result in a minus sign.

One might be tempted to break the irritating asymmetry of (5.30) by adding to the list

$$\mathbf{M}_4 = \mathbf{Z}_0\mathbf{X}_1\mathbf{X}_2\mathbf{Z}_3, \tag{5.30}$$

but it is not independent of the other four. Every $\mathbf{X}_i$ and $\mathbf{Z}_i$ appears exactly twice in the product of all five $\mathbf{M}_i$, so the product must be either 1 or -1. One easily checks that

$$\mathbf{M}_0\mathbf{M}_1\mathbf{M}_2\mathbf{M}_3\mathbf{M}_4 = \mathbf{1}, \tag{5.31}$$

and therefore

$$\mathbf{M}_4 = \mathbf{M}_0\mathbf{M}_1\mathbf{M}_2\mathbf{M}_3. \tag{5.32}$$

The 5-Qbit codewords are most clearly and usefully defined in terms of the $\mathbf{M}_i$ (rather than writing out their lengthy explicit expansions in computational-basis states):

$$\begin{aligned} |\bar{0}\rangle &= \tfrac{1}{4}(\mathbf{1}+\mathbf{M}_0)(\mathbf{1}+\mathbf{M}_1)(\mathbf{1}+\mathbf{M}_2)(\mathbf{1}+\mathbf{M}_3)|00000\rangle, \\ |\bar{1}\rangle &= \tfrac{1}{4}(\mathbf{1}+\mathbf{M}_0)(\mathbf{1}+\mathbf{M}_1)(\mathbf{1}+\mathbf{M}_2)(\mathbf{1}+\mathbf{M}_3)|11111\rangle. \end{aligned} \tag{5.33}$$

Before examining how one might produce five Qbits in either of these states, we discuss how the states work to correct 1-Qbit errors.

Since each $\mathbf{M}$ flips two Qbits, $|\bar{0}\rangle$ is a superposition of computational-basis states with an odd number of zeros (and an even number of ones), while $|\bar{1}\rangle$ is a superposition of states with an odd number of ones (and an even number of zeros). Consequently the two codeword states are orthogonal. They are also normalized to unity. Since $\mathbf{M}_i^2 = 1$,

$$(\mathbf{1}+\mathbf{M}_i)^2 = 2(\mathbf{1}+\mathbf{M}_i). \tag{5.34}$$

So we have

$$\begin{aligned} \langle\bar{0}|\bar{0}\rangle &= \langle 00000|(\mathbf{1}+\mathbf{M}_0)(\mathbf{1}+\mathbf{M}_1)(\mathbf{1}+\mathbf{M}_2)(\mathbf{1}+\mathbf{M}_3)|00000\rangle, \\ \langle\bar{1}|\bar{1}\rangle &= \langle 11111|(\mathbf{1}+\mathbf{M}_0)(\mathbf{1}+\mathbf{M}_1)(\mathbf{1}+\mathbf{M}_2)(\mathbf{1}+\mathbf{M}_3)|11111\rangle. \end{aligned} \tag{5.35}$$

If we expand the products of $\mathbf{1}+\mathbf{M}_i$ into 16 terms, the term $\mathbf{1}$ contributes 1 to $\langle\bar{0}|\bar{0}\rangle$ and to $\langle\bar{1}|\bar{1}\rangle$. Each of the remaining 15 terms can be reduced, using (5.31) (and the fact that each $\mathbf{M}_i^2 = \mathbf{1}$), to either a single $\mathbf{M}_i$ or a product of two ($i = 0, \ldots, 4$). So each of the 15 terms flips either two or four Qbits and contributes 0 to the inner products.

Because the $\mathbf{M}_i$ all commute and because

$$\mathbf{M}_i(\mathbf{1}+\mathbf{M}_i) = \mathbf{1}+\mathbf{M}_i, \tag{5.36}$$

the states $|\bar{0}\rangle$, $|\bar{1}\rangle$, and their superpositions

$$|\Psi\rangle = \alpha|\bar{0}\rangle + \beta|\bar{1}\rangle \tag{5.37}$$

are all eigenstates of each of the $\mathbf{M}_i$ with eigenvalue 1.

The 15 possible corruptions of (5.37) appearing in the corrupted state (5.18) are also eigenstates of the $\mathbf{M}_i$, distinguished by the

Table 5.2. The four error-syndrome operators $\mathbf{M}_i$ for the 5-Qbit code, and whether each of them commutes (+) or anticommutes (−) with each of the 15 operators $\mathbf{X}_i$, $\mathbf{Y}_i$, and $\mathbf{Z}_i$, $i = 1, \ldots, 5$, associated with the 15 different terms in the corrupted codeword. Note that each of the 15 columns, and the 16th column associated with **1** (no error), has a unique pattern of + and − signs.

	$\mathbf{X}_0\mathbf{Y}_0\mathbf{Z}_0$	$\mathbf{X}_1\mathbf{Y}_1\mathbf{Z}_1$	$\mathbf{X}_2\mathbf{Y}_2\mathbf{Z}_2$	$\mathbf{X}_3\mathbf{Y}_3\mathbf{Z}_3$	$\mathbf{X}_4\mathbf{Y}_4\mathbf{Z}_4$	**1**
$\mathbf{M}_0 = \mathbf{Z}_1\mathbf{X}_2\mathbf{X}_3\mathbf{Z}_4$	$+ + +$	$- - +$	$+ - -$	$+ - -$	$- - +$	$+$
$\mathbf{M}_1 = \mathbf{Z}_2\mathbf{X}_3\mathbf{X}_4\mathbf{Z}_0$	$- - +$	$+ + +$	$- - +$	$+ - -$	$+ - -$	$+$
$\mathbf{M}_2 = \mathbf{Z}_3\mathbf{X}_4\mathbf{X}_0\mathbf{Z}_1$	$+ - -$	$- - +$	$+ + +$	$- - +$	$+ - -$	$+$
$\mathbf{M}_3 = \mathbf{Z}_4\mathbf{X}_0\mathbf{X}_1\mathbf{Z}_2$	$+ - -$	$+ - -$	$- - +$	$+ + +$	$- - +$	$+$

$15 = 2^4 - 1$ other possible sets of eigenvalues ± 1 that the four $\mathbf{M}_i$ ($i = 0, \ldots, 3$) can have. To see this, note first that each $\mathbf{X}_i$, $\mathbf{Y}_i$, and $\mathbf{Z}_i$ commutes or anticommutes with all four $\mathbf{M}_i$. Therefore each of the terms $\mathbf{X}_i|\Psi\rangle$, $\mathbf{Y}_i|\Psi\rangle$, and $\mathbf{Z}_i|\Psi\rangle$ appearing in (5.18) is indeed an eigenstate of each $\mathbf{M}_i$ with eigenvalue 1 or -1.

Table 5.2 indicates whether each $\mathbf{M}_i$ commutes (+) or anticommutes (−) with each of the $\mathbf{X}_i$, $\mathbf{Y}_i$, $\mathbf{Z}_i$, and (trivially) the unit operator **1**. Inspection of the table reveals that each of the 16 possible binary columns of four symbols (+ or −) appears in exactly one column. Therefore, when the four $\mathbf{M}_i$ are measured, the corrupted state (5.18) is projected back to its original form if all four eigenvalues are $+1$, or projected onto one of the 15 corrupted states $\mathbf{X}_0|\Psi\rangle, \ldots, \mathbf{Z}_4|\Psi\rangle$ depending on which column in the table describes the eigenvalues. In each corrupted case the original state can be restored by application of the corresponding unitary transformation $\mathbf{X}_i$, $-\mathbf{Y}_i = \mathbf{X}_i\mathbf{Z}_i$, or $\mathbf{Z}_i$ to the appropriate Qbit. A circuit that measures the four operators (5.29) is shown in Figure 5.8.

The perfect efficiency of the 5-Qbit code leads to a straightforward way to manufacture the two 5-Qbit codeword states (5.33). As noted above, the 16 distinct sets of eigenvalues for the four mutually commuting operators $\mathbf{M}_i$ decompose the 32-dimensional space of five Qbits into 16 mutually orthogonal two-dimensional subspaces, spanned by $|\overline{0}\rangle$ and $|\overline{1}\rangle$ and by each of their 15 pairs of 1-Qbit corruptions.

The two-fold degeneracy of the four $\mathbf{M}_i$ within each of these 16 subspaces is lifted by the operator

$$\overline{\mathbf{Z}} = \mathbf{Z}_0\mathbf{Z}_1\mathbf{Z}_2\mathbf{Z}_3\mathbf{Z}_4, \tag{5.38}$$

which commutes with all the $\mathbf{M}_i$. Since $|00000\rangle$ and $|11111\rangle$ are eigenstates of $\overline{\mathbf{Z}}$ with eigenvalues 1 and -1, and since $\overline{\mathbf{Z}}$ commutes with $\mathbf{Z}_i$,

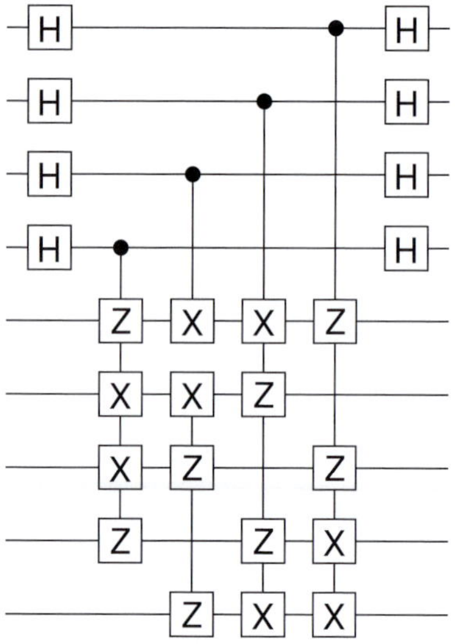

Fig 5.8 A circuit to measure the error syndrome for the 5-Qbit code. The five Qbits are the five lower wires. The four upper wires are the ancillas to be measured in the manner of Figure 5.7, associated with measuring the four commuting operators $\mathbf{Z}_1\mathbf{X}_2\mathbf{X}_3\mathbf{Z}_4$, $\mathbf{Z}_2\mathbf{X}_3\mathbf{X}_4\mathbf{Z}_0$, $\mathbf{Z}_3\mathbf{X}_4\mathbf{X}_0\mathbf{Z}_1$, and $\mathbf{Z}_4\mathbf{X}_0\mathbf{X}_1\mathbf{Z}_2$ of (5.29). When controlled-Z gates are present together with controlled-NOT gates, the figure is more readable if the cNOT gates are represented as controlled-X gates.

while anticommuting with $\mathbf{X}_i$ and $\mathbf{Y}_i$, it follows that

$$\begin{aligned} \overline{\mathbf{Z}}|\overline{0}\rangle &= |\overline{0}\rangle, & \overline{\mathbf{Z}}|\overline{1}\rangle &= -|\overline{1}\rangle, \\ \overline{\mathbf{Z}}\mathbf{Z}_i|\overline{0}\rangle &= \mathbf{Z}_i|\overline{0}\rangle, & \overline{\mathbf{Z}}\mathbf{Z}_i|\overline{1}\rangle &= -\mathbf{Z}_i|\overline{1}\rangle, \\ \overline{\mathbf{Z}}\mathbf{X}_i|\overline{0}\rangle &= -\mathbf{X}_i|\overline{0}\rangle, & \overline{\mathbf{Z}}\mathbf{X}_i|\overline{1}\rangle &= \mathbf{X}_i|\overline{1}\rangle, \\ \overline{\mathbf{Z}}\mathbf{Y}_i|\overline{0}\rangle &= -\mathbf{Y}_i|\overline{0}\rangle, & \overline{\mathbf{Z}}\mathbf{Y}_i|\overline{1}\rangle &= \mathbf{Y}_i|\overline{1}\rangle. \end{aligned} \tag{5.39}$$

Consequently if one takes five Qbits in any state you like (perhaps most conveniently $|00000\rangle$) and measures the four $\mathbf{M}_i$ together with $\overline{\mathbf{Z}}$, one projects the Qbits into one of the 32 states

$$|\overline{0}\rangle,\ \ \mathbf{X}_i|\overline{0}\rangle,\ \ \mathbf{Y}_i|\overline{0}\rangle,\ \ \mathbf{Z}_i|\overline{0}\rangle,\ \ |\overline{1}\rangle,\ \ \mathbf{X}_i|\overline{1}\rangle,\ \ \mathbf{Y}_i|\overline{1}\rangle,\ \ \mathbf{Z}_i|\overline{1}\rangle, \tag{5.40}$$

and learns from the results of the measurement which it is. Just as in the error-correction procedure, if the state is not $|\overline{0}\rangle$ or $|\overline{1}\rangle$ we can restore it to either of these forms by applying the appropriate $\mathbf{X}_i$, $\mathbf{Y}_i$, or $\mathbf{Z}_i$. If we wish to initialize the five Qbits to $|\overline{0}\rangle$ we can apply $\overline{\mathbf{X}}$, where

$$\overline{\mathbf{X}} = \mathbf{X}_0\mathbf{X}_1\mathbf{X}_2\mathbf{X}_3\mathbf{X}_4, \tag{5.41}$$

should the measurement indicate that the error-corrected state is $|\overline{1}\rangle$. This process of using a generalized measurement to produce five Qbits in the state $|\overline{0}\rangle$ is analogous to the procedure of using an ordinary measurement to produce a single Qbit in the state $|0\rangle$ described in Section 1.10.

There is quite a different way to construct the 5-Qbit codewords, by applying a set of 1- and 2-Qbit unitary gates to an uncoded 1-Qbit state and four ancillary Qbits all initially in the state $|0\rangle$. This is described in Section 5.9.

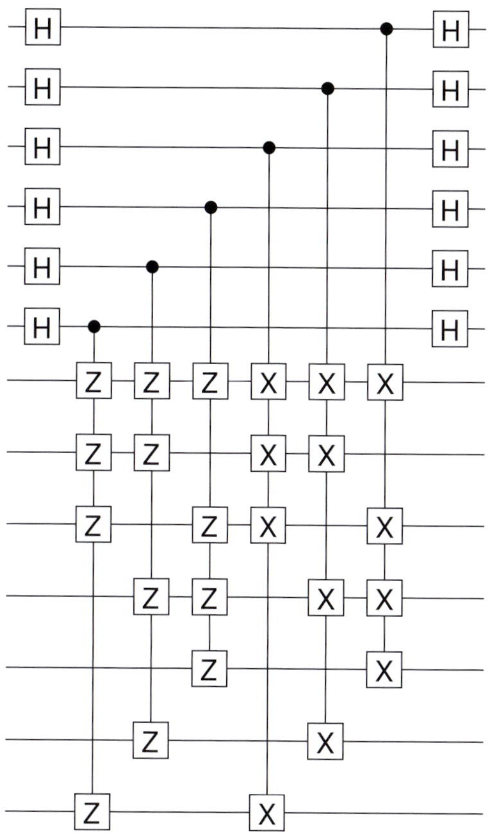

Fig 5.9 A circuit to measure the error syndrome for the 7-Qbit code. The seven Qbits are the seven lower wires. The six upper wires are the ancillas to be measured, resulting in a measurement of the six commuting operators $\mathbf{Z}_0\mathbf{Z}_4\mathbf{Z}_5\mathbf{Z}_6$, $\mathbf{Z}_1\mathbf{Z}_3\mathbf{Z}_5\mathbf{Z}_6$, $\mathbf{Z}_2\mathbf{Z}_3\mathbf{Z}_4\mathbf{Z}_6$, $\mathbf{X}_0\mathbf{X}_4\mathbf{X}_5\mathbf{X}_6$, $\mathbf{X}_1\mathbf{X}_3\mathbf{X}_5\mathbf{X}_6$, and $\mathbf{X}_2\mathbf{X}_3\mathbf{X}_4\mathbf{X}_6$ of (5.42).

5.6 The 7-Qbit error-correcting code

The 5-Qbit code is theoretically ideal but suffers from the problem that circuits performing extensions of many of the basic 1- and 2-Qbit operations to the 5-Qbit codewords are cumbersome. The current favorite is a 7-Qbit code, devised by Andrew Steane, which permits implementations of many basic operations on codewords, which are not only quite simple but also themselves susceptible to error correction.

The Steane code uses six mutually commuting operators to diagnose the error syndrome:

$$\begin{aligned} \mathbf{M}_0 &= \mathbf{X}_0\mathbf{X}_4\mathbf{X}_5\mathbf{X}_6, & \mathbf{N}_0 &= \mathbf{Z}_0\mathbf{Z}_4\mathbf{Z}_5\mathbf{Z}_6, \\ \mathbf{M}_1 &= \mathbf{X}_1\mathbf{X}_3\mathbf{X}_5\mathbf{X}_6, & \mathbf{N}_1 &= \mathbf{Z}_1\mathbf{Z}_3\mathbf{Z}_5\mathbf{Z}_6, \\ \mathbf{M}_2 &= \mathbf{X}_2\mathbf{X}_3\mathbf{X}_4\mathbf{X}_6, & \mathbf{N}_2 &= \mathbf{Z}_2\mathbf{Z}_3\mathbf{Z}_4\mathbf{Z}_6. \end{aligned} \tag{5.42}$$

The six operators in (5.42) clearly square to give the unit operator. The $\mathbf{M}_i$ trivially commute among themselves as do the $\mathbf{N}_i$, and each $\mathbf{M}_i$ commutes with each $\mathbf{N}_j$, in spite of the anticommutation of each $\mathbf{X}_k$ with the corresponding $\mathbf{Z}_k$, because in every case they share an even number of such pairs. A circuit that measures the six operators (5.42) is shown in Figure 5.9.

The 7-Qbit codewords are defined by

$$|\bar{0}\rangle = 2^{-3/2}(\mathbf{1} + \mathsf{M}_0)(\mathbf{1} + \mathsf{M}_1)(\mathbf{1} + \mathsf{M}_2)|0\rangle_7,$$
$$|\bar{1}\rangle = 2^{-3/2}(\mathbf{1} + \mathsf{M}_0)(\mathbf{1} + \mathsf{M}_1)(\mathbf{1} + \mathsf{M}_2)\overline{\mathsf{X}}|0\rangle_7, \tag{5.43}$$

where

$$\overline{\mathsf{X}} = \mathsf{X}_0\mathsf{X}_1\mathsf{X}_2\mathsf{X}_3\mathsf{X}_4\mathsf{X}_5\mathsf{X}_6, \tag{5.44}$$

so that

$$|1111111\rangle = \overline{\mathsf{X}}|0000000\rangle. \tag{5.45}$$

We again defer our discussion of how to produce these states until after our discussion of how they are used in error correction.

The two states in (5.43) are orthogonal, since each M flips four Qbits while $\overline{\mathsf{X}}$ flips all seven of them, so the first state is a superposition of 7-Qbit states with an odd number of zeros while the second is a superposition with an even number of zeros. They are normalized to unity, for essentially the same reasons as in the case of 5-Qbit code.

Since $\overline{\mathsf{X}}$ commutes with all the M_i, a general superposition of the two codewords can be written as

$$|\Psi\rangle = \alpha|\bar{0}\rangle + \beta|\bar{1}\rangle = \big(\alpha\mathbf{1} + \beta\overline{\mathsf{X}}\big)|\bar{0}\rangle, \tag{5.46}$$

and its corruption (5.18) assumes the form

$$|e\rangle|\Psi\rangle \to \left(|d\rangle\mathbf{1} + \sum_{i=1}^{7}\big[\,|a_i\rangle\mathsf{X}_i + |b_i\rangle\mathsf{Y}_i + |c_i\rangle\mathsf{Z}_i\big]\right)|\Psi\rangle. \tag{5.47}$$

Because the M_i all commute and $\mathsf{M}_i(\mathbf{1} + \mathsf{M}_i) = \mathbf{1} + \mathsf{M}_i$, and because the N_j commute with the the M_i and with $\overline{\mathsf{X}}$ and have $|0000000\rangle$ as an eigenstate with eigenvalue 1, it follows that $|\bar{0}\rangle$, $|\bar{1}\rangle$, and the general superposition (5.46) are eigenstates of each of the M_i and N_i with eigenvalue 1. The 21 possible corruptions of (5.46) appearing in (5.47) are also eigenstates, distinguished by the possible sets of eigenvalues ± 1 that the three M_i and three N_i can have. As in the 5-Qbit case, this is because each X_i, Y_i, and Z_i commutes or anticommutes with each of the M_i and N_i, so each state appearing in (5.47) is indeed an eigenstate of each M_i and N_i with eigenvalue 1 or -1.

To see why the results of the six measurements of the M_i and N_i determine a unique one of the 22 terms in (5.47), examine Table 5.3, which indicates by a bullet (•) whether an X_i appears in each of the M_i and whether a Z_i appears in each of the N_i. Each M_i commutes with every X_j; it anticommutes with Y_j and Z_j if a bullet appears in the column associated with X_j and commutes if there is no bullet; each N_i commutes with every Z_j; it anticommutes with X_j and Y_j if a bullet appears in the column associated with Z_j and commutes if there is no bullet.

Table 5.3. The six error-syndrome operators $\mathbf{M}_i$ and $\mathbf{N}_i$, $i = 0, 1, 2$, for the 7-Qbit code. A bullet (•) indicates whether a given $\mathbf{X}_i$ appears in each $\mathbf{M}_i$ and whether a given $\mathbf{Z}_i$ appears in each $\mathbf{N}_i$.

	$\mathbf{X}_0$	$\mathbf{X}_1$	$\mathbf{X}_2$	$\mathbf{X}_3$	$\mathbf{X}_4$	$\mathbf{X}_5$	$\mathbf{X}_6$
$\mathbf{M}_0$	•				•	•	•
$\mathbf{M}_1$		•		•		•	•
$\mathbf{M}_2$			•	•	•		•
	$\mathbf{Z}_0$	$\mathbf{Z}_1$	$\mathbf{Z}_2$	$\mathbf{Z}_3$	$\mathbf{Z}_4$	$\mathbf{Z}_5$	$\mathbf{Z}_6$
$\mathbf{N}_0$	•				•	•	•
$\mathbf{N}_1$		•		•		•	•
$\mathbf{N}_2$			•	•	•		•

The signature of an $\mathbf{X}_i$ error (or no error) is that all three $\mathbf{M}_i$ measurements give $+1$. The pattern of -1 eigenvalues in the $\mathbf{N}_i$ measurements then determines which of the seven possible $\mathbf{X}_i$ characterize the error. (If all three $\mathbf{N}_i$ measurements also give $+1$ there is no error.)

In the same way, the signature of a $\mathbf{Z}_i$ error (or no error) is that all three $\mathbf{N}_i$ measurements give $+1$ and then the pattern of -1 eigenvalues in the $\mathbf{M}_i$ measurements determines which of the seven possible $\mathbf{Z}_i$ characterize the error.

Finally, the signature of a $\mathbf{Y}_i$ error is that at least some of both the $\mathbf{M}_i$ and the $\mathbf{N}_i$ measurements give -1. The resulting pattern of -1 eigenvalues (which will be the same for both the $\mathbf{M}_i$ and the $\mathbf{N}_i$ measurements) then determines which of the seven possible $\mathbf{Y}_i$ characterize the error.

So the six measurements project the corrupted state into a unique one of the 22 terms in (5.47) and establish which term it is. One can then undo the corruption by applying the appropriate one of the 22 operators $\mathbf{1}$, $\mathbf{X}_0, \ldots, \mathbf{Z}_6$.

To produce the 7-Qbit codewords one cannot immediately extend the method we used in Section 5.5 to produce the 5-Qbit codewords, because the two 7-Qbit codewords and their 21 1-Qbit corruptions constitute only 44 mutually orthogonal states, while the space of seven Qbits has dimension $2^7 = 128$. One can, however, provide the missing 84 dimensions by noting the following.

The $2 \times 7 \times 6 = 84$ states given by

$$\mathbf{X}_i\mathbf{Z}_j|\overline{0}\rangle \quad \text{and} \quad \mathbf{X}_i\mathbf{Z}_j|\overline{1}\rangle, \quad i \neq j, \tag{5.48}$$

are also easily verified to be eigenstates of all the $\mathbf{M}_i$ and $\mathbf{N}_i$. These states can be associated with some of the possible 2-Qbit errors, but this is not pertinent to the use to which we put them here. Like the 1-Qbit $\mathbf{Y}_i$ errors, these states result in at least some of both the $\mathbf{M}_i$ and the $\mathbf{N}_i$ measurements giving -1, but unlike the $\mathbf{Y}_i$ errors, the resulting pattern of -1 eigenvalues will *not* be the same for both the $\mathbf{M}_i$ and the $\mathbf{N}_i$ measurements, since $i \neq j$. Each of the $7 \times 6 = 42$ possibilities for $\mathbf{X}_i\mathbf{Z}_j$ leads to its own characteristic pattern of $+1$ and -1 eigenvalues.

This gets us back to the situation we encountered in the 5-Qbit case. By measuring the seven mutually commuting operators $\mathbf{M}_i$, $\mathbf{N}_i$, and

$$\overline{\mathbf{Z}} = \mathbf{Z}_0\mathbf{Z}_1\mathbf{Z}_2\mathbf{Z}_3\mathbf{Z}_4\mathbf{Z}_5\mathbf{Z}_6, \tag{5.49}$$

we can produce from seven Qbits in an arbitrarily chosen state a unique one of the 128 mutually orthogonal states given by $|\overline{0}\rangle$, $|\overline{1}\rangle$, their 42 different 1-Qbit corruptions, and their 84 different special kinds of 2-Qbit corruptions. The results of the measurement tell us the character (if any) of the corruption, from which we know what operators ($\mathbf{X}_i$, $\mathbf{Y}_i$, $\mathbf{Z}_i$, or $\mathbf{X}_i\mathbf{Z}_j$, possibly combined with $\overline{\mathbf{X}}$) we must apply to the post-measurement state to convert it into $|\overline{0}\rangle$.

A simpler way to produce 7-Qbit codewords is to start with seven Qbits in the standard initial state $|0\rangle_7$, and then measure $\mathbf{M}_0$, $\mathbf{M}_1$, and $\mathbf{M}_2$. The resulting state will be one of the eight states

$$2^{-3/2}(\mathbf{1} \pm \mathbf{M}_0)(\mathbf{1} \pm \mathbf{M}_1)(\mathbf{1} \pm \mathbf{M}_2)|0\rangle_7, \tag{5.50}$$

with the specific pattern of $+$ and $-$ signs being revealed by the measurement. The upper part of Table 5.3 now permits one to choose a unique $\mathbf{Z}_i$ that commutes or anticommutes with each $\mathbf{M}_i$ depending on whether it appears in (5.50) with a $+$ or a $-$ sign. Since $\mathbf{Z}_i|0\rangle_7 = |0\rangle_7$, acting on the seven Qbits with that particular $\mathbf{Z}_i$ converts their state to

$$2^{-3/2}(\mathbf{1} + \mathbf{M}_0)(\mathbf{1} + \mathbf{M}_1)(\mathbf{1} + \mathbf{M}_2)|0\rangle_7 = |\overline{0}\rangle. \tag{5.51}$$

In Section 5.8 we examine a surprisingly simple circuit that encodes a general 1-Qbit state into a 7-Qbit codeword state in the manner of Figure 5.1, without using any measurement gates.

5.7 Operations on 7-Qbit codewords

The virtue of the 7-Qbit code, that makes it preferable to the 5-Qbit code in spite of its greater expenditure of Qbits, is that many of the fundamental 1- and 2-Qbit gates are trivially extended to 7- and 14-Qbit gates acting on the codewords. Because, for example, $\overline{\mathbf{X}}$ commutes with the $\mathbf{M}_i$ and flips all seven Qbits, it implements the logical NOT on the codewords (5.43):

$$\overline{\mathbf{X}}|\overline{0}\rangle = |\overline{1}\rangle, \qquad \overline{\mathbf{X}}|\overline{1}\rangle = |\overline{0}\rangle. \tag{5.52}$$

Similarly, $\overline{\mathbf{Z}}$ commutes with the $\mathbf{M}_i$, anticommutes with $\overline{\mathbf{X}}$, and leaves $|0\rangle_7$ invariant, so it implements the logical $\mathbf{Z}$ on the codewords:

$$\overline{\mathbf{Z}}|\overline{0}\rangle = |\overline{0}\rangle, \qquad \overline{\mathbf{Z}}|\overline{1}\rangle = -|\overline{1}\rangle. \tag{5.53}$$

This much works equally well for the 5-Qbit code. More remarkably, for the 7-Qbit code the bitwise Hadamard transformation,

$$\overline{\mathbf{H}} = \mathbf{H}_0\mathbf{H}_1\mathbf{H}_2\mathbf{H}_3\mathbf{H}_4\mathbf{H}_5\mathbf{H}_6, \tag{5.54}$$

also implements the logical Hadamard transformation on the codewords:

$$\overline{\mathbf{H}}|\overline{0}\rangle = \tfrac{1}{\sqrt{2}}\big(|\overline{0}\rangle + |\overline{1}\rangle\big), \qquad \overline{\mathbf{H}}|\overline{1}\rangle = \tfrac{1}{\sqrt{2}}\big(|\overline{0}\rangle - |\overline{1}\rangle\big). \tag{5.55}$$

(This does not hold for the 5-Qbit code.)

To see this, note first that two normalized states $|\phi\rangle$ and $|\psi\rangle$ are identical if and only if their inner product is 1. (For one can always express $|\psi\rangle$ in the form $|\psi\rangle = \alpha|\phi\rangle + \beta|\chi\rangle$, where $|\chi\rangle$ is orthogonal to $|\phi\rangle$ and $|\alpha|^2 + |\beta|^2 = 1$. We then have $\langle\phi|\psi\rangle = \alpha$, so if $\langle\phi|\psi\rangle = 1$, then $\alpha = 1$ and $\beta = 0$.) Since $|\overline{0}\rangle$ and $|\overline{1}\rangle$ are normalized and orthogonal and since $\overline{\mathbf{H}}$ is unitary and therefore preserves the normalization of $|\overline{0}\rangle$ and $|\overline{1}\rangle$, the four states appearing in the two equalities in (5.55) are all normalized. Therefore, to establish those equalities it suffices to show that

$$1 = \tfrac{1}{\sqrt{2}}\big(\langle\overline{0}|\overline{\mathbf{H}}|\overline{0}\rangle + \langle\overline{0}|\overline{\mathbf{H}}|\overline{1}\rangle\big), \qquad 1 = \tfrac{1}{\sqrt{2}}\big(\langle\overline{1}|\overline{\mathbf{H}}|\overline{0}\rangle - \langle\overline{1}|\overline{\mathbf{H}}|\overline{1}\rangle\big). \tag{5.56}$$

This in turn would follow if we could show that the matrix of the encoded Hadamard in the encoded states is the same as the matrix of the 1-Qbit Hadamard in the 1-Qbit states:

$$\langle\overline{0}|\overline{\mathbf{H}}|\overline{0}\rangle = \langle\overline{0}|\overline{\mathbf{H}}|\overline{1}\rangle = \langle\overline{1}|\overline{\mathbf{H}}|\overline{0}\rangle = \tfrac{1}{\sqrt{2}}, \qquad \langle\overline{1}|\overline{\mathbf{H}}|\overline{1}\rangle = -\tfrac{1}{\sqrt{2}}. \tag{5.57}$$

To establish (5.57), note that it follows from the definition (5.43) of the codewords $|\overline{0}\rangle$ and $|\overline{1}\rangle$ that the four matrix elements appearing in (5.57) are

$$\begin{aligned}\langle\overline{x}|\overline{\mathbf{H}}|\overline{y}\rangle = 2^{-3}\,{}_7\langle 0|\overline{\mathbf{X}}^x(\mathbf{1}+\mathbf{M}_0)(\mathbf{1}+\mathbf{M}_1)(\mathbf{1}+\mathbf{M}_2)\overline{\mathbf{H}}(\mathbf{1}+\mathbf{M}_0)\\ \times(\mathbf{1}+\mathbf{M}_1)(\mathbf{1}+\mathbf{M}_2)\overline{\mathbf{X}}^y|0\rangle_7.\end{aligned} \tag{5.58}$$

Since $\mathbf{HX} = \mathbf{ZH}$ and $\mathbf{XH} = \mathbf{HZ}$, and since each $\mathbf{N}_i$ differs from $\mathbf{M}_i$ only by the replacement of each $\mathbf{X}$ by the corresponding $\mathbf{Z}$, it follows that

$$\overline{\mathbf{H}}\,\mathbf{M}_i = \mathbf{N}_i\,\overline{\mathbf{H}}, \qquad \mathbf{M}_i\,\overline{\mathbf{H}} = \overline{\mathbf{H}}\,\mathbf{N}_i. \tag{5.59}$$

So we can bring all three terms $\mathbf{1}+\mathbf{M}_i$ in (5.58) on the right of $\overline{\mathbf{H}}$ over to the left if we replace each by $\mathbf{1}+\mathbf{N}_i$. But since the $\mathbf{M}$s and $\mathbf{N}$s all commute we can then bring all three terms $\mathbf{1}+\mathbf{M}_i$ on the left of $\overline{\mathbf{H}}$ over to the right if we again replace each by $\mathbf{1}+\mathbf{N}_i$. The effect of these interchanges is simply to change all the $\mathbf{M}$s in (5.58) into $\mathbf{N}$s:

$$\begin{aligned}\langle\overline{x}|\overline{\mathbf{H}}|\overline{y}\rangle = 2^{-3}\,{}_7\langle 0|\overline{\mathbf{X}}^x(\mathbf{1}+\mathbf{N}_0)(\mathbf{1}+\mathbf{N}_1)(\mathbf{1}+\mathbf{N}_2)\overline{\mathbf{H}}(\mathbf{1}+\mathbf{N}_0)\\ \times(\mathbf{1}+\mathbf{N}_1)(\mathbf{1}+\mathbf{N}_2)\overline{\mathbf{X}}^y|0\rangle_7.\end{aligned} \tag{5.60}$$

Since each $\mathbf{N}_i$ commutes with $\overline{\mathbf{X}}$ (there are four anticommutations) we have

$$\langle\overline{x}|\overline{\mathbf{H}}|\overline{y}\rangle = 2^{-3}\,{}_7\langle 0|(\mathbf{1}+\mathbf{N}_0)(\mathbf{1}+\mathbf{N}_1)(\mathbf{1}+\mathbf{N}_2)\overline{\mathbf{X}}^x\,\overline{\mathbf{H}}\,\overline{\mathbf{X}}^y(\mathbf{1}+\mathbf{N}_0) \times(\mathbf{1}+\mathbf{N}_1)(\mathbf{1}+\mathbf{N}_2)|0\rangle_7, \tag{5.61}$$

but since each $\mathbf{N}_i$ acts as the identity on $|0\rangle_7$, each of the six $\mathbf{1}+\mathbf{N}_i$ can be replaced by a factor of 2, reducing (5.61) simply to

$$\langle\overline{x}|\overline{\mathbf{H}}|\overline{y}\rangle = 2^3\,{}_7\langle 0|\overline{\mathbf{X}}^x\,\overline{\mathbf{H}}\,\overline{\mathbf{X}}^y|0\rangle_7. \tag{5.62}$$

Since $\overline{\mathbf{X}}$, $\overline{\mathbf{H}}$, and $|0\rangle_7$ are tensor products of the seven 1-Qbit quantities $\mathbf{X}$, $\mathbf{H}$, and $|0\rangle$, (5.62) is just

$$\langle\overline{x}|\overline{\mathbf{H}}|\overline{y}\rangle = 2^3\,\langle x|\mathbf{H}|y\rangle^7. \tag{5.63}$$

But since

$$\langle 0|\mathbf{H}|0\rangle = \langle 0|\mathbf{H}|1\rangle = \langle 1|\mathbf{H}|0\rangle = \tfrac{1}{\sqrt{2}}, \qquad \langle 1|\mathbf{H}|1\rangle = -\tfrac{1}{\sqrt{2}}, \tag{5.64}$$

(5.63) does indeed reduce to (5.57), establishing that $\overline{\mathbf{H}} = \mathbf{H}^{\otimes 7}$ does indeed act as a logical Hadamard gate on the codewords.

Nor is it difficult to make a 14-Qbit logical cNOT gate that takes the pair of codewords $|\overline{x}\rangle|\overline{y}\rangle$ into $|\overline{x}\rangle|\overline{x\oplus y}\rangle$. One simply applies ordinary cNOT gates to each of the seven pairs of corresponding Qbits in the two codewords. This works because each of the codewords in (5.43) is left invariant by each of the $\mathbf{M}_i$. If the control codeword is in the state $|\overline{0}\rangle$ then the pattern of flips applied to the target codeword for each of the eight terms in the expansion of the control codeword

$$|\overline{0}\rangle = 2^{-3/2}\big(\mathbf{1}+\mathbf{M}_0+\mathbf{M}_1+\mathbf{M}_2+\mathbf{M}_1\mathbf{M}_2+\mathbf{M}_2\mathbf{M}_0 +\mathbf{M}_0\mathbf{M}_1+\mathbf{M}_0\mathbf{M}_1\mathbf{M}_2\big)|0\rangle_7 \tag{5.65}$$

is simply given by the corresponding product of $\mathbf{M}_i$. Since each $\mathbf{M}_i$ acts as the identity on both $|\overline{0}\rangle$ and $|\overline{1}\rangle$, the target codeword is unchanged. On the other hand, if the control codeword is in the state $|\overline{1}\rangle$ then the pattern of flips applied to the target codeword differs from this by an additional application of $\overline{\mathbf{X}}$, which has precisely the effect of interchanging $|\overline{0}\rangle$ and $|\overline{1}\rangle$.

Because of the simplicity of all these encoded gates, one can use error correction to eliminate malfunctions of the elementary gates themselves, if the rate of malfunctioning is so low that only a single one of the seven elementary gates is likely to malfunction. In the case of the 1-Qbit encoded gates, their elementary components act only on single Qbits in the codeword, so if only a single one of them malfunctions then only a single Qbit in the codeword will be corrupted and the error-correction procedure described above will restore the correct output. But this works as well for the encoded cNOT gate, since if only a single one of the elementary 2-Qbit cNOT gates malfunctions,

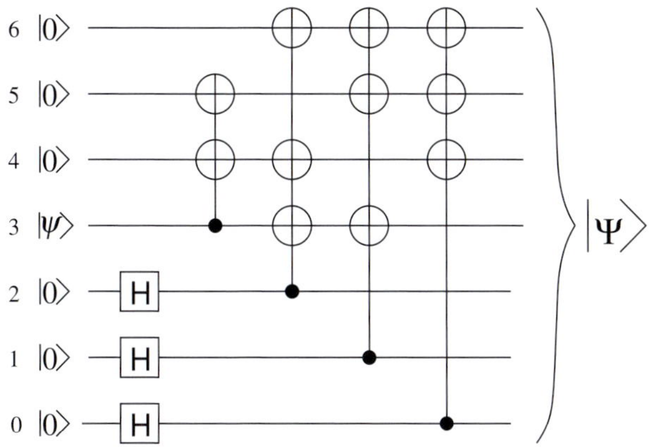

Fig 5.10 A 7-Qbit encoding circuit (a) that takes $|\psi\rangle = \alpha|0\rangle + \beta|1\rangle$ into the corresponding superposition of the two 7-Qbit codewords given in (5.43), $|\Psi\rangle = \alpha|\overline{0}\rangle + \beta|\overline{1}\rangle$. The numbering of the Qbits from 6 to 0 is made explicit to facilitate comparison with the form (5.42)–(5.44) of the codewords.

this will affect only single Qbits in each of the two encoded 7-Qbit words, and the correct output will again be restored by applying error correction to both of the codewords.

Another virtue of codeword gates that can be constructed as tensor products of uncoded gates is that they cannot (when functioning correctly) convert single-Qbit errors to multiple-Qbit errors, as more elaborate constructions of codeword gates might do. This highly desirable property is called *fault tolerance*. The great advantage of the 7-Qbit code is that many of the most important logical gates can be implemented in a fault-tolerant way.

5.8 A 7-Qbit encoding circuit

The circuit in Figure 5.10 encodes a general 1-Qbit state into a 7-Qbit codeword without using any measurement gates, in a manner analogous to the way Figure 5.1 produces 3-Qbit codewords.

Since the circuit is unitary and therefore linear, it is enough to show that it works when $|\psi\rangle = |0\rangle$ and when $|\psi\rangle = |1\rangle$. This follows from the fact that if the $(n+1)$-Qbit gate $\mathbf{C}^U$ is a controlled n-Qbit unitary $\mathbf{U}$, then

$$\begin{aligned}\mathbf{C}^U\big(\mathbf{H}|0\rangle\big) \otimes |\Phi\rangle_n &= \mathbf{C}^U \tfrac{1}{\sqrt{2}}\big(|0\rangle + |1\rangle\big) \otimes |\Phi\rangle_n \\ &= \tfrac{1}{\sqrt{2}}\big(\mathbf{1} + \mathbf{X} \otimes \mathbf{U}\big)|0\rangle \otimes |\Phi\rangle_n, \qquad (5.66)\end{aligned}$$

where the control Qbit is on the left. If this is applied to the three controlled triple-NOT gates in Figure 5.10 then, reading from left to right, the resulting operations are $(1/\sqrt{2})(\mathbf{1} + \mathbf{X}_2\mathbf{X}_3\mathbf{X}_4\mathbf{X}_6) =$

$(1/\sqrt{2})(\mathbf{1}+\mathbf{M}_2)$, $(1/\sqrt{2})(\mathbf{1}+\mathbf{X}_1\mathbf{X}_3\mathbf{X}_5\mathbf{X}_6) = (1/\sqrt{2})(\mathbf{1}+\mathbf{M}_1)$, and $(1/\sqrt{2})(\mathbf{1}+\mathbf{X}_0\mathbf{X}_4\mathbf{X}_5\mathbf{X}_6) = (1/\sqrt{2})(\mathbf{1}+\mathbf{M}_0)$.

When $|\psi\rangle = |0\rangle$ the controlled double-NOT on the left acts as the identity, so the circuit does indeed produce the codeword $|\overline{0}\rangle$ in (5.43). When $|\psi\rangle = |1\rangle$, the controlled double-NOT on the left acts as $\mathbf{X}_4\mathbf{X}_5$. The circuit after that action is exactly the same as when $|\psi\rangle = |0\rangle$, except that the initial state of Qbits 3, 4, and 5 on the left is $|1\rangle$ rather than $|0\rangle$. Since all $\mathbf{X}_i$ commute, the state that results is not $|\overline{0}\rangle$ but $\mathbf{X}_3\mathbf{X}_4\mathbf{X}_5|\overline{0}\rangle$. But

$$\mathbf{X}_3\mathbf{X}_4\mathbf{X}_5 = \mathbf{X}_0\mathbf{X}_1\mathbf{X}_2\mathbf{X}_3\mathbf{X}_4\mathbf{X}_5\mathbf{X}_6\mathbf{M}_0\mathbf{M}_1\mathbf{M}_2 = \overline{\mathbf{X}}\mathbf{M}_0\mathbf{M}_1\mathbf{M}_2. \quad (5.67)$$

Since $\mathbf{M}_0\mathbf{M}_1\mathbf{M}_2$ acts as the identity on $|\overline{0}\rangle$, the resulting state is indeed $|\overline{1}\rangle = \overline{\mathbf{X}}|\overline{0}\rangle$.

A less direct method to confirm that Figure 5.10 produces the 7-Qbit encoding, analogous to the method described in Section 5.9 for the 5-Qbit encoding, is given in Appendix O.

5.9 A 5-Qbit encoding circuit

The circuit in Figure 5.11 encodes a general 1-Qbit state into a 5-Qbit codeword without using any measurement gates.

The circuit differs from one reported by David DiVincenzo[2] only by the presence of the 1-Qbit gates **ZHZ** on the left. When $|\psi\rangle = |x\rangle$ DiVincenzo's circuit produces two orthogonal linear combinations of the codewords (5.43), which are, of course, equally valid choices. But to get the codewords in (5.43) one needs these additional gates. (I have written them in the symmetric form **ZHZ** rather than in the simpler equivalent form **YH** both to spare the reader from having to remember that $\mathbf{Y} = \mathbf{ZX}$ and not $\mathbf{XZ}$, and also to spare her the confusion of having to reverse the order of gates when going from a circuit diagram to the corresponding equation.)

In contrast to the superficially similar circuit for the 7-Qbit code in Figure 5.10, there does not seem to be a transparently simple way to demonstrate that the circuit in Figure 5.11 does produce the 5-Qbit codewords. One can always, of course, write down the action of each successive gate in the circuit, and check that the resulting unwieldy expressions are identical to the explicit expansions of the codewords (5.33) in computational-basis states. A less clumsy proof follows from the fact that $|\overline{0}\rangle$ is the unique (to within an overall phase factor $e^{i\varphi}$) joint eigenvector with all eigenvalues 1 of the five mutually commuting operators consisting of the four error-syndrome operators $\mathbf{M}_0, \ldots, \mathbf{M}_3$

2 David P. DiVincenzo, "Quantum Gates and Circuits," *Proceedings of the Royal Society of London* A **454**, 261–276 (1998), `http://arxiv.org/abs/quant-ph/9705009`.

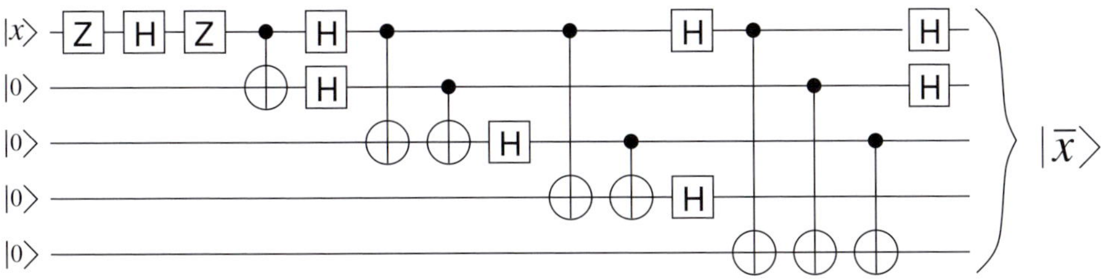

Fig 5.11 A 5-Qbit encoding circuit. If the initial state of the Qbit on the top wire is $|\psi\rangle = \alpha|0\rangle + \beta|1\rangle$, then the circuit produces the corresponding superposition of the two 5-Qbit codewords given in (5.33), $|\Psi\rangle = \alpha|\overline{0}\rangle + \beta|\overline{1}\rangle$. This fact is established in Figures 5.12–5.20. The figure illustrates this for the states $|0\rangle$ and $|1\rangle$ ($x = 0$ or 1) on the upper wire. Since a product of unitary gates is linear, the circuit encodes arbitrary superpositions of these states.

of Equation (5.29) and the operator $\overline{\mathbf{Z}}$ of Equation (5.38). So if we can establish that the state $|\overline{x}\rangle$ produced in Figure 5.11 is invariant under the four $\mathbf{M}_i$, that it is invariant under $\overline{\mathbf{Z}}$ when $x = 0$, and that applying the 5-Qbit $\overline{\mathbf{X}}$ to $|\overline{x}\rangle$ is the same as applying the 1-Qbit $\mathbf{X}$ to $|x\rangle$, then we will have shown that the circuit produces the 5-Qbit encoding to within an overall phase factor $e^{i\varphi}$. Having done this, we can then confirm that $e^{i\varphi} = 1$ by evaluating the projection on $|0\rangle_5$ of the state produced by the circuit when $x = 0$.

To learn the actions of various products of 1-Qbit $\mathbf{X}$s and $\mathbf{Z}$s on the state produced by the circuit in Figure 5.11, we apply them on the right side of the diagram, and then bring them to the left through the cNOT gates and 1-Qbit gates that make up the circuit, until they act directly on the input state on the left. In doing this we must use the fact that bringing an $\mathbf{X}$ (or a $\mathbf{Z}$) through a Hadamard converts it to a $\mathbf{Z}$ (or an $\mathbf{X}$), bringing an $\mathbf{X}$ through a $\mathbf{Z}$ introduces a factor of -1, and bringing an $\mathbf{X}$ or a $\mathbf{Z}$ through a cNOT has the results shown in Figure 5.12: a $\mathbf{Z}$ on the control Qbit (or an $\mathbf{X}$ on the target Qbit) commutes with a cNOT, while bringing a $\mathbf{Z}$ through the target Qbit (or an $\mathbf{X}$ through the control Qbit) introduces an additional $\mathbf{Z}$ on the control Qbit (or $\mathbf{X}$ on the target Qbit).

Figure 5.13 uses these elementary facts to show that $\mathbf{M}_0 = \mathbf{Z}_1\mathbf{X}_2\mathbf{X}_3\mathbf{Z}_4$ leaves both codewords invariant, by demonstrating that it can be brought to the left through all the gates in the circuit to act on the input state $|x0000\rangle$ as $\mathbf{Z}_2$. Figures 5.14–5.16 show similar things for $\mathbf{M}_1 = \mathbf{Z}_2\mathbf{X}_3\mathbf{X}_4\mathbf{Z}_0$, $\mathbf{M}_2 = \mathbf{Z}_3\mathbf{X}_4\mathbf{X}_0\mathbf{Z}_1$, and $\mathbf{M}_3 = \mathbf{Z}_4\mathbf{X}_0\mathbf{X}_1\mathbf{Z}_2$, which can be brought to the left through all the gates to act on the input state as $\mathbf{Z}_0$, $\mathbf{Z}_3$, and $\mathbf{Z}_1$. Figure 5.17 shows that $\overline{\mathbf{X}} = \mathbf{X}_0\mathbf{X}_1\mathbf{X}_2\mathbf{X}_3\mathbf{X}_4$ can be brought to the left through all the gates of the circuit to act on the input state $|x0000\rangle$ as $\mathbf{X}_4\mathbf{Z}_2\mathbf{Z}_1$, which simply interchanges $x = 0$ and

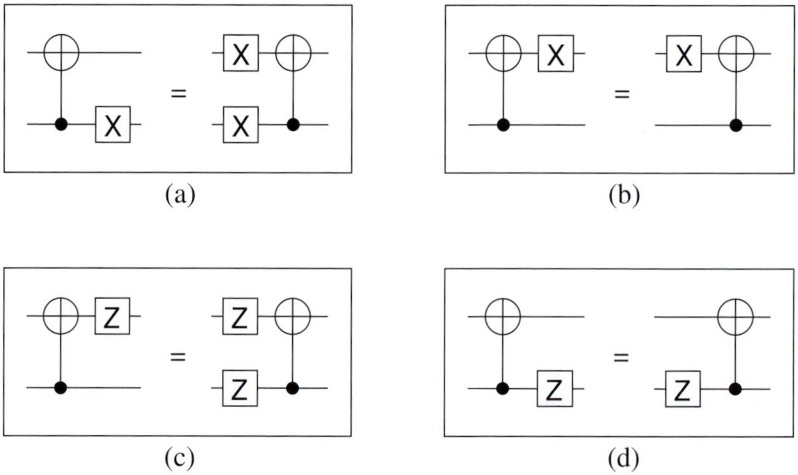

Fig 5.12 Easily verifiable identities useful in determining how various products of $\mathbf{X}$s and $\mathbf{Z}$s act on the circuit of Figure 5.11. (a) A cNOT can be interchanged with an $\mathbf{X}$ acting on the control Qbit, if another $\mathbf{X}$ acting on the target Qbit is introduced. (b) A cNOT commutes with an $\mathbf{X}$ acting on the target Qbit. (c) A cNOT can be interchanged with a $\mathbf{Z}$ acting on the target Qbit, if another $\mathbf{Z}$ acting on the control Qbit is introduced. (d) A cNOT commutes with a $\mathbf{Z}$ acting on the control Qbit.

$x = 1$, thereby demonstrating that $\overline{\mathbf{X}}$ acts as logical X on the codewords. Figure 5.18 shows the analogous property for $\overline{\mathbf{Z}} = \mathbf{Z}_0\mathbf{Z}_1\mathbf{Z}_2\mathbf{Z}_3\mathbf{Z}_4$, which can be brought to the left through all the gates of the circuit to act on the input state $|x0000\rangle$ as $\mathbf{Z}_4\mathbf{Z}_3\mathbf{Z}_0$, which multiplies it by $(-1)^x$, thereby demonstrating that $\overline{\mathbf{Z}}$ acts as logical Z on the codewords. Finally Figures 5.19 and 5.20 show that the inner product of the codeword state $|\overline{0}\rangle$ with the computational-basis state $|00000\rangle$ is $\frac{1}{4}$, thereby demonstrating that the circuit produces the codewords (5.33) with the right phase.

In Appendix O this circuit-theoretic approach is used to give a second (more complicated, but instructive) demonstration of the validity of the 7-Qbit encoding circuit of Figure 5.10.

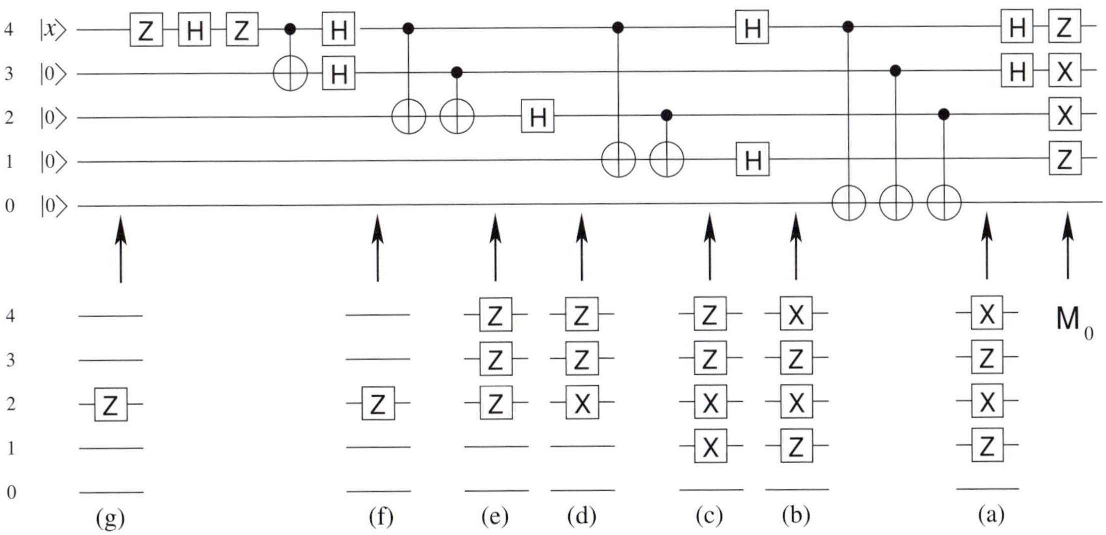

Fig 5.13 Demonstration that $\mathbf{M}_0 = \mathbf{Z}_1\mathbf{X}_2\mathbf{X}_3\mathbf{Z}_4$ acting on the output of the encoding circuit in Figure 5.11 is the same as $\mathbf{Z}_2$ acting on the input, which leaves the input invariant. On the extreme left $\mathbf{M}_0$ is applied to the output of the circuit. The insets (a)–(g) show what happens as the $\mathbf{X}$ and $\mathbf{Z}$ gates making up $\mathbf{M}_0$ are moved to the left through the gates of the circuit. (a) $\mathbf{Z}_4$ and $\mathbf{X}_3$ are changed to $\mathbf{X}_4$ and $\mathbf{Z}_3$ as a result of having been brought through Hadamard gates. (b) Bringing the two $\mathbf{X}$ gates through the control Qbits of cNOT gates produces a pair of cancelling $\mathbf{X}$ gates on the common target Qbit, so the set of gates in (a) is unchanged when it is moved to (b). (c) The Hadamard gates convert $\mathbf{X}_4$ and $\mathbf{Z}_1$ to $\mathbf{Z}_4$ and $\mathbf{X}_1$. (d) Bringing $\mathbf{X}_2$ through the control Qbit of the cNOT produces an $\mathbf{X}$ on its target Qbit which cancels the $\mathbf{X}$ already there. (e) The Hadamard on Qbit 2 converts the $\mathbf{X}$ to a $\mathbf{Z}$. (f) Moving the $\mathbf{Z}_2$ through the targets of the two cNOTs produces $\mathbf{Z}$ gates on their control Qbits which cancel the two $\mathbf{Z}$ gates already there. (g) The resulting $\mathbf{Z}_2$ can be moved all the way to the left.

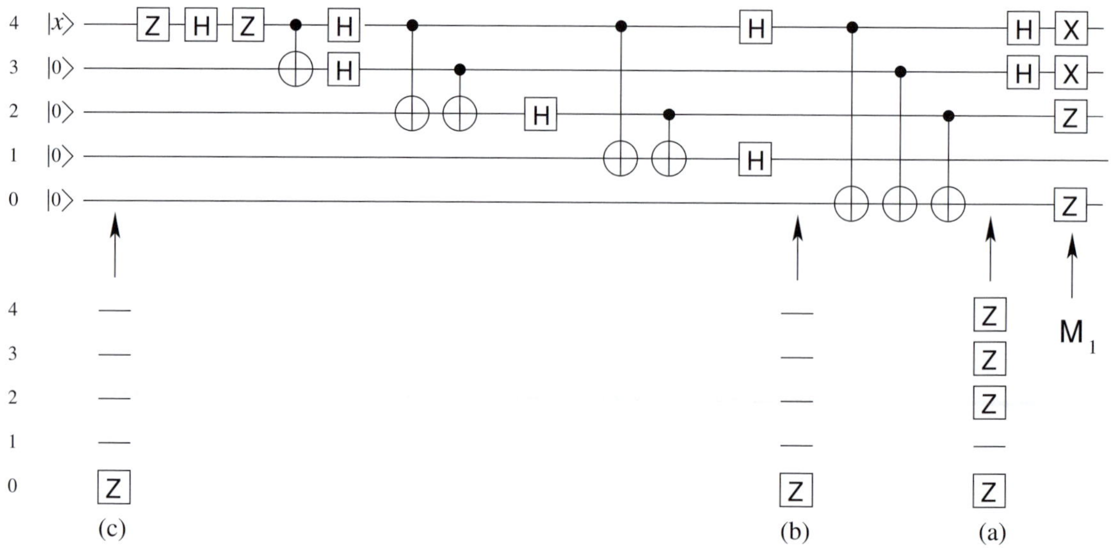

Fig 5.14 Using the identities in Figure 5.12 and the fact that bringing a **Z** through a Hadamard converts it to an **X** and vice versa establishes that $\mathbf{M}_1$ can be brought to the left through the gates of the encoding circuit to act directly on $|x0000\rangle$ as $\mathbf{Z}_0$.

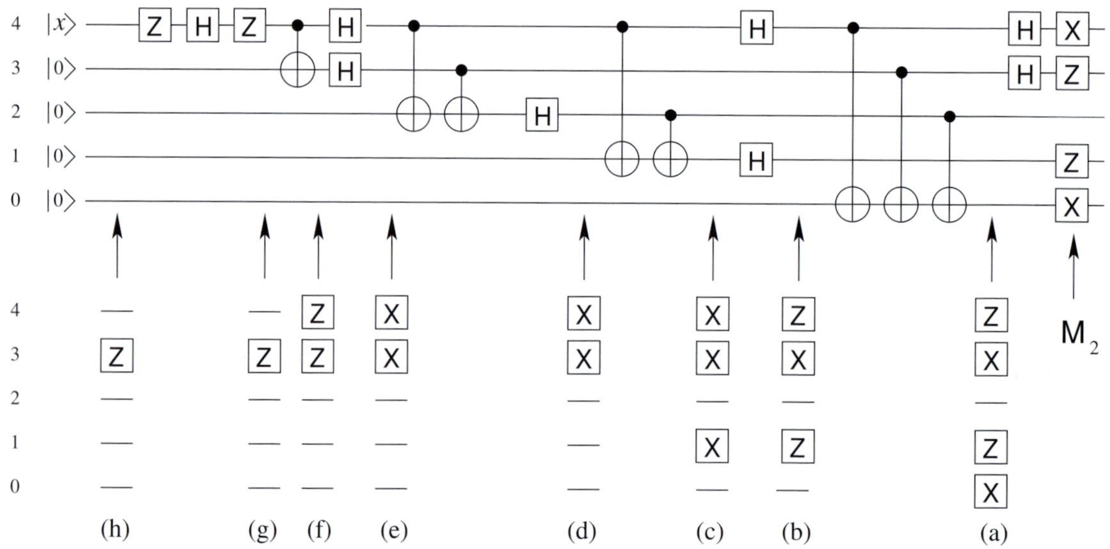

Fig 5.15 $\mathbf{M}_2$ can be brought to the left through the gates of the encoding circuit to act directly on $|x0000\rangle$ as $\mathbf{Z}_3$.

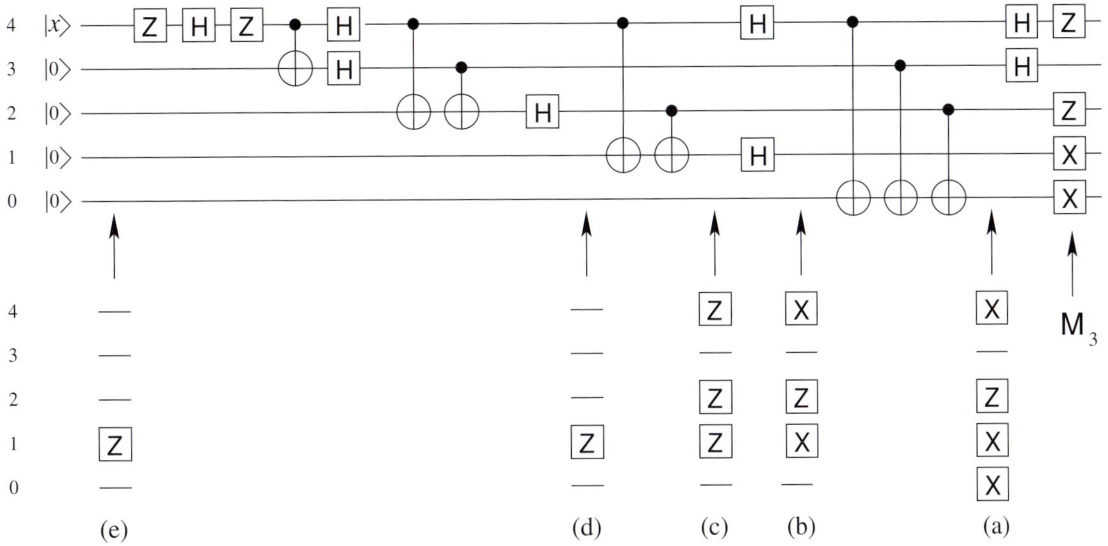

Fig 5.16 $\mathbf{M}_3$ can be brought to the left through the gates of the encoding circuit to act directly on $|x0000\rangle$ as $\mathbf{Z}_1$.

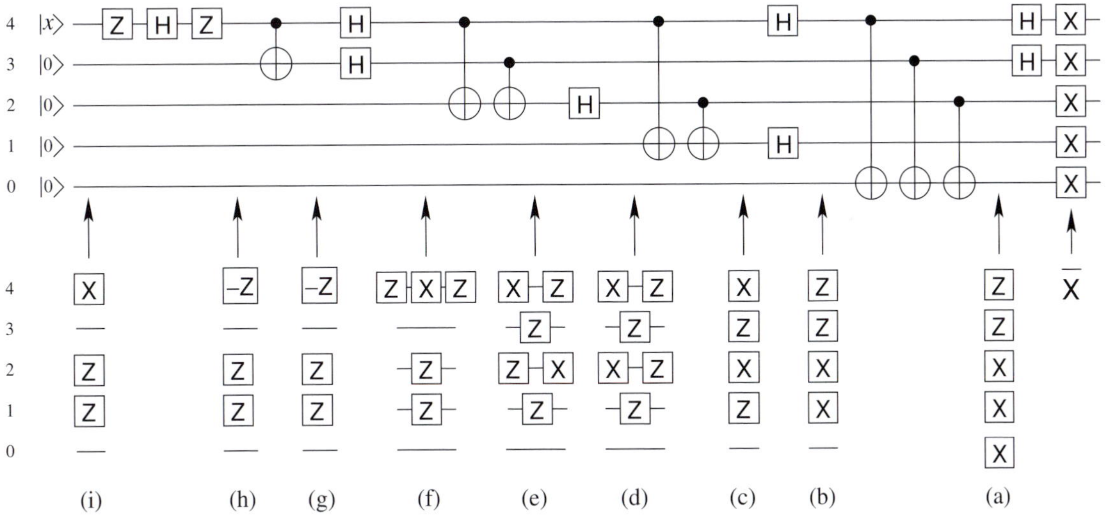

Fig 5.17 Demonstration that $\overline{\mathbf{X}} = \mathbf{X}_0\mathbf{X}_1\mathbf{X}_2\mathbf{X}_3\mathbf{X}_4$ acting on the output of the encoding circuit in Figure 5.11 is the same as $\mathbf{X}_4\mathbf{Z}_2\mathbf{Z}_1$ acting on the input, which interchanges $|00000\rangle$ and $|10000\rangle$. (a) Bringing $\mathbf{X}_4$ and $\mathbf{X}_3$ through the Hadamards converts them to $\mathbf{Z}_4$ and $\mathbf{Z}_3$. (b) Bringing $\mathbf{X}_2$ through the cNOT controlled by Qbit 2 produces an $\mathbf{X}$ on the target Qbit 0, which cancels the $\mathbf{X}$ already there. (c) The Hadamards convert $\mathbf{Z}_4$ and $\mathbf{X}_1$ to $\mathbf{X}_4$ and $\mathbf{Z}_1$. (d) Bringing $\mathbf{X}_4$ and $\mathbf{X}_2$ to the left produces two $\mathbf{X}_1$ gates which cancel; bringing $\mathbf{Z}_1$ to the left then produces additional $\mathbf{Z}_4$ and $\mathbf{Z}_2$ gates. (e) The Hadamard $\mathbf{H}_2$ interchanges the $\mathbf{X}_2$ and $\mathbf{Z}_2$ gates. (f) First bring to the left the $\mathbf{Z}_2$ gate, then the $\mathbf{X}_4$ gate. (g) The $\mathbf{H}_4$ converts $\mathbf{ZXZ}$ to $\mathbf{XZX} = -\mathbf{Z}$. (h) No further changes. (i) $\mathbf{Z}$ commutes with itself, is changed to $\mathbf{X}$ on passing through $\mathbf{H}$, and acquires another minus sign on passage through $\mathbf{Z}$.

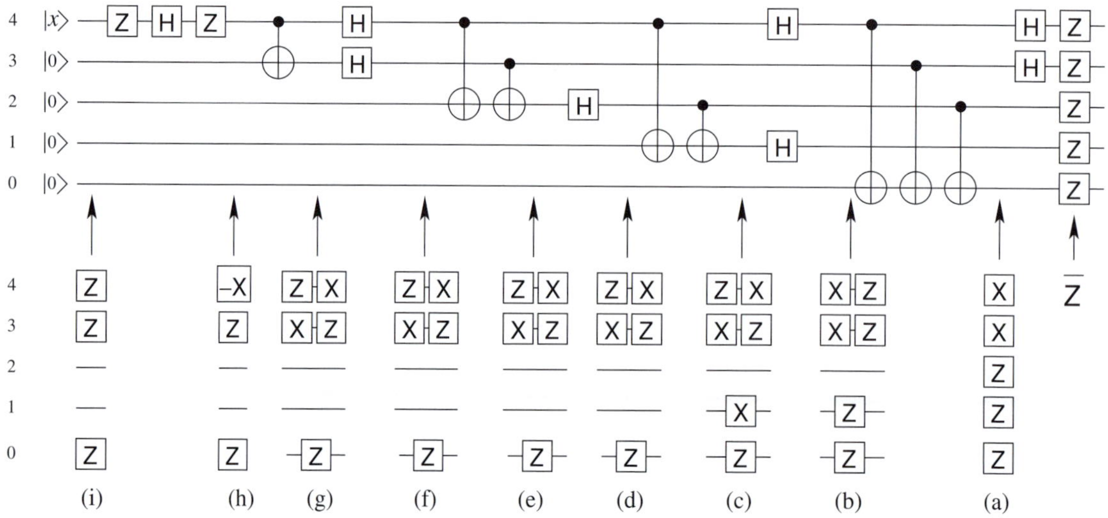

Fig 5.18 Demonstration that $\overline{Z} = Z_0 Z_1 Z_2 Z_3 Z_4$ acting on the output of the encoding circuit in Figure 5.11 is the same as $Z_4 Z_3 Z_0$ acting on the input, which takes $|x0000\rangle$ into $(-1)^x |x0000\rangle$.

$$|\Phi\rangle \quad -\boxed{A}-\boxed{B}- \quad |\Psi\rangle \quad = \quad BA\,|\Phi\rangle$$

(a)

$$\langle\Psi|\Omega\rangle \; = \; \langle\Phi|A^\dagger B^\dagger|\Omega\rangle \; = \; \langle\Phi|-\boxed{A^\dagger}-\boxed{B^\dagger}-|\Omega\rangle$$

(b)

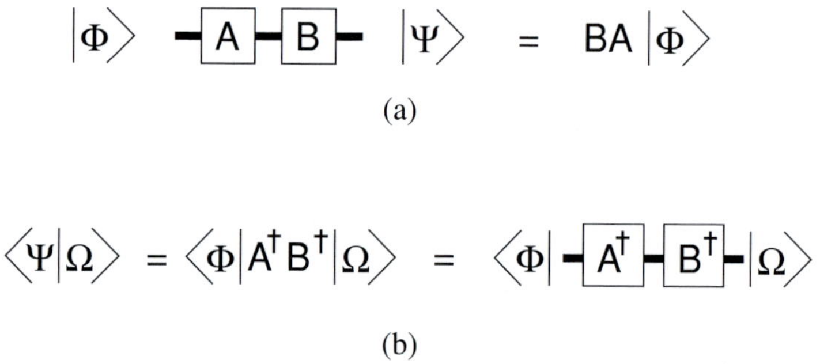

Fig 5.19 A circuit-theoretic way to evaluate inner products. (a) A circuit taking the input $|\Phi\rangle$ into the output $|\Psi\rangle = BA|\Phi\rangle$. The inner product $\langle\Psi|\Omega\rangle$ of the output state Ψ with some other state $|\Omega\rangle$ is given by $\langle\Phi|A^\dagger B^\dagger|\Omega\rangle$. The diagram on the right in (b) shows this inner product being evaluated by first letting $B^\dagger$ act on $|\Omega\rangle$, then letting $A^\dagger$ act on the result, and then taking the inner product with the input state $|\Phi\rangle$. Evidently this generalizes to the product of many gates. If the gates are all Hermitian, as they are in the circuit of Figure 5.11, then the circuit on the right of (b) is identical to the circuit on the left of (a). The resulting evaluation of the inner product of $|\Omega\rangle = |0\rangle_5$ with the state $|\overline{0}\rangle$ produced by letting the circuit of Figure 5.11 act on $|\Phi\rangle = |0\rangle_5$ is carried out in Figure 5.20.

$$= \frac{1}{4} \quad \langle 1| - \boxed{H} - \bullet - |0\rangle \quad + \quad \frac{1}{4} \quad \langle 1| - \boxed{H} - \bullet - |1\rangle$$
$$\langle 0| - \boxed{H} - \boxed{Z} - |0\rangle \qquad \langle 0| - \boxed{H} - \boxed{Z} - |1\rangle$$

(d)

$$= \frac{1}{4} \quad \langle 1| - \boxed{H} - |0\rangle \quad - \quad \frac{1}{4} \quad \langle 1| - \boxed{H} - |1\rangle \quad = \quad \frac{1}{4}$$
$$\langle 0| - \boxed{H} - |0\rangle \qquad \langle 0| - \boxed{H} - |1\rangle$$

(e)

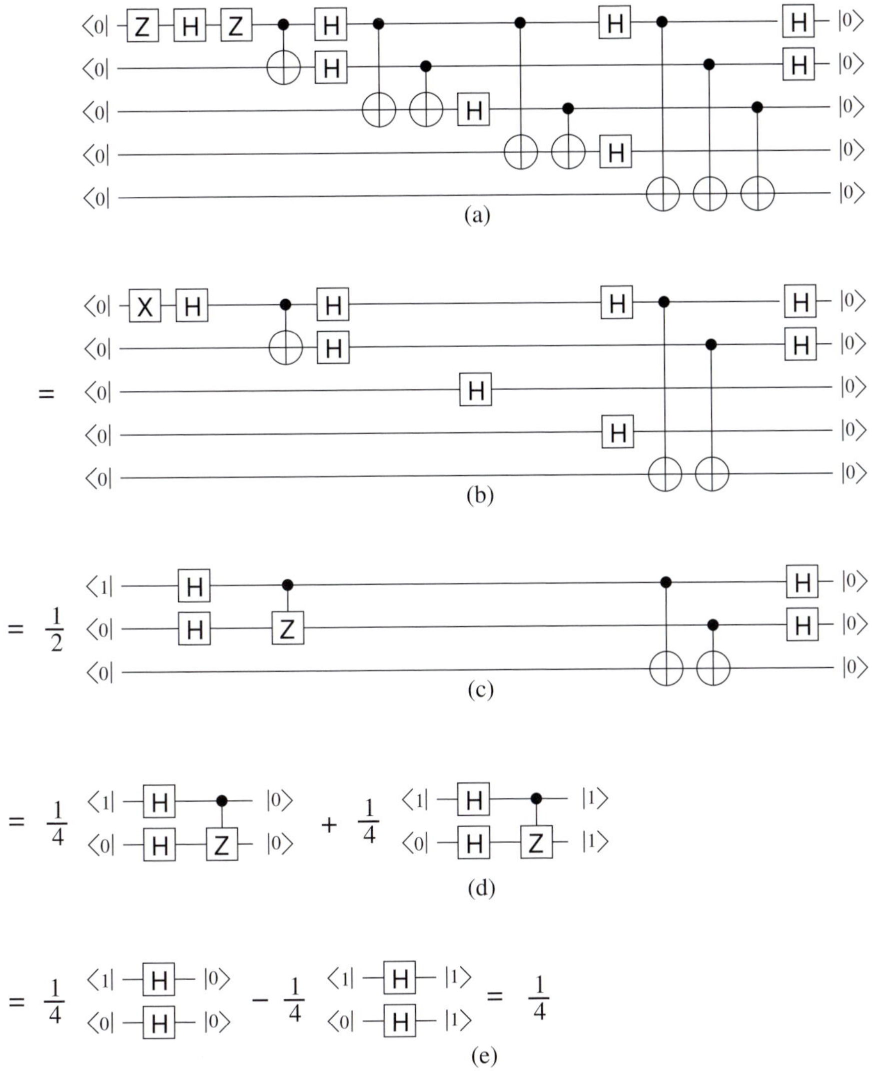

Fig 5.20 Demonstration that the state produced by the encoding circuit in Figure 5.11 when $x = 0$ has an inner product with the state $|0\rangle_5$ that is $\frac{1}{4}$, thereby establishing that the phase factor $e^{i\varphi} = 1$ – i.e. that the state is precisely $|\bar{0}\rangle$ without any additional phase factor. (a) Circuit-theoretic representation of the inner product, following the procedure developed in Figure 5.19; all gates now act to the right. (b) Elimination of operations in (a) that act as the identity: the cNOT on the extreme right of (a) can be dropped since its control Qbit is in the state $|0\rangle$; since $\mathbf{H}|0\rangle$ is invariant under $\mathbf{X}$, the pair of cNOT gates targeting Qbit 1 can be dropped, as can the pair targeting Qbit 2. (c) A pair of Hadamards on Qbit 4 in (b) cancel; a Hadamard on Qbit 3 in (b) is moved to the left converting a cNOT to a controlled-Z; Qbits 2 and 1 in (b) simply give the matrix element $\langle 0|\mathbf{H}|0\rangle = \frac{1}{\sqrt{2}}$, resulting in an overall factor of $\frac{1}{2}$. (d) Expanding both states $\mathbf{H}|0\rangle = \frac{1}{\sqrt{2}}(|0\rangle + |1\rangle)$ on the right of (c), the effect of the two cNOT gates in (c) is that only the terms in $|0\rangle|0\rangle$ and $|1\rangle|1\rangle$ give nonzero contributions. (e) The action of the controlled-Z gates in (d) has been carried out, leaving a sum of products of matrix elements of $\mathbf{H}$.

Chapter 6

Protocols that use just a few Qbits

6.1 Bell states

In this chapter we examine some elementary quantum information-theoretic protocols which are often encountered in the context of quantum computation, though they also have applications in the broader area of quantum information processing. Because they use only a small number of Qbits, they have all been carried out in at least one laboratory, unlike any but the most trivial and atypical examples of the protocols we have considered in earlier chapters.

Most of these examples make use of the 2-Qbit entangled state,

$$|\psi_{00}\rangle = \tfrac{1}{\sqrt{2}}\big(|00\rangle + |11\rangle\big). \tag{6.1}$$

This state can be assigned to two Qbits, each in the state $|0\rangle$, by applying a Hadamard to one of them, and then using it as the control Qbit for a cNOT that targets the other (Figure 6.1(a)):

$$|\psi_{00}\rangle = \mathbf{C}_{10}\mathbf{H}_1|00\rangle. \tag{6.2}$$

We generalize (6.2) by letting the original pair of unentangled Qbits be in any of the four 2-Qbit computational-basis states $|00\rangle$, $|01\rangle$, $|10\rangle$, and $|11\rangle$ (Figure 6.1(b)):

$$|\psi_{xy}\rangle = \mathbf{C}_{10}\mathbf{H}_1|xy\rangle. \tag{6.3}$$

Since the four states $|xy\rangle$ are an orthonormal set and the Hadamard and cNOT gates are unitary, the four entangled states $|\psi_{xy}\rangle$ are also an orthonormal set, called the *Bell basis* to honor the memory of the physicist John S. Bell, who discovered in 1964 one of the most extraordinary facts about 2-Qbit entangled states. We examine a powerful 3-Qbit version of Bell's theorem in Section 6.6.

If we rewrite (6.3) as

$$|\psi_{xy}\rangle = \mathbf{C}_{10}\mathbf{H}_1\mathbf{X}_1^x\mathbf{X}_0^y|00\rangle, \tag{6.4}$$

and recall that $\mathbf{HX} = \mathbf{ZH}$ and that either a $\mathbf{Z}$ on the control Qbit or an $\mathbf{X}$ on the target Qbit commutes with a cNOT, then we have

$$|\psi_{xy}\rangle = \mathbf{Z}_1^x\mathbf{X}_0^y\mathbf{C}_{10}\mathbf{H}_1|00\rangle = \mathbf{Z}_1^x\mathbf{X}_0^y\tfrac{1}{\sqrt{2}}\big(|00\rangle + |11\rangle\big), \tag{6.5}$$

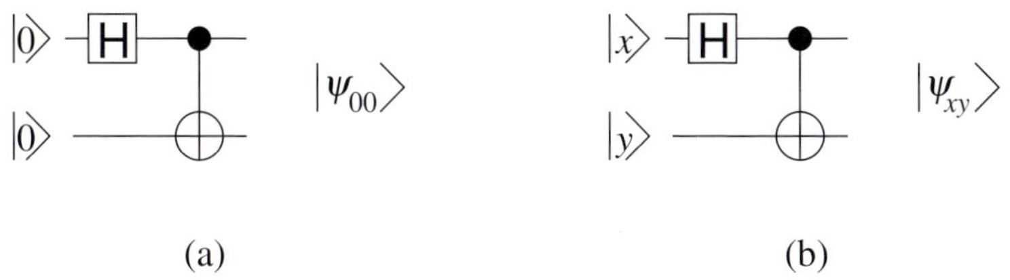

Fig 6.1 (a) A circuit that creates the entangled state $|\psi_{00}\rangle = \frac{1}{\sqrt{2}}\big(|00\rangle + |11\rangle\big)$ from the unentangled computational-basis state $|00\rangle$. (b) A circuit that creates the four orthonormal entangled Bell states $|\psi_{xy}\rangle$ from the unentangled computational-basis state $|xy\rangle$.

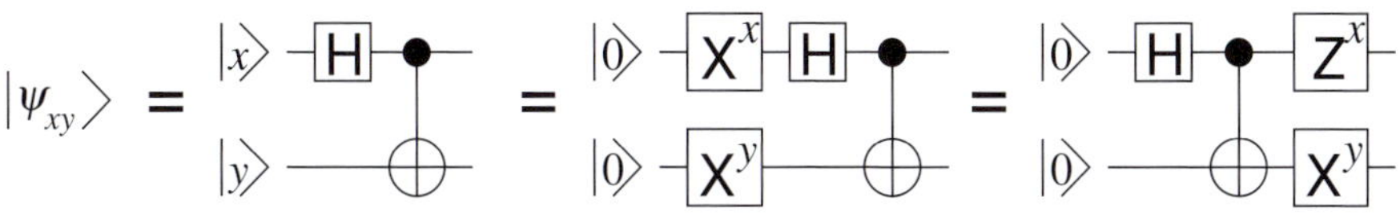

Fig 6.2 The Bell states $|\psi_{xy}\rangle$ can be constructed from $|\psi_{00}\rangle = \frac{1}{\sqrt{2}}\big(|00\rangle + |11\rangle\big)$ by flipping a single Qbit, changing the sign from + to −, or doing both of these.

as illustrated in Figure 6.2. This shows that the other Bell states are obtained from $(1/\sqrt{2})(|00\rangle + |11\rangle)$ by flipping one of the Qbits, by changing the + to a −, or by doing both. This, of course, can also be derived directly from (6.3) by letting the Hadamard and cNOT act for each of the four choices for the pair xy.

We now examine a few simple protocols in which some or all of the Bell states (or, in Section 6.6, their 3-Qbit generalizations) play an important role.

6.2 Quantum cryptography

A decade before Shor's discovery that quantum computation posed a threat to the security of RSA encryption, it was pointed out that Qbits (though the term did not exist at the time) offered a quite different and demonstrably secure basis for the exchange of secret messages.

Of all the various possible applications of quantum mechanics to information processing, quantum cryptography arguably holds the most promise for becoming a practical technology. There are several reasons for this. First of all, it works Qbit by Qbit. The only relevant gates are a small number of simple 1-Qbit gates. Interactions between pairs of Qbits like those mediated by cNOT gates play no role, at least in the most straightforward versions of the protocol.

Furthermore, in actual realizations of quantum cryptography the physical Qbits are extremely simple. Each Qbit is a single photon of light. The state of the Qbit is the linear polarization state of the photon. If the states $|0\rangle$ and $|1\rangle$ describe photons with vertical and horizontal polarization, then the states $\mathbf{H}|0\rangle = (1/\sqrt{2})\big(|0\rangle + |1\rangle\big)$ and $\mathbf{H}|1\rangle = (1/\sqrt{2})\big(|0\rangle - |1\rangle\big)$ describe photons diagonally polarized, either at 45° or at -45° to the vertical. Photons in any of these four polarization states can be prepared in any number of ways, most simply (if not most efficiently) by sending a weak beam of light through an appropriately oriented polaroid filter. Once a photon has been prepared in its initial polarization state it does not have to be manipulated any further beyond eventually measuring either its horizontal–vertical or its diagonal polarization by, for example, sending it through an appropriately oriented birefringent crystal and seeing which beam it emerges in, or seeing whether it does or does not get through another appropriately oriented polaroid filter. Photons can effectively be shielded from extraneous interactions by sending them through optical fibers, where they can travel in a polarization-preserving manner at the speed of light.

This procedure can be viewed as the simplest possible quantum computation. First the Qbit is assigned an initial state by sending it through a 1-Qbit measurement gate. Then a 1-Qbit unitary gate is or is not applied (depending on whether a subsequent polarization measurement is to be along the same direction as the first). And finally the Qbit is sent through a second 1-Qbit measurement gate.

The usefulness of easily transportable single Qbits for secret communication stems from one important cryptographic fact: Alice and Bob can have an unbreakable code if they share newly created identical strings of random bits, called *one-time codepads*. If they both have such identical random strings, then Alice can take her message, in the form of a long string of zeros and ones, and transform it into its bitwise modulo-2 sum (also called the *exclusive or* or XOR) with a random string of zeros and ones of the same length taken from her one-time codepad. Flipping or not flipping each bit of a coherent message according to whether the corresponding bit of a random string is 0 or 1 converts the message into another random string. (If this is not obvious, think of the process as flipping or not flipping each bit of the random string, according to whether the corresponding bit of the coherent message is 0 or 1.) Nobody can reconstruct the original string without knowing the random string used to encode it, so only Bob can decode the message. He does this by taking the XOR of the now meaningless string of zeros and ones, received from Alice, with his own copy of the random string that she used to do the encoding. The string he gets in this way is $M \oplus S \oplus S$, where M is the message, S is the random string, and $M \oplus S$ is the encoded message from Alice. Since $S \oplus S = 0$, Bob recovers the original message.

The problem with one-time codepads is that they can be used only once. If an eavesdropper (Eve) picks up two messages encoded with the same pad, she can take the XOR of the two encoded messages. The random string used to encode the two messages drops out of the process, leaving the XOR of the two unencoded messages. But the XOR of two meaningful messages, combined with the usual code-breaking tricks based on letter frequencies, can be used (with more subtlety than would be required for a single message) to separate and decode both texts. So to be perfectly secure Alice and Bob must continually refresh their one-time codepad with new identical random strings of bits.

The problem of exchanging such random strings in a secure way might appear to be identical to the original problem of exchanging meaningful messages in a secure way. But at this point quantum mechanics comes to the rescue and provides an entirely secure means for exchanging identical sequences of random bits. Pause to savor this situation. Nobody has figured out how to exploit quantum mechanics to provide a secure means for directly exchanging meaningful messages. The secure exchange is possible only because the bit sequences are random. On the face of it one would think nothing could be more useless than such a transmission of noise. What is bizarre is that human ingenuity combined with human perversity has succeeded in inventing a context in which the need to hide information from a third party actually provides a purpose for such an otherwise useless exchange of random strings of bits.

The scheme for doing this is known as BB84 after its inventors, Charles Bennett and Gilles Brassard, who published the idea in 1984. Alice sends Bob a long sequence of photons. For each photon Alice randomly chooses a polarization type for the photon (horizontal–vertical or diagonal) and within each type she randomly chooses a polarization state for the photon – one of the two orthogonal states associated with that type of polarization. In Qbit language Alice sends Bob a long sequence of Qbits randomly chosen to be in one of four states: $|0\rangle$ (polarized horizontally), $|1\rangle$ (polarized vertically), $\mathbf{H}|0\rangle = (1/\sqrt{2})(|0\rangle + |1\rangle)$ (polarized diagonally along 45°), or $\mathbf{H}|1\rangle = (1/\sqrt{2})(|0\rangle - |1\rangle)$ (polarized diagonally along −45°).

Reverting from photon-polarization language to our more familiar quantum-computational language, we divide the four equally likely types of Qbits that Alice sends to Bob into two categories: those with state $|0\rangle$ or $|1\rangle$, which we call type-1 Qbits, and those with state $\mathbf{H}|0\rangle$ or $\mathbf{H}|1\rangle$, which we call type-H Qbits. As each Qbit arrives Bob randomly decides whether to send it directly through a measurement gate, or to apply a Hadamard and only then send it through a measurement gate. We call these two options type-1 and type-H measurements. The Qbits must be individually identifiable – for example by the sequence in which they arrive – so that Alice and Bob can compare what each of them knows about each one.

Fig 6.3 Quantum cryptography. For each Qbit she sends to Bob, Alice randomly decides which type of state to prepare it in (type **1** means $|x\rangle$ and type **H** means $\mathbf{H}|x\rangle$) and which state of that type ($x = 0$ or 1) to prepare. For each Qbit he receives from Alice, Bob randomly decides whether (**H**) or not (**1**) to apply a Hadamard gate before measuring it. In those cases (about half, enclosed in rectangular boxes) for which Bob's choice of measurement type is the same as Alice's choice of state, they acquire identical random bits. When their choices differ they acquire no useful information.

		1	2	3	4	5	6	7	8	9 . . .
Alice:	**Type:**	1	H	H	H	1	1	H	1	H . . .
	State:	0	1	0	1	1	0	1	0	0 . . .
Bob:	**Measure-ment type:**	H	H	H	1	1	H	1	1	1 . . .
	Outcome:	1	1	0	0	1	1	1	0	0 . . .

When Bob has measured all the Qbits in this way, Alice tells him over an insecure channel which of the Qbits she sent him were type-1 and which were type-H. But she does not reveal which of the two possible states she prepared within each type: $|0\rangle$ or $|1\rangle$ for type-1 Qbits and $\mathbf{H}|0\rangle$ or $\mathbf{H}|1\rangle$ for type-H. For those Qbits (about half of them) for which Bob's random choice of measurement type agrees with Alice's random choice of which type to send, Bob learns from the result of his measurement the actual random bit – 0 or 1 – that Alice chose to send. For those Qbits (the other half) for which Bob's choice of which type to measure disagrees with Alice's choice of which type to send, the result of his measurement is completely uncorrelated with Alice's choice of bit, and reveals nothing about it. This is illustrated in Figure 6.3.

Finally, Bob tells Alice, over an insecure channel, which of the Qbits he subjected to a type of measurement that agreed with her choice of which type to prepare – i.e. which Qbits were of the kind that provides them with identical random bits. They discard the useless half of their data for which Bob's type of measurement differed from Alice's type of preparation. They are then able to construct their one-time codepads from the identical strings of random bits they have acquired.

You might wonder why Bob doesn't wait to decide what type of measurement to make on each Qbit until he learns Alice's choice of type for that photon, thereby doubling the number of shared random bits. This would indeed be a sensible strategy if Bob could store the Qbits he received from Alice. However, storing individual photons in a polarization-preserving manner is difficult. For feasible quantum cryptography today, Bob must make his decision and measure the polarization of each photon as it arrives.

The reason Alice randomly varies the type of Qbit she sends to Bob is to provide security against eavesdroppers. If Alice sent all Qbits of the same type, then an eavesdropper, Eve, could acquire the same information as Bob without being detected. If, for example, Alice and Bob had agreed that all the Qbits would be type-1 and Eve learned of this, then she could intercept each Qbit before it reached Bob and send it directly through a measurement gate without altering its state, subsequently sending it (or another Qbit she prepared in the state she just learned) on to Bob. In this way she too could acquire the random bit that Alice sends out and that Bob subsequently acquires when he makes his own type-1 measurement. Nothing in the protocol would give Bob a clue that Eve was listening in. But by making each Qbit secretly and randomly of type 1 or type H Alice deprives Eve of this strategy.

The best Eve can do, like Bob, is to make type-1 or type-H measurements randomly. In doing so she necessarily reveals her presence. Bob and Alice can determine that Eve has compromised the security of their bits by sacrificing some of the supposedly identical random bits they extracted from the Qbits they both ended up treating in the same way. They take a sample of these bits and check (over an insecure channel) to see whether they actually do agree, as they would in the absence of eavesdropping. If Eve intercepts the Qbits, randomly making type-1 or type-H measurements of her own before sending them on to Bob, then for about half of the useful Qbits her choice will differ from the common choice of Alice and Bob. In about half of *those* cases, Eve's intervention will result in the outcome of Bob's measurement disagreeing with what Alice sent him. If, for example, Eve makes a type-1 measurement of a Qbit that Alice has prepared in the state $\mathbf{H}|0\rangle$, then she will necessarily change its state to one or the other of the two states $|0\rangle$ or $|1\rangle$. In either case if Bob then applies a Hadamard before measuring he will get the result 0 only half the time.

So if Eve is systematically intercepting Qbits, Bob's result will fail to agree with Alice's preparation for about a quarter of their sample. This warns them that the transmission was insecure. If all the sample data agree except for a tiny fraction, then they can set an upper limit to the fraction of bits that Eve might have picked up, enabling them to make an informed judgment of the security with which they can use the remaining ones.

Can Eve do better by a more sophisticated attack, that involved capturing each of Alice's Qbits and processing it in a quantum computer that restored it to its initial state, before sending it on to Bob? This would eliminate the possibility of her eavesdropping being revealed to Bob. But the requirement that Alice's Qbit be returned to its initial state also eliminates the possibility of Eve learning anything useful, for reasons rather like our earlier proof of the no-cloning theorem.

Let $|\phi_\mu\rangle$, $\mu = 0, \ldots, 3$, be the four possible states of Alice's Qbit: $|0\rangle$, $|1\rangle$, $\mathbf{H}|0\rangle$, and $\mathbf{H}|1\rangle$. Let $|\Phi\rangle$ be the initial state of the n Qbits in Eve's computer, and let $\mathbf{U}$ be the $(n+1)$-Qbit unitary transformation the computer executes on its own Qbits and Alice's. Since Alice's Qbit must emerge in its original state, we have

$$\mathbf{U}\Big(|\phi_\mu\rangle \otimes |\Phi\rangle\Big) = |\phi_\mu\rangle \otimes |\Psi_\mu\rangle. \tag{6.6}$$

Eve's hope is to devise a $\mathbf{U}$ that yields four $|\Psi_\mu\rangle$ whose differences enable her, by subsequent processing and measurement, to extract useful information about which of the four possible states $|\phi_\mu\rangle$ was. But unitary transformations preserve inner products, so

$$\langle\phi_\nu|\phi_\mu\rangle\langle\Phi|\Phi\rangle = \langle\phi_\nu|\phi_\mu\rangle\langle\Psi_\nu|\Psi_\mu\rangle. \tag{6.7}$$

Because $\langle\Phi|\Phi\rangle = 1$ and because $\langle\phi_\nu|\phi_\mu\rangle \neq 0$ for $\mu\nu = 02, 03, 12, 13$, it follows that

$$\langle\Psi_\nu|\Psi_\mu\rangle = 1, \quad \mu\nu = 02, 03, 12, 13. \tag{6.8}$$

Since the inner product of two normalized states can be 1 only if they are identical, it follows from (6.8) that

$$|\Psi_0\rangle = |\Psi_1\rangle = |\Psi_2\rangle = |\Psi_3\rangle. \tag{6.9}$$

The price Eve pays for eliminating all traces of her eavesdropping is that the resulting state of her quantum computer can teach her nothing whatever about the four possible states of Alice's Qbit.

There is a less practical version of this cryptographic protocol that appears, at first sight, to be different, but turns out to be exactly the same. Suppose that there were some central source that produced pairs of Qbits in the entangled state

$$|\Psi\rangle = \tfrac{1}{\sqrt{2}}\big(|00\rangle + |11\rangle\big), \tag{6.10}$$

and then sent one member of each pair to Alice and the other to Bob. One easily verifies that

$$\big(\mathbf{H} \otimes \mathbf{H}\big)\tfrac{1}{\sqrt{2}}\big(|00\rangle + |11\rangle\big) = \tfrac{1}{\sqrt{2}}\big(|00\rangle + |11\rangle\big), \tag{6.11}$$

so if Alice and Bob make measurements of the same type, they will get identical random results.

This might seem even more secure than the first protocol, since the Qbits are in an entangled state until Alice or Bob actually makes a measurement. The correlated bits – the outcomes of the measurement – do not even *exist* until a measurement has been made, and that does not happen until both Qbits are safely in Alice's and Bob's separate possession. But this is only the case if Eve does not intercept a Qbit. If she does measure one before it gets to Bob or Alice, then the correlated bits do come into existence at the moment of her own measurement. This is later than in the first protocol (when each bit exists from the moment Alice performs her measurement) but early enough to help Eve in the same way as before.

If Alice and Bob decided to produce their perfectly correlated random bits by always making type-1 measurements then if Eve finds this out she can intercept one member of the pair with type-1 measurements of her own, disentangling the state prematurely, but in a way that enables her to learn what each random bit is, while not altering the perfect correlations between the values Alice and Bob will subsequently measure. Alice and Bob can guard against this possibility by each randomly (and, necessarily, independently) alternating between type-1 and type-H measurements, and then following a procedure identical to the one they used when Alice sent Bob Qbits in definite states.

This returns us to the original protocol that made no use of entangled pairs. Indeed, if Alice measures her member of the entangled pair (making either a type-1 or a type-H measurement) before Bob measures his, this is equivalent to her sending Bob a Qbit with a randomly selected state that she knows. The only difference is that now the random choice of which of the two states to send within each type is not made by Alice tossing a coin, but by the basic laws of quantum mechanics that guarantee that the outcome of her own measurement is random.

6.3 Bit commitment

One can try to formulate a similar protocol for a procedure called bit commitment. Suppose that Alice wishes to assure Bob that she has made a binary decision by a certain date, but does not wish to reveal that decision until some future time. She can do this by writing YES or NO on a card, putting the card in a box, locking the box, and sending the box, but not the key, to Bob. Once the box is in Bob's possession he can be sure that Alice has not altered her decision, but while the key is in Alice's possession she can be sure that Bob has not learned what that decision was. When it is time for her to reveal the decision she sends the key to Bob who opens the box and learns what it was.

Of course Alice might worry about Bob breaking into the box by other means. Quantum mechanics offers a more secure procedure (but with an exotic loophole, which we return to momentarily). Alice prepares a large number n of labeled Qbits. If her answer is YES, she takes each Qbit to be randomly in the state $|0\rangle$ or the state $|1\rangle$. If her answer is NO, she prepares each Qbit randomly in the state $\mathbf{H}|0\rangle$ or $\mathbf{H}|1\rangle$. In either case she notes which Qbits are in which state, and then sends them all off to Bob, who stores them in a way that preserves both their state and their labels. (As noted above, such storage is beyond the range of current technology for polarized photons.)

If Bob has a collection of n Qbits, each of which has been chosen with equal probability to be in one of two orthogonal states $|\phi\rangle$ and $|\psi\rangle$, then there is no way for Bob to get any hint of what the two orthogonal states are. If, for example, he measures every Qbit, then the probability of getting 0 is

$$p(0) = \tfrac{1}{2}|\langle 0|\phi\rangle|^2 + \tfrac{1}{2}|\langle 0|\psi\rangle|^2. \tag{6.12}$$

But

$$|\langle 0|\phi\rangle|^2 + |\langle 0|\psi\rangle|^2 = 1, \tag{6.13}$$

since this is the sum of the squared moduli of the amplitudes of the expansion of $|0\rangle$ in the orthonormal basis given by $|\phi\rangle$ and $|\psi\rangle$:

$$|0\rangle = |\phi\rangle\langle\phi|0\rangle + |\psi\rangle\langle\psi|0\rangle. \tag{6.14}$$

So $p(0) = \frac{1}{2}$. Bob's measurement outcomes are completely random, regardless of what the orthogonal pair of states actually is.

In Appendix P it is shown, more generally, that no information Bob can extract from his collection of Qbits can distinguish between the case in which each has a 50–50 chance of being in the state $|0\rangle$ or $|1\rangle$ and the case in which each has a 50–50 chance of being in the state $\mathbf{H}|0\rangle$ or $\mathbf{H}|1\rangle$. There is no way Bob can learn Alice's choice from the Qbits that Alice has sent him. He cannot break into the locked box.

(It is crucial for Bob's inability to learn Alice's choice that, regardless of what that choice is, she sends him a collection of Qbits each of whose two possible states is picked randomly. If, for example, she sent him *exactly* $\frac{1}{2}n$ Qbits in the state $|0\rangle$ and $\frac{1}{2}n$ in the state $|1\rangle$, in some random order, then with probability 1 Bob would get an equal number of zeros and ones if he measured in the computational basis. But if he applied $\mathbf{H}$ before measuring, the outcome of each measurement would be random, and the probability of getting equal numbers of zeros and ones for his measurements would be quite small (asymptotically $\sqrt{2/(\pi n)}$) for large n. So if he got equal numbers of zeros and ones he could be rather sure that Alice had sent him photons in the states $|0\rangle$ and $|1\rangle$ rather than in the states $\mathbf{H}|0\rangle$ and $\mathbf{H}|1\rangle$.)

When the time comes for Alice to reveal her choice for the pair of orthogonal states, she says to Bob something like this: "My answer was YES, so each of the Qbits I sent you was either in the state $|0\rangle$ or in the state $|1\rangle$. To prove this I now tell you that I put Qbits 1, 2, 4, 6, 7, 11, ... into the state $|0\rangle$ and I put Qbits 3, 5, 8, 9, 10, 12, ... into the state $|1\rangle$. You can confirm that I'm telling the truth by measuring each Qbit directly."

Bob makes the direct measurements and gets every one of Alice's predicted outcomes. If instead Alice had sent him Qbits whose states were randomly $\mathbf{H}|0\rangle$ or $\mathbf{H}|1\rangle$ she could do the same trick by telling Bob exactly what he would find if he preceded each of his measurements with a Hadamard gate. But there is no way she could do the trick for measurements preceded by Hadamard gates in the first case or for direct measurements in the second. The best she could do if she wanted to deceive Bob would be to make random guesses for each outcome, and with n Qbits she would succeed in fooling him only with probability $1/2^n$. So this works perfectly well, and without the worry of Bob possessing unexpected safe-cracking skills.

But, as noted above, there is a loophole – in fact, a fatal problem. The technological skills required to take advantage of the loophole are spectacularly greater than those required for the naive protocol, so one could imagine a stretch of years, decades, or even centuries during which the naive protocol might actually be useful. But ultimately it will be insecure. Suppose that Alice, unknown to Bob, has actually prepared n labeled pairs in the entangled state (6.10), sending one member to Bob while retaining the other for herself. Then the Qbits Bob receives will have no states of their own, being entangled with the Qbits Alice keeps for herself. Nevertheless, if Bob chooses to test some of them with measurements, (6.11) insures that the results he gets will be indistinguishable from the random outcomes he would have got if Alice had been playing the game honestly. No hint of her deception will be revealed by any test Bob can perform.

But now when the time comes for Alice to reveal her choice, if she wants to prove to Bob that it was YES, she makes a direct measurement on each of the Qbits she has kept and correctly informs Bob what he will get if he makes a direct measurement on each of the paired Qbits. But if she wants to prove that it was NO, she instead applies Hadamards before measuring each of her Qbits, enabling her, because of the identity (6.11), to tell Bob what he will find if he also applies Hadamards before measuring his own Qbits. So she can use entangled pairs of Qbits to cheat at what would otherwise be a perfectly secure bit-commitment protocol.

Alice can cheat in the same way even if Bob measures his Qbits (randomly applying or not applying a Hadamard before each measurement) before she "reveals" her commitment. If she wants to "prove" to Bob she had sent him YES she directly measures each of her Qbits and tells Bob all her results. He notes that they do indeed agree with all the results he found for his direct measurements, and is persuaded that she had indeed sent him YES. To "prove" she sent him NO she applies Hadamards before measuring each of her Qbits.

Of course the success of Alice's cheating depends crucially on Bob's knowing all about 1-Qbit states, but never having taken the kind of course in quantum mechanics that would have taught him anything about entangled 2-Qbit states. If Bob is as sophisticated a student of the quantum theory as Alice, they will both realize that the protocol is fatally flawed, since it can be defeated by entanglement.

It is in this context that Einstein's famous complaint about spooky actions at a distance ("*spukhafte Fernwirkungen*") seems pertinent. By finally measuring her members of the entangled pairs, Alice seems to convert the distant Qbits in Bob's possession into the kind she deceptively said she had sent him long ago, while retaining until the last minute the option of which of the two kinds to pick. But of course Alice's action is not so much on the Qbits in Bob's possession as it is on what it is possible for her to *tell him* about what he can learn from those Qbits. It is this peculiar tension between what is objective (ontology) and what is known (epistemology) that makes quantum mechanics such a source of delight (or anguish) to the philosophically inclined.

Something like Alice's discovery of the value of entanglement for cheating actually happened in the historical development of these ideas about quantum information processing. When the bit-commitment protocol described above was first put forth it was realized that entangled pairs could be used to thwart it, but more sophisticated versions were proposed that were believed to be immune to cheating with entanglement. There developed a controversy over whether some form of bit commitment could or could not be devised that would be secure even if entanglement were fully exploitable. The current consensus is that there is no way to use Qbits in a bit-commitment protocol that cannot be defeated by using entangled states. Indeed, it has even been suggested that the structure of quantum mechanics might be uniquely determined by requiring it to enable the secure exchange of random strings of bits, as in quantum cryptography, but not to enable bit commitment. Nobody has managed to show this. It does seem implausible that God would have taken as a fundamental principle of design that certain kinds of covert activity should be possible while others should be forbidden.

6.4 Quantum dense coding

Although an infinite amount of information is needed to specify the state $|\psi\rangle = \alpha|0\rangle + \beta|1\rangle$ of a single Qbit, there is no way for somebody who has acquired possession of the Qbit to learn what that state is, as we have often noted. If Alice prepares a Qbit in the state $|\psi\rangle$ and sends it to Bob, all he can do is apply a unitary transformation of his choice and then measure the Qbit, getting the value 0 or 1. After that the state of the Qbit is either $|0\rangle$ or $|1\rangle$ and no further measurement can teach him anything about its original state $|\psi\rangle$. The most Alice can communicate to Bob by sending him a single Qbit is a single bit of information.

If, however, Alice has one member of an entangled pair of Qbits in the state

$$|\Psi\rangle = \tfrac{1}{\sqrt{2}}\big(|0\rangle|0\rangle + |1\rangle|1\rangle\big) \tag{6.15}$$

and Bob has the other, then by suitably preparing her member of the pair and then sending it to Bob, she can convey to him *two* bits of information. She does this by first applying the transformation $\mathbf{1}$, $\mathbf{X}$, $\mathbf{Z}$, or $\mathbf{ZX}$ to her Qbit, depending on whether she wants to send Bob the message 00, 01, 10, or 11. If hers is the Qbit on the left in (6.15) these transform the state of the pair into one of the four mutually orthogonal Bell states (6.5),

$$\begin{aligned}
\mathbf{1}_a|\Psi\rangle &= \tfrac{1}{\sqrt{2}}\big(|0\rangle|0\rangle + |1\rangle|1\rangle\big),\\
\mathbf{X}_a|\Psi\rangle &= \tfrac{1}{\sqrt{2}}\big(|1\rangle|0\rangle + |0\rangle|1\rangle\big),\\
\mathbf{Z}_a|\Psi\rangle &= \tfrac{1}{\sqrt{2}}\big(|0\rangle|0\rangle - |1\rangle|1\rangle\big),\\
\mathbf{Z}_a\mathbf{X}_a|\Psi\rangle &= \tfrac{1}{\sqrt{2}}\big(|0\rangle|1\rangle - |1\rangle|0\rangle\big).
\end{aligned} \tag{6.16}$$

She then sends her Qbit over to Bob. He sends the pair through the controlled-NOT gate $\mathbf{C}_{ab}$, using the Qbit he received from Alice as control, to get

$$\begin{aligned}
\mathbf{C}_{ab}\mathbf{1}_a|\Psi\rangle &= \tfrac{1}{\sqrt{2}}\big(|0\rangle + |1\rangle\big)|0\rangle, \\
\mathbf{C}_{ab}\mathbf{X}_a|\Psi\rangle &= \tfrac{1}{\sqrt{2}}\big(|0\rangle + |1\rangle\big)|1\rangle, \\
\mathbf{C}_{ab}\mathbf{Z}_a|\Psi\rangle &= \tfrac{1}{\sqrt{2}}\big(|0\rangle - |1\rangle\big)|0\rangle, \\
\mathbf{C}_{ab}\mathbf{Z}_a\mathbf{X}_a|\Psi\rangle &= \tfrac{1}{\sqrt{2}}\big(|0\rangle - |1\rangle\big)|1\rangle,
\end{aligned} \tag{6.17}$$

and then he applies a Hadamard transform to get

$$\begin{aligned}
\mathbf{H}_a\mathbf{C}_{ab}\mathbf{1}_a|\Psi\rangle &= |0\rangle|0\rangle, \\
\mathbf{H}_a\mathbf{C}_{ab}\mathbf{X}_a|\Psi\rangle &= |0\rangle|1\rangle, \\
\mathbf{H}_a\mathbf{C}_{ab}\mathbf{Z}_a|\Psi\rangle &= |1\rangle|0\rangle, \\
\mathbf{H}_a\mathbf{C}_{ab}\mathbf{Z}_a\mathbf{X}_a|\Psi\rangle &= |1\rangle|1\rangle.
\end{aligned} \tag{6.18}$$

Measuring the two Qbits then gives him 00, 01, 10, or 11 – precisely the two-bit message Alice wished to send.

This process of transforming the Bell basis back into the computational basis – i.e. undoing the process (6.3) by which the Bell basis was constructed from the computational basis – and then measuring is called "measuring in the Bell basis."

One can directly demonstrate that this works with circuit diagrams, without going through any of the analysis in (6.15)–(6.18). Suppose that Alice represents the two bits x and y she wishes to transmit to Bob as the computational-basis state $|x\rangle|y\rangle$ of two Qbits (the top two wires, Figure 6.4(a)). If Bob has two Qbits initially in the state $|0\rangle|0\rangle$ (the bottom two wires in Figure 6.4(a)), then the circuit in Figure 6.4(a) gets the two bits to Bob in a straightforward classical way, transforming the state $|x\rangle|y\rangle|0\rangle|0\rangle$ on the right to $|x\rangle|y\rangle|x\rangle|y\rangle$ on the left by means of direct Qbit-to-Qbit coupling via two cNOT gates. The procedure involves only classical operations on classically meaningful states. It gets the two bits from Alice to Bob by explicit interactions between her Qbits and his. It would work equally well for Cbits.

One can transform this direct classical procedure into the more exotic quantum protocol by expanding the cNOT gates into products of quantum gates. One first expands one of the $\mathbf{C}$ gates into $\mathbf{H}\mathbf{C}^Z\mathbf{H}$ in Figure 6.4(b). Because $\mathbf{Z}$ acting on the control Qbit commutes with $\mathbf{C}$ and because $\mathbf{C}$ is its own inverse, we can further expand Figure 6.4(b) to Figure 6.4(c). We can then bring the $\mathbf{H}$ and $\mathbf{C}$ gates on either side of the $\mathbf{C}^Z$ to the extreme left and right to get Figure 6.4(d). We can also expand the two $\mathbf{C}$ gates on the left of Figure 6.4(d) into the three $\mathbf{C}$ gates on the left of Figure 6.4(e), since the action of either set is to flip the target Qbit if and only if the computational-basis states of the two control Qbits are different, while leaving the states of the control

Fig 6.4 A circuit-theoretic derivation of the quantum dense-coding protocol.

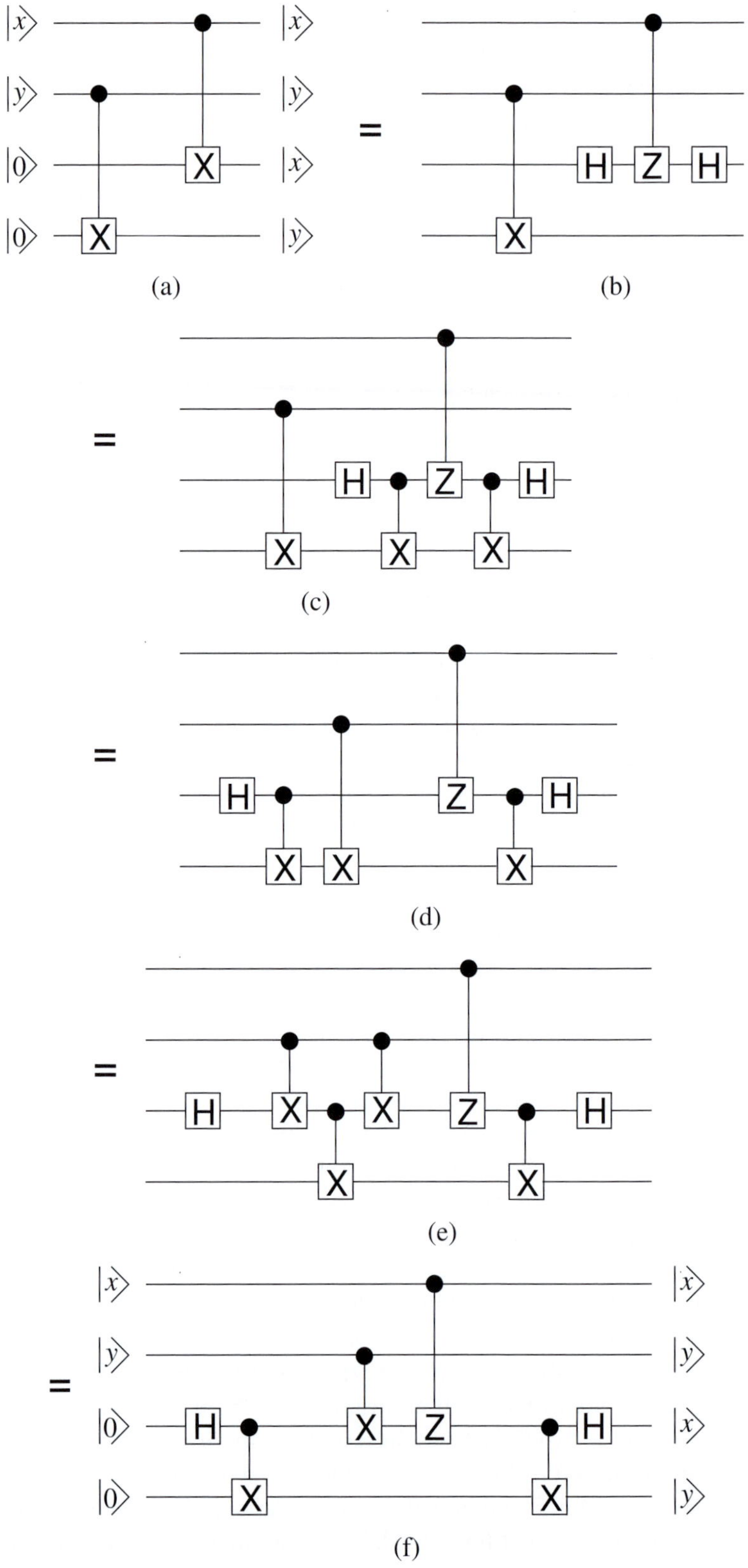

Qbits unaltered. Because the state $\mathbf{H}|0\rangle = (1/\sqrt{2})(|0\rangle + |1\rangle)$ is invariant under the action of $\mathbf{X}$, the $\mathbf{C}$ on the extreme left of Figure 6.4(e) acts as the identity, and Figure 6.4(e) simplifies to Figure 6.4(f).

The fact that Figure 6.4(f) has the same action as Figure 6.4(a) contains all the content of the dense-coding protocol. The pair of gates $\mathbf{C}_{10}\mathbf{H}_1$ on the left of Figure 6.4(f) acts on the state $|0\rangle|0\rangle$ to produce the entangled state (6.15). The bottom Qbit of the pair, Qbit 0, is given to Bob and the upper one, Qbit 1, is given to Alice, who also possesses the upper two, Qbits 2 and 3. The pair of gates $\mathbf{C}^Z_{31}\,\mathbf{C}_{21}$ acts as $\mathbf{1}$, $\mathbf{X}$, $\mathbf{Z}$, or $\mathbf{ZX}$ on Qbit 1 depending on whether the states of Qbits 3 and 2 are $|0\rangle|0\rangle$, $|0\rangle|1\rangle$, $|1\rangle|0\rangle$, or $|1\rangle|1\rangle$. This reproduces the transformation Alice applies to the member of the entangled pair in her possession, depending on the values of the two bits she wishes to transmit to Bob. Alice then sends Qbit 1 to Bob. The final pair $\mathbf{H}_1\mathbf{C}_{10}$ on the right is precisely the transformation (6.18) that Bob performs on the reunited entangled pair before making his measurement, which yields the values x, y that Alice wished to transmit.

Like dense coding, many tricks of quantum information theory, including the one we examine next, teleportation, rely on two or more people sharing entangled Qbits, prepared some time ago, carefully stored in their remote locations awaiting an occasion for their use. While the preparation of entangled Qbits (in the form of photons) and their transmission to distant places has been achieved, putting them into entanglement-preserving, local, long-term storage remains a difficult challenge.

6.5 Teleportation

Suppose that Alice has a Qbit in a state

$$|\psi\rangle = \alpha|0\rangle + \beta|1\rangle, \tag{6.19}$$

but she does not know the amplitudes α and β. Carol, for example, may have prepared the Qbit for Alice by taking a Qbit initially assigned the standard state $|0\rangle$, applying a specific unitary transformation to it, and then giving it to Alice, without telling her what unitary transformation she applied.

Alice would like to reassign that precise state to another Qbit possessed by Bob. Neither Alice nor Bob (who could be far away from Alice) has any access to the other's Qbit. Alice is, however, allowed to send "classical information" to Bob – e.g. she can talk to him over the telephone. And, crucially, Bob's Qbit shares with a second Qbit of Alice the 2-Qbit entangled state

$$|\Phi\rangle = \frac{1}{\sqrt{2}}\big(|0\rangle|0\rangle + |1\rangle|1\rangle\big). \tag{6.20}$$

The no-cloning theorem prohibits duplicating the unknown state of Alice's first Qbit, either far away from her or nearby. But it turns out to be possible for Alice and Bob to cooperate over the telephone in assigning the state $|\psi\rangle$ to Bob's member of the entangled pair. The no-cloning theorem is not violated because in doing so Alice obliterates all traces of the state $|\psi\rangle$ from either of her own Qbits. The process – called teleporting the state from Alice to Bob – also eliminates the entanglement Alice and Bob formerly shared. For each shared entangled pair, they can teleport just a single 1-Qbit state. The term "teleportation" emphasizes that the state assignment acquired by Bob's Qbit no longer applies to Alice's; it has been transported from her Qbit to his.

Here is how teleportation works. Alice's first Qbit and the entangled pair she shares with Bob are characterized by the 3-Qbit state

$$|\psi\rangle_a|\Phi\rangle_{ab} = \big(\alpha|0\rangle_a + \beta|1\rangle_a\big)\tfrac{1}{\sqrt{2}}\big(|0\rangle_a|0\rangle_b + |1\rangle_a|1\rangle_b\big), \tag{6.21}$$

where I have given the state symbols for the Qbits in Alice's and Bob's possession subscripts a and b. To teleport the unknown state of her Qbit to Bob's member of the entangled pair, Alice first applies a cNOT gate, using her first Qbit in the state $|\psi\rangle$ as the control and her member of the shared entangled pair as the target. This produces the 3-Qbit state

$$\alpha|0\rangle_a\tfrac{1}{\sqrt{2}}\big(|0\rangle_a|0\rangle_b + |1\rangle_a|1\rangle_b\big) + \beta|1\rangle_a\tfrac{1}{\sqrt{2}}\big(|1\rangle_a|0\rangle_b + |0\rangle_a|1\rangle_b\big). \tag{6.22}$$

Next she applies a Hadamard transformation **H** to her first Qbit, giving all three Qbits the state

$$\begin{aligned}
&\alpha\tfrac{1}{\sqrt{2}}\big(|0\rangle_a + |1\rangle_a\big)\tfrac{1}{\sqrt{2}}\big(|0\rangle_a|0\rangle_b + |1\rangle_a|1\rangle_b\big)\\
&\qquad\qquad + \beta\tfrac{1}{\sqrt{2}}\big(|0\rangle_a - |1\rangle_a\big)\tfrac{1}{\sqrt{2}}\big(|1\rangle_a|0\rangle_b + |0\rangle_a|1\rangle_b\big)\\
&= \tfrac{1}{2}|0\rangle_a|0\rangle_a\big(\alpha|0\rangle_b + \beta|1\rangle_b\big) + \tfrac{1}{2}|1\rangle_a|0\rangle_a\big(\alpha|0\rangle_b - \beta|1\rangle_b\big)\\
&\quad + \tfrac{1}{2}|0\rangle_a|1\rangle_a\big(\alpha|1\rangle_b + \beta|0\rangle_b\big) + \tfrac{1}{2}|1\rangle_a|1\rangle_a\big(\alpha|1\rangle_b - \beta|0\rangle_b\big).
\end{aligned} \tag{6.23}$$

Now Alice measures both Qbits in her possession. (As remarked in Section 6.4, such an application of cNOT and Hadamard gates, immediately followed by measurement gates, is called "measuring in the Bell basis.") If the result is 00, Bob's Qbit will indeed acquire the state $|\psi\rangle$ originally possessed by Alice's first Qbit (whose state would then be reduced to $|0\rangle$). But if the result of Alice's measurement is 10, 01, or 11 then the state of Bob's Qbit becomes

$$\alpha|0\rangle_b - \beta|1\rangle_b, \quad \alpha|1\rangle_b + \beta|0\rangle_b, \quad \text{or} \quad \alpha|1\rangle_b - \beta|0\rangle_b. \tag{6.24}$$

In each of these three cases there is a unitary transformation that restores the state of Bob's Qbit to Alice's original state $|\psi\rangle$. In the first case we can apply **Z** (which leaves $|0\rangle$ alone but changes the sign of $|1\rangle$), in the second case, **X** (which interchanges $|0\rangle$ and $|1\rangle$), and in the third case, **ZX**.

So all Alice need do to transfer the state of her Qbit to Bob's member of their entangled pair is to telephone Bob and report to him the results of her two measurements. He then knows whether the state has already been transferred (if Alice's result is 00) or what unitary transformation he must apply to his member of the entangled pair in order to complete the transfer (if Alice's result is one of the other three.) Note the resemblance to quantum error correction: by making a measurement Alice acquires the information needed for Bob to reconstruct a particular quantum state, without anybody acquiring any information about what the state actually is.

This appears to be remarkable. A general state of a Qbit is described by two complex numbers α and β that take on a continuum of values, constrained only by the requirement that $|\alpha|^2 + |\beta|^2 = 1$. Yet, with the aid of a standard entangled pair, whose state does not depend on α and β, Alice is able to provide Bob with a Qbit described by the unknown state, at the price of only two bits of classical information (giving the results of her two measurements) and the loss of the entanglement of their pair.

But of course the teleportation process does not communicate to Bob the information that can be encoded in α and β. Bob is no more able to learn the values of α and β from manipulating his Qbit, now assigned the state $|\psi\rangle$, than Alice was able to do when it was her Qbit that was assigned the same state $|\psi\rangle$. On the other hand Alice's state could be produced at a crucial stage of an elaborate quantum computation, and its transfer to Bob could enable him to continue with the computation on his own far-away quantum computer, so one can achieve a nontrivial objective by such teleportations.

Like dense coding, teleportation can also be constructed by manipulating an elementary classical circuit diagram, without going through any of the analysis in (6.21)–(6.24). Figure 6.5(a) shows a circuit that exchanges the state $|\psi\rangle = |x\rangle$ of Alice's Cbit with the state $|0\rangle$ of Bob's Cbit, regardless of whether $x = 0$ or 1. The transfer is achieved by direct physical coupling between the two Cbits. As a linear quantum circuit it continues to perform this exchange for arbitrary superpositions, $|\psi\rangle = \alpha|0\rangle + \beta|1\rangle$. The entire teleportation protocol can be constructed by appropriately expanding the two gates in Figure 6.5(a), with the aid of an ancillary Qbit. The aim of the expansion is to eliminate the direct interaction between Alice's and Bob's Qbits through the two cNOT gates in Figure 6.5(a), in favor of the telephoned message from Alice to Bob, *and* the interaction necessary to produce their shared pair of entangled Qbits (which can take place well before Alice has even acquired her Qbit in the state $|\psi\rangle$).

In Figure 6.5(b) an ancillary Qbit, not acted upon throughout the process, is introduced in the state

$$|\phi\rangle = \mathbf{H}|0\rangle = \tfrac{1}{\sqrt{2}}\big(|0\rangle + |1\rangle\big). \tag{6.25}$$

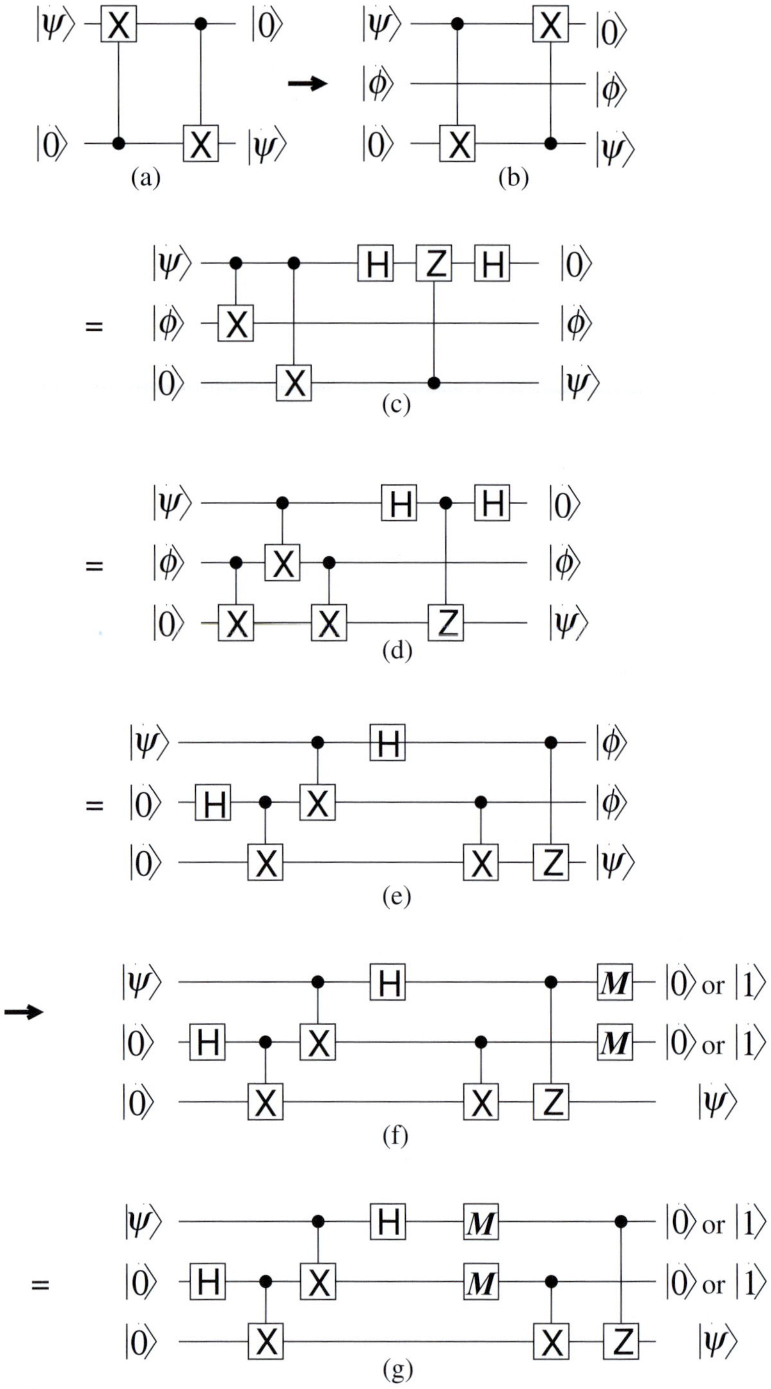

Fig 6.5 A circuit-theoretic derivation of the quantum teleportation protocol.

In Figure 6.5(c) the identities $\mathbf{X} = \mathbf{HZH}$ and $\mathbf{1} = \mathbf{HH}$ have been used to rewrite the cNOT gate on the right of Figure 6.5(b), and an additional cNOT gate has been added on the left, which acts as the identity, since $\mathbf{X}$ acts as the identity on the state $\mathbf{H}|0\rangle$.

Figure 6.5(d) follows from Figure 6.5(c) because the action of $\mathbf{C}^Z$ is independent of which Qbit is the control and which the target, and because the two cNOT gates on the left of Figure 6.5(c) have exactly the same action as the three cNOT gates on the left of Figure 6.5(d): acting on the computational basis, both sets of gates apply $\mathbf{X}$ on both of the bottom two wires if the state of the top wire is $|1\rangle$ and act as the identity if the state of the top wire is $|0\rangle$.

Figure 6.5(e) follows from Figure 6.5(d) if we write the $|\phi\rangle$ on the left of Figure 6.5(d) as $\mathbf{H}|0\rangle$ and explicitly write the $|0\rangle$ on the right of Figure 6.5(d) as $\mathbf{H}|\phi\rangle$. But Figure 6.5(e) is an automated version of teleportation. To relate it to ordinary teleportation, introduce measurements of the upper two Qbits after the circuit of Figure 6.5(e) has acted, as in Figure 6.5(f). Their effect is to collapse the states of each of the two upper wires randomly and independently to $|0\rangle$ or $|1\rangle$. But as noted in Section 3.6, measurement of a control Qbit commutes with any operation controlled by that Qbit, so the measurement gates can be moved to the positions they occupy in Figure 6.5(g).

Figure 6.5(g) is precisely the teleportation protocol. The two gates on the left transform the two lower Qbits into the entangled state (6.20). The subsequent applications to the top two Qbits of cNOT followed by $\mathbf{H}$ followed by two measurement gates are precisely Alice's "measurement in the Bell basis." Since Alice knows the outcomes of the measurements, she knows whether the subsequent cNOT and $\mathbf{C}^Z$ gates will or will not act, and she can replace these physical couplings by a phone call to Bob telling him whether or not to apply $\mathbf{X}$ and/or $\mathbf{Z}$ directly to his own Qbit.

Figure 6.6 demonstrates that entanglement can also be teleported. The figure reproduces parts (b), (e), and (g) of Figure 6.5 with three changes. (1) A bar representing n Qbits in the n-Qbit state $|\Phi\rangle_i$ has been added above each part of the figure. No operations act on these additional Qbits. (2) The state to be teleported has been given a subscript i so it is now one of several possible states $|\psi_i\rangle$. (3) Because of the linearity of the unitary gates we may sum over the index i. The effect of the circuit is to transfer participation in the entangled state $\sum_i |\Phi_i\rangle|\psi_i\rangle$ from the third wire from the bottom to the bottom wire.

So even if Alice's Qbit has no state of its own but is entangled with other Qbits, Alice can use the same protocol to teleport its role in the entangled state over to Bob's Qbit. The result is that Bob's Qbit becomes entangled in exactly the same way Alice's was, and Alice's Qbit becomes entirely unentangled.

Fig 6.6 Figure 6.6. A demonstration that entanglement can be teleported.

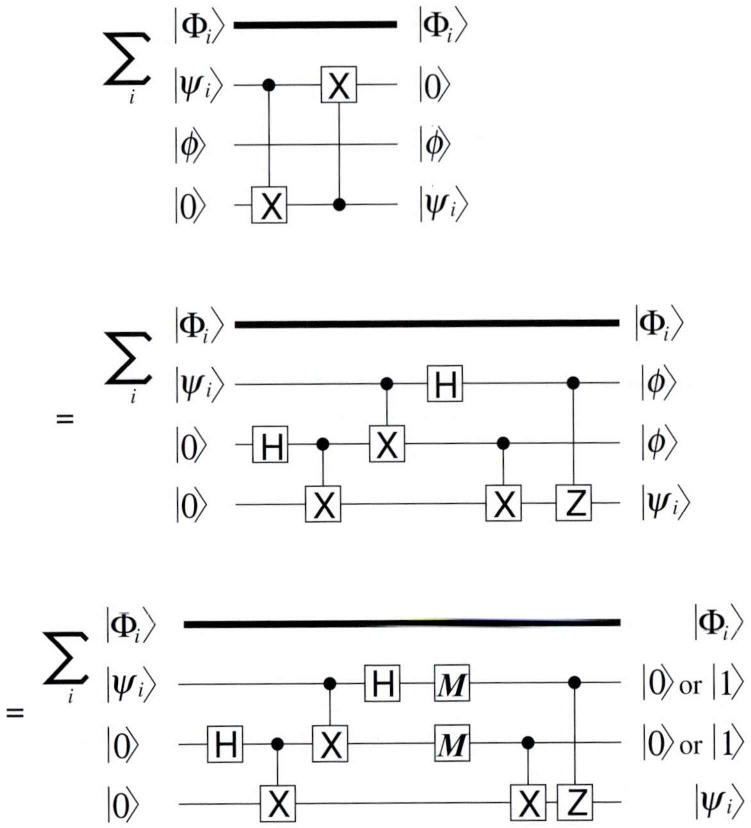

6.6 The GHZ puzzle

We conclude with another illustration of just how strange the behavior of Qbits can be. The situation described below is a 3-Qbit version of one first noticed by Daniel Greenberger, Michael Horne, and Anton Zeilinger ("GHZ") in the late 1980s, which gives a very striking version of Bell's theorem. An alternative version, discovered by Lucien Hardy in the early 1990s, is given in Appendix D.

Consider the 3-Qbit state

$$|\Psi\rangle = \tfrac{1}{2}\big(|000\rangle - |110\rangle - |011\rangle - |101\rangle\big). \tag{6.26}$$

Note that the form of $|\Psi\rangle$ is explicitly invariant under any permutation of the three Qbits. Numbering the Qbits from left to right 2, 1, and 0, we have

$$|\Psi\rangle = \mathbf{C}_{21}\mathbf{H}_2\mathbf{X}_2\tfrac{1}{\sqrt{2}}\big(|000\rangle - |111\rangle\big). \tag{6.27}$$

Since

$$\tfrac{1}{\sqrt{2}}\big(|000\rangle - |111\rangle\big) = \mathbf{C}_{21}\mathbf{C}_{20}\mathbf{H}_2\mathbf{X}_2|000\rangle, \tag{6.28}$$

(6.27) and (6.28) provide an explicit construction of $|\Psi\rangle$ from elementary 1- and 2-Qbit gates acting on the standard state $|0\rangle_3$.

Because $|\Psi\rangle$ in the form (6.26) and the state $(1/\sqrt{2})(|000\rangle - |111\rangle)$ appearing in (6.27) are both invariant under permutations of the Qbits 0, 1, and 2, any of the other five forms of (6.27) associated with permutations of the subscripts 0, 1, and 2 are equally valid. In particular

$$|\Psi\rangle = \mathbf{C}_{12}\mathbf{H}_1\mathbf{X}_1\tfrac{1}{\sqrt{2}}\big(|000\rangle - |111\rangle\big). \tag{6.29}$$

It follows from (6.29) that

$$\begin{aligned}\mathbf{H}_2\mathbf{H}_1|\Psi\rangle &= \mathbf{H}_2\mathbf{H}_1\mathbf{C}_{12}\mathbf{H}_1\mathbf{X}_1\tfrac{1}{\sqrt{2}}\big(|000\rangle - |111\rangle\big)\\ &= \big(\mathbf{H}_2\mathbf{H}_1\mathbf{C}_{12}\mathbf{H}_1\mathbf{H}_2\big)\mathbf{H}_2\mathbf{X}_1\tfrac{1}{\sqrt{2}}\big(|000\rangle - |111\rangle\big)\\ &= \mathbf{C}_{21}\mathbf{H}_2\mathbf{X}_1\big(|000\rangle - |111\rangle\big)\end{aligned} \tag{6.30}$$

(since sandwiching a cNOT between Hadamards exchanges target and control Qbits). Comparing the last expression in (6.30) with the form of $|\Psi\rangle$ in (6.27) reveals that

$$\mathbf{H}_2\mathbf{H}_1|\Psi\rangle = \mathbf{Z}_2\mathbf{X}_1|\Psi\rangle \tag{6.31}$$

(which can, of course, be confirmed more clumsily directly from the definition (6.26) of $|\Psi\rangle$.) Because of the invariance of $|\Psi\rangle$ under permutation of the three Qbits we also have

$$\mathbf{H}_2\mathbf{H}_0|\Psi\rangle = \mathbf{Z}_2\mathbf{X}_0|\Psi\rangle, \tag{6.32}$$

$$\mathbf{H}_1\mathbf{H}_0|\Psi\rangle = \mathbf{Z}_1\mathbf{X}_0|\Psi\rangle. \tag{6.33}$$

Now suppose that we have prepared three Qbits in the state $|\Psi\rangle$ and then allowed no further interactions among them. If we measure each Qbit, it follows from the form (6.26) that because $|\Psi\rangle$ is a superposition of computational-basis states having either none or two of the Qbits in the state $|1\rangle$, the three outcomes are constrained to satisfy

$$x_2 \oplus x_1 \oplus x_0 = 0 \tag{6.34}$$

(where $\oplus$, as usual, denotes addition modulo 2).

Suppose, on the other hand, that we apply Hadamards to Qbits 2 and 1 before measuring all three. According to (6.31) this has the effect of flipping the state of Qbit 1 in each term of the superposition (6.26) (and changing the signs of some of the terms). As a result the 3-Qbit state (6.26) is changed into a superposition of computational-basis states having either one or three of the Qbits in the state $|1\rangle$. So if the outcomes are x_2^H, x_1^H, and x_0, we must have

$$x_2^H \oplus x_1^H \oplus x_0 = 1. \tag{6.35}$$

Similarly, if we apply Hadamards to Qbits 2 and 0 before measuring all three, then (6.32) requires that the outcomes must obey

$$x_2^H \oplus x_1 \oplus x_0^H = 1, \tag{6.36}$$

and if Hadamards are applied to Qbits 1 and 0 then according to (6.33) if all three are measured we will have

$$x_2 \oplus x_1^H \oplus x_0^H = 1. \tag{6.37}$$

Consider now the following question. If we are talking about a single trio of Qbits, assigned the state $|\Psi\rangle$, must the x_0 appearing in (6.34) be the same as the x_0 appearing in (6.35)? A little reflection reveals that this question makes no sense. After all, (6.34) describes the outcomes of immediately measuring the three Qbits, whereas (6.35) describes the outcomes of measuring them after Hadamards have been applied to Qbits 2 and 1. Since only one of these two possibilities can actually be carried out, there is no way to compare the results of measuring Qbit 0 in the two cases. You can't compare the x_0 you found in the case you actually carried out with the x_0 you might have found in the case you didn't carry out. It's just a stupid question.

Or is it? Suppose that Qbits 2 and 1 are measured *before* Qbit 0 is measured. If no Hadamards were applied before the measurements of 2 and 1, then (6.34) assures us that when 0 is finally measured the result will be

$$x_0 = x_1 \oplus x_2. \tag{6.38}$$

So the outcome of measuring Qbit 0 is predetermined by the outcomes of the earlier measurements of Qbits 2 and 1. Since all interactions among the Qbits ceased after the state $|\Psi\rangle$ had been prepared, subjecting Qbits 2 and 1 to measurement gates can have no effect on Qbit 0. Since the outcomes of the measurements of Qbits 2 and 1 determine in advance the outcome of the subsequent measurement of Qbit 0, it would seem that Qbit 0 was already predisposed to give the result (6.38) upon being measured. Because the Qbits did not interact after their initial state was prepared, it would seem that Qbit 0 must have had that predisposition even before Qbits 2 and 1 were actually measured to reveal what the result of measuring Qbit 0 would have to be.

This is a bit disconcerting, since prior to any measurements the state of the Qbits was (6.26), in which none of them was individually predisposed to reveal any particular value. Indeed, it would seem that the 3-Qbit state (6.26) gives an *incomplete* description of the Qbits. The omitted predisposition of Qbit 0 seems to be an additional *element of reality* that a more complete description than that afforded by the quantum theory would take into account.

But if Qbit 0 did indeed have a predetermined predisposition to give x_0 when measured, even before Qbits 1 and 2 were measured to reveal what x_0 actually was, then the value of x_0 surely would not be altered if Hadamards were applied to Qbits 1 and 2 before they were measured, since the Qbits have ceased to interact, and the predisposition to give x_0 was present before the decision to apply Hadamards or not had been made. This means that the value x_0 appearing in (6.34) must indeed be identical to the value of x_0 appearing in (6.35). So our question is not meaningless. The answer is Yes!

Such an argument for elements of reality – predetermined values – was put forth in 1935 (in a different context) by Albert Einstein, Boris Podolsky, and Nathan Rosen (EPR). The controversy and discussion it has given rise to has steadily increased over the past seven decades. The terms "incomplete" and "element of reality" originated with EPR. Today it is Einstein's most cited paper.

The wonderful thing about three Qbits in the state (6.26) is that they not only provide a beautiful illustration of the EPR argument, but also, when examined further, reveal that the appealing argument establishing predetermined measurement outcomes cannot be correct. To see this, note that exactly the same reasoning establishes that the values of x_1 appearing in (6.34) and (6.36) must be the same, as well as the values of x_2 appearing in (6.34) and (6.37). And the same line of thought establishes that the values of x_0^H in (6.37) and (6.36) must be the same, as well as the values of x_1^H in (6.37) and (6.35) and the values of x_2^H in (6.36) and (6.35).

If all this is true, then adding together the left sides of (6.34)–(6.37) must give 0 modulo 2, since each of x_2, x_1, x_0, x_2^H, x_1^H, and x_0^H appears in exactly two of the equations. But the modulo 2 sum of the right sides is $0 \oplus 1 \oplus 1 \oplus 1 = 1$.

So the appealing EPR argument must be wrong. There are no elements of reality – no predetermined measurement outcomes that a more complete theory would take into account. The answer to what is mistaken in the simple and persuasive reasoning that led Einstein, Podolsky, and Rosen to the existence of elements of reality is still a matter of debate more than 70 years later. How, after all, can Qbit 0 and its measurement gate "know" that if they interact only after Qbits 1 and 2 have gone through their own measurement gates (and no Hadamards were applied) then the result of the measurement of Qbit 0 *must* be given by (6.38)?

The best explanation anybody has come up with to this day is to insist that no explanation is needed beyond what one can infer from the laws of quantum mechanics. Those laws are correct. Quantum mechanics works. There is no controversy about that. What fail to work are attempts to provide underlying mechanisms, that go beyond the quantum-mechanical rules, for how certain strong quantum correlations can actually operate. One gets puzzled only if one tries to understand how the rules can work not only for the actual situation in which they are applied, but also in alternative situations that might have been chosen but were not.

By concluding with this "paradoxical" state of affairs, I am not suggesting that there is anything wrong with the quantum-theoretic description of Qbits and the gates that act on them. On the contrary, the quantum theory has to be regarded as the must accurate and successful theory in the history of physics, and there is no doubt whatever among physicists that if the formidable technological obstacles standing in the way of building a quantum computer can be overcome, then the computer will behave exactly as described in the preceding chapters.

But I cannot, in good conscience, leave you without a warning that the simple theory of Qbits developed here, though correct, is in some respects exceedingly strange. The strangeness emerges only when one seeks to go beyond the straightforward rules enunciated in Chapter 1. In particular one must not ask for an underlying mechanism that accounts not only for the behavior of the circuit actually applied to a particular collection of Qbits, but also for the possible behavior of other circuits that might have been applied to the very same collection of Qbits, but were not.

A good motto for the quantum physicist and for future quantum computer scientists might be "What didn't happen *didn't happen*." On that firm note I conclude (except for the 16 appendices that follow).

Appendix A

Vector spaces: basic properties and Dirac notation

In quantum computation the integers from 0 to N are associated with $N+1$ orthogonal unit vectors in a vector space of $D = N+1$ dimensions over the complex numbers. The nature of this association is the subject of Chapter 1. Here we review some of the basic properties of such a vector space, while relating conventional vector-space notation to the Dirac notation used in quantum computer science. Usually the dimension D is a power of 2, but this does not matter for our summary of the basic facts and nomenclature.

In conventional notation such a set of $D = N+1$ orthonormal vectors might be denoted by symbols such as $\phi_0, \phi_1, \phi_2, \ldots, \phi_N$. The orthogonality and normalization conditions are expressed in terms of the inner products (ϕ_x, ϕ_y):

$$(\phi_x, \phi_y) = \begin{cases} 0, & x \neq y; \\ 1, & x = y. \end{cases} \tag{A.1}$$

In quantum computation the indices x and y describing the integers associated with the vectors play too important a role to be relegated to tiny fonts in subscripts. Fortunately quantum mechanics employs a notation for vectors, invented by the physicist Paul Dirac, which is well suited for representing such information more prominently. One replaces the symbols ϕ_x and ϕ_y by $|x\rangle$ and $|y\rangle$, and represents the inner product (ϕ_x, ϕ_y) by the symbol $\langle x|y\rangle$. The orthonormality condition (A.1) becomes

$$\langle x|y\rangle = \begin{cases} 0, & x \neq y; \\ 1, & x = y. \end{cases} \tag{A.2}$$

Vectorial character is conveyed by the symbol $|\;\rangle$, with the specific vector being identified by whatever it is that goes between the bent line $\rangle$ and the vertical line $|$. This notational strategy is reminiscent of the notation for vectors in ordinary three-dimensional physical space (which we will use here for such vectors) in which vectorial character is indicated by a horizontal arrow above a symbol denoting the specific vector being referred to: $\vec{r}$.

Symbols like ϕ and ψ remain useful in the notation of quantum computation for representing generic vectors, but for consistency with the notation for vectors associated with specific integers, and to emphasize their vectorial character, they too are enclosed between a bent line $\rangle$

and a vertical line |, becoming $|\phi\rangle$ and $|\psi\rangle$. Some mathematicians disapprove of this practice. Why write $|\psi\rangle$, introducing the spurious symbols $\rangle$ and |, when ψ by itself does the job perfectly well? This gets it backwards. The real point is that the important information – for example the number 7798 – is easier to read in the form $|7798\rangle$ than when presented in small print in the form ϕ_{7798}. Why introduce in a normal font the often uninformative symbol ϕ, at the price of demoting the most important information to a mere subscript?

The vector space that describes the operation of a quantum computer consists of all linear combinations $|\psi\rangle$ of the $N+1$ orthonormal vectors $|x\rangle$, $x = 0, \ldots, N$, with coefficients α_x taken from the complex numbers:

$$|\psi\rangle = \alpha_0|0\rangle + \alpha_1|1\rangle + \cdots + \alpha_N|N\rangle = \sum_{x=0}^{N} \alpha_x|x\rangle, \qquad \text{(A.3)}$$

where $\alpha_x = u_x + iv_x$, u_x and v_x are real numbers, and $i = \sqrt{-1}$.

The mathematicians' preference for writing ψ instead of $|\psi\rangle$ for generic vectors is explicitly acknowledged in the useful convention that $|\alpha\psi + \beta\phi\rangle$ is nothing more than an alternative way of writing the vector $\alpha|\psi\rangle + \beta|\phi\rangle$:

$$|\alpha\psi + \beta\phi\rangle = \alpha|\psi\rangle + \beta|\phi\rangle. \qquad \text{(A.4)}$$

In a vector space over the complex numbers the inner product of two general vectors is a complex number satisfying

$$\langle\psi|\phi\rangle = \langle\phi|\psi\rangle^*, \qquad \text{(A.5)}$$

where * denotes complex conjugation:

$$(u + iv)^* = u - iv, \quad u, v \text{ real}. \qquad \text{(A.6)}$$

The inner product is linear in the right-hand vector,

$$\langle\phi|\alpha\psi_1 + \beta\psi_2\rangle = \alpha\langle\phi|\psi_1\rangle + \beta\langle\phi|\psi_2\rangle, \qquad \text{(A.7)}$$

and therefore, from (A.5), "anti-linear" in the left-hand vector,

$$\langle\alpha\phi_1 + \beta\phi_2|\psi\rangle = \alpha^*\langle\phi_1|\psi\rangle + \beta^*\langle\phi_2|\psi\rangle. \qquad \text{(A.8)}$$

The inner product of a vector with itself is a real number satisfying

$$\langle\phi|\phi\rangle > 0, \quad |\phi\rangle \neq 0. \qquad \text{(A.9)}$$

It follows from the orthonormality condition (A.2) that the inner product of the vector $|\psi\rangle$ in (A.3) with another vector

$$|\phi\rangle = \beta_0|0\rangle + \beta_1|1\rangle + \cdots + \beta_N|N\rangle = \sum_x \beta_x|x\rangle \qquad \text{(A.10)}$$

is given in terms of the expansion coefficients α_x and β_x (called *amplitudes* in quantum computation) by

$$\langle\phi|\psi\rangle = \sum_x \beta_x^* \alpha_x. \tag{A.11}$$

The squared magnitude of a vector is its inner product with itself, so (A.11) gives for the squared magnitude

$$\langle\psi|\psi\rangle = \sum_x |\alpha_x|^2, \tag{A.12}$$

where

$$|u + iv|^2 = u^2 + v^2, \quad u, v \text{ real.} \tag{A.13}$$

The form (A.12) gives an explicit confirmation of the rule (A.9).

A linear transformation $\mathbf{A}$ associates with every vector $|\psi\rangle$ another vector, called $\mathbf{A}|\psi\rangle$, subject to the rule (linearity)

$$\mathbf{A}\big(\alpha|\psi\rangle + \beta|\phi\rangle\big) = \alpha\mathbf{A}|\psi\rangle + \beta\mathbf{A}|\phi\rangle. \tag{A.14}$$

With a nod to the mathematicians, it is notationally useful to define

$$|\mathbf{A}\psi\rangle = \mathbf{A}|\psi\rangle. \tag{A.15}$$

A linear transformation that preserves the magnitudes of all vectors is called *unitary*, because it follows from linearity that all magnitudes will be preserved if and only if unit vectors (vectors of magnitude 1) are taken into unit vectors. It also follows from linearity that if a linear transformation $\mathbf{U}$ is unitary then it must preserve not only the inner products of arbitrary vectors with themselves, but also the inner products of arbitrary pairs of vectors. This follows straightforwardly for two general vectors $|\phi\rangle$ and $|\psi\rangle$ from the fact that $\mathbf{U}$ preserves the magnitudes of both of them, as well as the magnitudes of the vectors $|\phi\rangle + |\psi\rangle$ and $|\phi\rangle + i|\psi\rangle$.

One can associate with any given vector $|\phi\rangle$ the *linear functional* that takes every vector $|\psi\rangle$ into the number $\langle\phi|\psi\rangle$. Linearity follows from property (A.7) of the inner product. The set of all such linear functionals is itself a vector space, called the *dual space* of the original space. The functional associated with the vector $\alpha|\phi\rangle + \beta|\psi\rangle$ is the sum of α^* times the functional associated with $|\phi\rangle$ and β^* times the functional associated with $|\psi\rangle$. It is an easy exercise to show that *any* linear functional on the original space is associated with some vector in the dual space. Dirac called vectors in the original space *ket vectors* and vectors in the dual space *bra vectors*. He denoted the bra associated with the ket $|\phi\rangle$ by the symbol $\langle\phi|$, so that the symbol $\langle\phi|\psi\rangle$ can equally well be viewed as the inner product of the two kets $|\phi\rangle$ and $|\psi\rangle$ or as a compact way of expressing the action $\langle\phi|(|\psi\rangle)$ of the associated linear functional $\langle\phi|$ on the vector $|\psi\rangle$. Note that one has

$$\langle\alpha\phi + \beta\psi| = \alpha^*\langle\phi| + \beta^*\langle\psi|. \tag{A.16}$$

A linear transformation $\mathbf{A}$ on the space of ket vectors induces a linear transformation $\mathbf{A}^\dagger$ (called "A-adjoint") on the dual space of bra vectors, according to the rule

$$\langle\mathbf{A}\psi| = \langle\psi|\mathbf{A}^\dagger. \tag{A.17}$$

The operation adjoint to the trivial linear transformation that multiplies by a given complex number is multiplication by the complex conjugate of that number.

It is convenient to extend the dagger notation to the vectors themselves, defining

$$\big(|\psi\rangle\big)^\dagger = \langle\psi|, \tag{A.18}$$

so that the bra dual to a given ket is viewed as adjoint to that ket. The definition (A.17) of $\mathbf{A}^\dagger$ then becomes

$$\big(|\mathbf{A}\psi\rangle\big)^\dagger = \langle\psi|\mathbf{A}^\dagger, \tag{A.19}$$

or, with (A.15),

$$\big(\mathbf{A}|\psi\rangle\big)^\dagger = \langle\psi|\mathbf{A}^\dagger, \tag{A.20}$$

which provides a simple example of a very general rule that the adjoint of a product of quantities is the product of their adjoints taken in the opposite order. Another instance of the rule which follows from (A.20) is that

$$\langle\phi|\big(\mathbf{AB}\big)^\dagger = \langle\mathbf{AB}\phi| = \langle\mathbf{B}\phi|\mathbf{A}^\dagger = \langle\phi|\mathbf{B}^\dagger\mathbf{A}^\dagger. \tag{A.21}$$

Since this holds for arbitrary $\langle\phi|$ we have

$$\big(\mathbf{AB}\big)^\dagger = \mathbf{B}^\dagger\mathbf{A}^\dagger. \tag{A.22}$$

Although the adjoint $\mathbf{A}^\dagger$ of a linear transformation $\mathbf{A}$ on kets is a linear transformation on bras, one can also define its action on kets. One does so by requiring that the action of $\langle\phi|$ on $\mathbf{A}^\dagger|\psi\rangle$ should be equal to the action of $\langle\phi|\mathbf{A}^\dagger$ on $|\psi\rangle$. This amounts to stipulating that the symbol $\langle\phi|\mathbf{A}^\dagger|\psi\rangle$ should be unambiguous; it does not matter whether it is read as $(\langle\phi|\mathbf{A}^\dagger)|\psi\rangle$ or as $\langle\phi|(\mathbf{A}^\dagger|\psi\rangle)$. Implicit in this definition is the fact that a vector is completely defined by giving its inner product with all vectors. This in turn follows from the fact that a vector $|\psi\rangle$ can be defined by giving all the amplitudes α_x in its expansion (A.3) in the complete orthonormal set $|x\rangle$. But $\alpha_x = \langle x|\psi\rangle$. Similarly, a linear operator $\mathbf{A}$ is completely defined by giving its *matrix elements* $\langle\phi|\mathbf{A}|\psi\rangle$ for arbitrary pairs of vectors, since the subset $\langle x|\mathbf{A}|y\rangle$ is already enough to determine its action on a general vector (A.3).

Note that any matrix element of $\mathbf{A}^\dagger$ is equal to the complex conjugate of the *transposed* (with ϕ and ψ exchanged) matrix element of $\mathbf{A}$:

$$\langle\phi|\mathbf{A}^\dagger|\psi\rangle = \langle\mathbf{A}\phi|\psi\rangle = \langle\psi|\mathbf{A}\phi\rangle^* = \langle\psi|\mathbf{A}|\phi\rangle^*. \tag{A.23}$$

It follows from this that

$$\left(\mathbf{A}^\dagger\right)^\dagger = \mathbf{A}. \tag{A.24}$$

Since a unitary transformation $\mathbf{U}$ preserves inner products, we have

$$\langle\phi|\psi\rangle = \langle\mathbf{U}\phi|\mathbf{U}\psi\rangle = \langle\phi|\mathbf{U}^\dagger\mathbf{U}|\psi\rangle, \tag{A.25}$$

and therefore

$$\mathbf{U}^\dagger\mathbf{U} = \mathbf{1}, \tag{A.26}$$

where $\mathbf{1}$ is the unit (identity) operator that takes every vector into itself. It follows from (A.26) that

$$\mathbf{U}\mathbf{U}^\dagger\mathbf{U} = \mathbf{U}. \tag{A.27}$$

In a finite-dimensional vector space a unitary transformation $\mathbf{U}$ always takes an orthonormal basis into another orthonormal basis, so any $\mathbf{U}$ clearly has a right inverse – the linear transformation that takes the second basis back into the first. Multiplying (A.27) on the right by that inverse tells us that

$$\mathbf{U}\mathbf{U}^\dagger = \mathbf{1}, \tag{A.28}$$

so $\mathbf{U}^\dagger$ and $\mathbf{U}$ are inverses regardless of the order in which they act.

The vector $|\psi\rangle$ is an *eigenvector* of the linear operator $\mathbf{A}$ if the action of $\mathbf{A}$ on $|\psi\rangle$ is simply to multiply it by a complex number a, called an *eigenvalue* of $\mathbf{A}$:

$$\mathbf{A}|\psi\rangle = a|\psi\rangle. \tag{A.29}$$

Since the number a can be expressed as $a = \langle\psi|\mathbf{A}|\psi\rangle/\langle\psi|\psi\rangle$, it follows from (A.23) that if $\mathbf{A} = \mathbf{A}^\dagger$ (such operators are said to be *self-adjoint* or *Hermitian*) then a is a real number. Eigenvalues of Hermitian operators are necessarily real.

Since $\mathbf{A}$ is Hermitian and a is a real number, it follows from (A.29) (by forming the adjoints of both sides) that

$$\langle\psi|\mathbf{A} = a\langle\psi|, \tag{A.30}$$

so the vector dual to an eigenket of a Hermitian operator is an eigenbra with the same eigenvalue. It follows immediately that if $|\phi\rangle$ is another eigenvector of $\mathbf{A}$ with eigenvalue a', then

$$a\langle\psi|\phi\rangle = \langle\psi|\mathbf{A}|\phi\rangle = a'\langle\psi|\phi\rangle, \tag{A.31}$$

so if $a' \neq a$ then $\langle\psi|\phi\rangle = 0$: eigenvectors of a Hermitian operator with different eigenvalues are orthogonal.

It can be shown that for any Hermitian operator $\mathbf{A}$, one can choose an orthonormal basis for the entire D-dimensional space whose members are eigenvectors of $\mathbf{A}$. The basis is unique if and only if all the D eigenvalues of $\mathbf{A}$ are distinct. In the contrary case (in which $\mathbf{A}$ is said to be degenerate) one can pick arbitrary orthonormal bases within each of the subspaces spanned by eigenvectors of $\mathbf{A}$ with the same eigenvalue. More generally, if $\mathbf{A}, \mathbf{B}, \mathbf{C}, \ldots$ are mutually commuting Hermitian operators then one can choose an orthonormal basis whose members are eigenstates of every one of them.

If $\mathbf{B}$ is any linear operator, then $\mathbf{A}_1 = \mathbf{B} + \mathbf{B}^\dagger$ and $\mathbf{A}_2 = i\big(\mathbf{B}^\dagger - \mathbf{B}\big)$ are both Hermitian, and commute if $\mathbf{B}$ and $\mathbf{B}^\dagger$ commute. Since a joint eigenvector of $\mathbf{A}_1$ and $\mathbf{A}_2$ is also a joint eigenvector of $\mathbf{B} = \mathbf{A}_1 + i\mathbf{A}_2$ and $\mathbf{B}^\dagger = \mathbf{A}_1 - i\mathbf{A}_2$, it follows that if $\mathbf{B}$ commutes with $\mathbf{B}^\dagger$ then one can choose an orthonormal basis of eigenvectors of $\mathbf{B}$. In particular, since a unitary transformation $\mathbf{U}$ satisfies $\mathbf{U}\mathbf{U}^\dagger = \mathbf{U}^\dagger\mathbf{U} = \mathbf{1}$, one can choose an orthonormal basis consisting of eigenvectors of $\mathbf{U}$. Since unitary transformations preserve the magnitudes of vectors, the eigenvalues of $\mathbf{U}$ must be complex numbers of modulus 1. In the quantum theory such complex numbers are often called *phase factors*.

Given two vector spaces of dimensions D_1 and D_2, and given any two vectors $|\psi_1\rangle$ and $|\psi_2\rangle$ in the two spaces, one associates with each such pair a *tensor product* $|\psi_1\rangle \otimes |\psi_2\rangle$ (often the tensor-product sign $\otimes$ is omitted) which is bilinear:

$$\begin{aligned} |\psi_1\rangle \otimes \big(\alpha|\psi_2\rangle + \beta|\phi_2\rangle\big) &= \alpha|\psi_1\rangle \otimes |\psi_2\rangle + \beta|\psi_1\rangle \otimes |\phi_2\rangle, \\ \big(\alpha|\psi_1\rangle + \beta|\phi_1\rangle\big) \otimes |\psi_2\rangle &= \alpha|\psi_1\rangle \otimes |\psi_2\rangle + \beta|\phi_1\rangle \otimes |\psi_2\rangle. \end{aligned} \tag{A.32}$$

With the further rule that $|\psi_1\rangle \otimes |\psi_2\rangle = |\phi_1\rangle \otimes |\phi_2\rangle$ only if $|\phi_1\rangle$ and $|\phi_2\rangle$ are scalar multiples of $|\psi_1\rangle$ and $|\psi_2\rangle$, one easily sees that the set of all tensor products of vectors from the two spaces forms a vector space of dimension $D_1 D_2$.

One defines the inner product of $|\psi_1\rangle \otimes |\psi_2\rangle$ with $|\phi_1\rangle \otimes |\phi_2\rangle$ to be the ordinary product $\langle\psi_1|\phi_1\rangle\langle\psi_2|\phi_2\rangle$ of the inner products in the two original spaces. Given orthonormal bases for each of the two spaces, the set of tensor products of all pairs of vectors from the two bases forms an orthonormal basis for the tensor-product space. If $\mathbf{A}_1$ and $\mathbf{A}_2$ are linear operators on the two spaces, one defines the tensor-product operator $\mathbf{A}_1 \otimes \mathbf{A}_2$ to satisfy

$$(\mathbf{A}_1 \otimes \mathbf{A}_2)\big(|\psi_1\rangle \otimes |\psi_2\rangle\big) = |\mathbf{A}_1\psi_1\rangle \otimes |\mathbf{A}_2\psi_2\rangle = \big(\mathbf{A}_1|\psi_1\rangle\big) \otimes \big(\mathbf{A}_2|\psi_2\rangle\big), \tag{A.33}$$

and easily shows that it can be extended to a linear operator on the entire tensor-product space.

All of this generalizes in the obvious way to n-fold tensor products of n vector spaces.

If $\mathbf{A}$ is a linear operator whose eigenvectors constitute an orthonormal basis – i.e. if $\mathbf{A}$ is Hermitian or, more generally, if $\mathbf{A}$ and $\mathbf{A}^\dagger$ commute – and if f is a function taking complex numbers to complex numbers, then one can define $f(\mathbf{A})$ by specifying that each eigenvector $|\phi\rangle$ of $\mathbf{A}$, in the basis with eigenvalue a, is also an eigenvector of $f(\mathbf{A})$ with eigenvalue $f(a)$. This defines $f(\mathbf{A})$ on a basis, and it can therefore be extended to arbitrary vectors by requiring it to be linear. It follows from this definition that if $f(z)$ is a polynomial or convergent power series in z then $f(\mathbf{A})$ is the corresponding polynomial or convergent power series in $\mathbf{A}$.

In Dirac notation one defines the *outer product* of two vectors $|\phi\rangle$ and $|\psi\rangle$ to be the linear operator, denoted by $|\phi\rangle\langle\psi|$, that takes any vector $|\gamma\rangle$ into $|\phi\rangle$ multiplied by the inner product $\langle\psi|\gamma\rangle$:

$$\big(|\phi\rangle\langle\psi|\big)|\gamma\rangle = |\phi\rangle\big(\langle\psi|\gamma\rangle\big). \tag{A.34}$$

As is always the case with Dirac notation, the point is to define things in such a way that the evaluation of an ambiguous expression such as $|\phi\rangle\langle\psi|\gamma\rangle$ does not depend on how you read it; the notation is designed always to enforce the associative law.

Note that $|\psi\rangle\langle\psi|$ is the projection operator onto the one-dimensional subspace spanned by the unit vector $|\psi\rangle$; i.e. any vector $|\gamma\rangle$ can be written as the sum of a vector $|\gamma\rangle_{\|}$ in the one-dimensional subspace and a vector $|\gamma\rangle_{\perp}$ perpendicular to the one-dimensional subspace, and

$$\big(|\psi\rangle\langle\psi|\big)|\gamma\rangle = |\gamma\rangle_{\|}. \tag{A.35}$$

Similarly, if one has a set of orthonormal vectors $|\psi_i\rangle$ then $\sum_i |\psi_i\rangle\langle\psi_i|$ projects onto the subspace spanned by all the $|\psi_i\rangle$. If the orthonormal set is a complete orthonormal set – for example $|x\rangle$, $x = 0, \ldots, N$ – then the set spans the entire vector space and the projection operator is the unit operator $\mathbf{1}$:

$$\sum_{x=0}^{N} |x\rangle\langle x| = \mathbf{1}. \tag{A.36}$$

This trivial identity can be surprisingly helpful. Any vector $|\psi\rangle$, for example, satisfies

$$|\psi\rangle = \mathbf{1}|\psi\rangle = \sum_x |x\rangle\langle x|\psi\rangle, \tag{A.37}$$

which tells us that the amplitudes α_x appearing in the expansion (A.3) of $|\psi\rangle$ are just the inner products $\langle x|\psi\rangle$. Similarly, any linear operator

$\mathbf{A}$ satisfies

$$|\mathbf{A}\rangle = \mathbf{1A1} = \left(\sum_x |x\rangle\langle x|\right)\mathbf{A}\left(\sum_y |y\rangle\langle y|\right) = \sum_{xy} |x\rangle\langle y|\big(\langle x|\mathbf{A}|y\rangle\big), \tag{A.38}$$

which reveals the matrix elements $\langle x|\mathbf{A}|y\rangle$ to be the expansion coefficients of the operator $\mathbf{A}$ in the "operator basis" $|x\rangle\langle y|$. And note that

$$\langle x|\mathbf{AB}|y\rangle = \langle x|\mathbf{A1B}|y\rangle = \sum_z \langle x|\mathbf{A}|z\rangle\langle z|\mathbf{B}|y\rangle, \tag{A.39}$$

which gives the familiar matrix-multiplication rule for constructing the matrix of a product out of the matrix elements of the individual operators.

If you prefer to think of vectors in terms of their components in a specific basis, then you might note that the (ket) vector $|\psi\rangle$, with the expansion (A.3) with amplitudes α_x in the orthonormal basis $|x\rangle$, can be represented by a column vector:

$$|\psi\rangle \longrightarrow \begin{pmatrix} \alpha_0 \\ \alpha_1 \\ \vdots \\ \alpha_N \end{pmatrix}. \tag{A.40}$$

The associated bra vector is then the row vector:

$$\langle\psi| \longrightarrow \begin{pmatrix} \alpha_0^* & \alpha_1^* & \dots & \alpha_N^* \end{pmatrix}. \tag{A.41}$$

If

$$|\phi\rangle \longrightarrow \begin{pmatrix} \beta_0 \\ \beta_1 \\ \vdots \\ \beta_N \end{pmatrix}, \tag{A.42}$$

then the inner product $\langle\phi|\psi\rangle$ is given by the ordinary matrix product of the row and column vectors:

$$\langle\phi|\psi\rangle = \begin{pmatrix} \beta_0^* & \beta_1^* & \dots & \beta_N^* \end{pmatrix} \begin{pmatrix} \alpha_0 \\ \alpha_1 \\ \vdots \\ \alpha_N \end{pmatrix}. \tag{A.43}$$

The outer product $|\psi\rangle\langle\phi|$ is also a matrix product:

$$|\psi\rangle\langle\phi| = \begin{pmatrix} \alpha_0 \\ \alpha_1 \\ \vdots \\ \alpha_N \end{pmatrix} \begin{pmatrix} \beta_0^* & \beta_1^* & \dots & \beta_N^* \end{pmatrix}. \tag{A.44}$$

Note that in Dirac notation (A.43) is nothing more than the statement that

$$\langle\phi|\psi\rangle = \langle\phi|\mathbf{1}|\psi\rangle = \sum_x \langle\phi|x\rangle\langle x|\psi\rangle = \sum_x \langle x|\phi\rangle^* \langle x|\psi\rangle, \qquad \text{(A.45)}$$

while (A.44) asserts that

$$\langle x|\Big(|\psi\rangle\langle\phi|\Big)|y\rangle = \langle x|\psi\rangle\langle\phi|y\rangle = \langle x|\psi\rangle\langle y|\phi\rangle^*. \qquad \text{(A.46)}$$

Appendix B

Structure of the general 1-Qbit unitary transformation

I describe here some relations among Pauli matrices, 1-Qbit unitary transformations, and rotations of real-space three-dimensional vectors. The relations are of fundamental importance in many applications of quantum mechanics, and are an essential part of the intellectual equipment of anybody wanting to understand the mathematical structure of three-dimensional rotations. The reason for mentioning them here is that they can also make certain circuit identities quite transparent. The quantum-computation literature contains some unnecessarily cumbersome derivations of many such identities, suggesting that these useful mathematical facts deserve to be more widely known in the field.

The two-dimensional unit matrix **1** and the three Pauli matrices form a basis,

$$\begin{aligned} \mathbf{1} &= \begin{pmatrix} 1 & 0 \\ 0 & 1 \end{pmatrix}, & \boldsymbol{\sigma}_x &= \begin{pmatrix} 0 & 1 \\ 1 & 0 \end{pmatrix}, \\ \boldsymbol{\sigma}_y &= \begin{pmatrix} 0 & -i \\ i & 0 \end{pmatrix}, & \boldsymbol{\sigma}_z &= \begin{pmatrix} 1 & 0 \\ 0 & -1 \end{pmatrix}, \end{aligned} \tag{B.1}$$

for the four-dimensional algebra of two-dimensional matrices: any two-dimensional matrix **u** has a unique expansion of the form

$$\mathbf{u} = u_0\mathbf{1} + \vec{u} \cdot \vec{\boldsymbol{\sigma}} \tag{B.2}$$

for some complex number u_0 and 3-vector $\vec{u}$ with complex components u_x, u_y, and u_z. Here $\vec{\boldsymbol{\sigma}}$ represents the "3-vector" whose components are the Pauli matrices $\boldsymbol{\sigma}_x$, $\boldsymbol{\sigma}_y$, and $\boldsymbol{\sigma}_z$, so in expanded form (B.2) reads

$$\mathbf{u} = u_0\mathbf{1} + u_x\boldsymbol{\sigma}_x + u_y\boldsymbol{\sigma}_y + u_z\boldsymbol{\sigma}_z = \begin{pmatrix} u_0 + u_z & u_x - iu_y \\ u_x + iu_y & u_0 - u_z \end{pmatrix}. \tag{B.3}$$

As what follows demonstrates, however, it is invariably simpler to use the form (B.2) together with the multiplication rule (see Section 1.4)

$$(\vec{a} \cdot \vec{\boldsymbol{\sigma}})(\vec{b} \cdot \vec{\boldsymbol{\sigma}}) = (\vec{a} \cdot \vec{b})\mathbf{1} + i(\vec{a} \times \vec{b}) \cdot \vec{\boldsymbol{\sigma}}, \tag{B.4}$$

rather than dealing explicitly with two-dimensional matrices.

Impose on (B.2) the condition

$$\mathbf{u}\mathbf{u}^\dagger = \mathbf{u}^\dagger\mathbf{u} = \mathbf{1} \tag{B.5}$$

that **u** be unitary. Since any unitary matrix remains unitary if it is multiplied by an overall multiplicative phase factor $e^{i\theta}$ with θ real, we can require u_0 to be real and arrive at a form which is general except for such an overall phase factor. Since the Pauli matrices are Hermitian, we then have

$$\mathbf{u}^\dagger = u_0\mathbf{1} + \vec{u}^* \cdot \vec{\boldsymbol{\sigma}}. \tag{B.6}$$

The rule (B.4) now tells us that for **u** to be unitary we must have

$$0 = \mathbf{1} - \mathbf{u}^\dagger\mathbf{u} = \big(1 - u_0^2 - \vec{u}^* \cdot \vec{u}\big)\mathbf{1} - \big(u_0(\vec{u} + \vec{u}^*) + i\,\vec{u}^* \times \vec{u}\big) \cdot \vec{\boldsymbol{\sigma}}. \tag{B.7}$$

Since $\mathbf{1}$, $\boldsymbol{\sigma}_x$, $\boldsymbol{\sigma}_y$, and $\boldsymbol{\sigma}_z$ are linearly independent in the four-dimensional algebra of 1-Qbit operators, the coefficients of all four of them in (B.7) must vanish and we have

$$1 = u_0^2 + \vec{u}^* \cdot \vec{u}, \qquad 0 = u_0(\vec{u} + \vec{u}^*) + i\,\vec{u}^* \times \vec{u}. \tag{B.8}$$

The second of these requires the real and imaginary parts of the vector $\vec{u}$ to satisfy

$$u_0 \,\mathrm{Re}\,\vec{u} = \mathrm{Re}\,\vec{u} \times \mathrm{Im}\,\vec{u}. \tag{B.9}$$

If $u_0 \neq 0$, it follows from (B.9) that $\mathrm{Re}\,\vec{u} \cdot \mathrm{Re}\,\vec{u} = 0$, so $\mathrm{Re}\,\vec{u} = 0$, and the vector $\vec{u}$ must be i times a real vector $\vec{v}$. On the other hand if $u_0 = 0$ then (B.9) requires the real and imaginary parts of $\vec{u}$ to be parallel vectors, so that $\vec{u}$ itself is just a complex multiple of a real vector. But if $u_0 = 0$ we retain the freedom to pick the overall phase of the operator **u**, which we can choose to make the vector $\vec{u}$ purely imaginary. So irrespective of whether or not $u_0 = 0$, the general form for a two-dimensional unitary **u** is, to within an overall phase factor,

$$\mathbf{u} = u_0\mathbf{1} + i\,\vec{v} \cdot \vec{\boldsymbol{\sigma}}, \tag{B.10}$$

where u_0 is a real number, $\vec{v}$ is a real vector, and, from the first of (B.8),

$$u_0^2 + \vec{v} \cdot \vec{v} = 1. \tag{B.11}$$

The identity (B.11) allows us to parametrize u_0 and $\vec{v}$ in terms of a real unit vector $\vec{n}$ parallel to $\vec{v}$ and a real angle γ so that

$$\mathbf{u} = \cos\gamma\,\mathbf{1} + i\sin\gamma(\vec{n} \cdot \vec{\boldsymbol{\sigma}}). \tag{B.12}$$

An alternative way of writing (B.12) is

$$\mathbf{u} = \exp(i\gamma\,\vec{n} \cdot \vec{\boldsymbol{\sigma}}). \tag{B.13}$$

This follows from the forms of the power-series expansions of the exponential, sine, and cosine, together with the fact that $(\vec{n} \cdot \vec{\boldsymbol{\sigma}})^2 = \mathbf{1}$

for any unit vector $\vec{n}$ as a special case of (B.4). (The argument is the same as the argument that $e^{i\varphi} = \cos\varphi + i\sin\varphi$ for any real number φ.)

A remarkable connection between these two-dimensional unitary matrices and ordinary three-dimensional rotations emerges from the fact that each of the three Pauli matrices in (B.1) has zero trace, and that the operator unitary transformation

$$\mathbf{A} \to \mathbf{u}\mathbf{A}\mathbf{u}^\dagger \tag{B.14}$$

preserves the trace of $\mathbf{A}$.[1]

Note first that if $\vec{a}$ is a real vector then $\mathbf{u}(\vec{a}\cdot\vec{\sigma})\mathbf{u}^\dagger$ is Hermitian and can therefore be expressed as a linear combination of **1** and the three Pauli matrices with real coefficients. Since $\boldsymbol{\sigma}_x$, $\boldsymbol{\sigma}_y$, and $\boldsymbol{\sigma}_z$ all have zero trace, so does $\vec{a}\cdot\vec{\sigma}$ and therefore so does $\mathbf{u}(\vec{a}\cdot\vec{\sigma})\mathbf{u}^\dagger$. Its expansion as a linear combination of **1** and the three Pauli matrices must therefore be of the form $\vec{a}\,'\cdot\vec{\sigma}$ for some real vector $\vec{a}\,'$ (since **1** alone among the four matrices has nonzero trace):

$$\mathbf{u}(\vec{a}\cdot\vec{\sigma})\mathbf{u}^\dagger = \vec{a}\,'\cdot\vec{\sigma}. \tag{B.15}$$

It follows that

$$\mathbf{u}(\vec{a}\cdot\sigma)(\vec{b}\cdot\vec{\sigma})\mathbf{u}^\dagger = \big(\mathbf{u}(\vec{a}\cdot\vec{\sigma})\mathbf{u}^\dagger\big)\big(\mathbf{u}(\vec{b}\cdot\vec{\sigma})\mathbf{u}^\dagger\big) = \big(\vec{a}\,'\cdot\vec{\sigma}\big)\big(\vec{b}\,'\cdot\vec{\sigma}\big). \tag{B.16}$$

Since unitary transformations preserve the trace,

$$\mathrm{Tr}(\vec{a}\cdot\vec{\sigma})(\vec{b}\cdot\vec{\sigma}) = \mathrm{Tr}(\vec{a}\,'\cdot\vec{\sigma})\big(\vec{b}\,'\cdot\vec{\sigma}\big). \tag{B.17}$$

Hence, from (B.4),

$$\vec{a}\,'\cdot\vec{b}\,' = \vec{a}\cdot\vec{b}. \tag{B.18}$$

It follows directly from the form (B.15) of the transformation from unprimed to primed vectors that $(\vec{a}+\vec{b})' = \vec{a}\,'+\vec{b}\,'$ and $(\lambda\vec{a})' = \lambda\vec{a}\,'$ – i.e. the transformation $\vec{a}\to\vec{a}\,'$ is linear. But the most general real, linear, inner-product-preserving transformation on real 3-vectors is a rotation. Consequently the transformation from real 3-vectors $\vec{a}$ to real 3-vectors $\vec{a}\,'$ induced by any two-dimensional unitary $\mathbf{u}$ through (B.15) is a rotation:

$$\vec{a}\,' = \mathbf{R}_{\mathbf{u}}\vec{a}. \tag{B.19}$$

Furthermore, by applying the (unitary) product $\mathbf{uv}$ of two unitary

1 The trace of a matrix is the sum of its diagonal elements. Recall also the (easily verified) fact that the trace of a product of two matrices is independent of the order in which the matrices are multiplied, even when the matrices do not commute.

transformations in two steps,

$$\begin{aligned}(\mathbf{uv})(\vec{a}\cdot\vec{\sigma})(\mathbf{uv})^\dagger &= \mathbf{u}\big(\mathbf{v}(\vec{a}\cdot\vec{\sigma})\mathbf{v}^\dagger\big)\mathbf{u}^\dagger = \mathbf{u}\big([\mathbf{R_v}\vec{a}]\cdot\vec{\sigma}\big)\mathbf{u}^\dagger \\ &= [\mathbf{R_u}\mathbf{R_v}\vec{a}]\cdot\vec{\sigma},\end{aligned} \tag{B.20}$$

we deduce that

$$\mathbf{R_{uv}} = \mathbf{R_u}\mathbf{R_v}. \tag{B.21}$$

Thus the association of three-dimensional rotations with two-dimensional unitary matrices preserves the multiplicative structure of the rotation group: the rotation associated with the product of two unitary transformations is the product of the two associated rotations.

Which rotation is associated with which unitary transformation? To answer this, note first that when the vector $\vec{a}$ in (B.15) is taken to be the vector $\vec{n}$ appearing in $\mathbf{u}$ (in (B.12) or (B.13)) then $\vec{n}' = \vec{n}$, since $\mathbf{u}$ then commutes with $\vec{n}\cdot\vec{\sigma}$. Therefore $\vec{n}$ is along the axis of the rotation associated with $\mathbf{u} = \exp(i\gamma\vec{n}\cdot\vec{\sigma})$. To determine the angle θ of that rotation, let $\vec{m}$ be any unit vector perpendicular to the axis $\vec{n}$, so that

$$\cos\theta = \vec{m}\cdot\vec{m}'. \tag{B.22}$$

We then have

$$\begin{aligned}\cos\theta &= \tfrac{1}{2}\,\mathrm{Tr}\big((\vec{m}\cdot\vec{\sigma})(\vec{m}'\cdot\vec{\sigma})\big) \\ &= \tfrac{1}{2}\,\mathrm{Tr}\big((\vec{m}\cdot\vec{\sigma})(\cos\gamma\,\mathbf{1} + i\sin\gamma\;\vec{n}\cdot\vec{\sigma})(\vec{m}\cdot\sigma) \\ &\qquad\times(\cos\gamma\,\mathbf{1} - i\sin\gamma\;\vec{n}\cdot\vec{\sigma})\big) \\ &= \tfrac{1}{2}\,\mathrm{Tr}\big((\cos\gamma\;\vec{m} - \sin\gamma\;\vec{m}\times\vec{n})\cdot\vec{\sigma}) \\ &\qquad\times(\cos\gamma\;\vec{m} + \sin\gamma\;\vec{m}\times\vec{n})\cdot\vec{\sigma})\big) \\ &= \cos^2\gamma - \sin^2\gamma = \cos(2\gamma),\end{aligned} \tag{B.23}$$

where we have made repeated use of (B.4) and the fact that $\vec{m}\cdot\vec{n} = 0$.

So the unitary matrix (B.13) is associated with a rotation about the axis $\vec{n}$ through the angle 2γ. Since the identity rotation is associated both with $\mathbf{u} = \mathbf{1}$ and with $\mathbf{u} = -\mathbf{1}$, the correspondence between these unitary matrices and three-dimensional proper rotations is 2-to-1. It is useful to introduce the notation $\mathbf{u}(\vec{n}, \theta)$ for the 1-Qbit unitary transformation associated with the rotation $\mathbf{R}(\vec{n}, \theta)$ about the axis $\vec{n}$ through the angle θ:

$$\mathbf{u}(\vec{n}, \theta) = \exp\big(i\tfrac{1}{2}\theta\vec{n}\cdot\vec{\sigma}\big) = \cos\big(\tfrac{1}{2}\theta\big) + i(\vec{n}\cdot\vec{\sigma})\sin\big(\tfrac{1}{2}\theta\big). \tag{B.24}$$

The three-dimensional rotations arrived at in this way are all *proper* (i.e. they preserve rather than invert handedness) because they can all be continuously connected to the identity. *Any* proper rotation can be associated with a $\mathbf{u}$, and in just two different ways ($\mathbf{u}$ and $-\mathbf{u}$ clearly being associated with the same rotation). The choice of phase leading to the general form (B.10) with real u_0 can be imposed by requiring that the determinant of **u** must be 1, so in mathematical language we have a 2-to-1 homomorphism from the group SU(2) of unimodular unitary two-dimensional matrices to the group SO(3) of proper three-dimensional rotations.

Although this may all seem tediously abstract, it is surprisingly useful at a very practical level. It can reduce some highly nontrivial three-dimensional geometry to elementary algebra, just as Euler's relation $e^{i\phi} = \cos\phi + i\sin\phi$ reduces some slightly nontrivial two-dimensional trigonometry to simple algebra. Suppose, for example, that you combine a rotation through an angle α about an axis given by the unit vector $\vec{a}$ with a rotation through β about $\vec{b}$. The result, of course, is a single rotation. What are its angle γ and axis $\vec{c}$? Answering this question can be a nasty exercise in three-dimensional geometry. But to answer it using the Pauli matrices you need only note that $\mathbf{u}(\vec{c},\gamma) = \mathbf{u}(\vec{a},\alpha)\mathbf{u}(\vec{b},\beta)$, i.e.

$$\cos\left(\tfrac{1}{2}\gamma\right)\mathbf{1}+i\sin\left(\tfrac{1}{2}\gamma\right)(\vec{c}\cdot\vec{\sigma}) = \left(\cos\left(\tfrac{1}{2}\alpha\right)\mathbf{1}+i\sin\left(\tfrac{1}{2}\alpha\right)(\vec{a}\cdot\vec{\sigma})\right) \times\left(\cos\left(\tfrac{1}{2}\beta\right)\mathbf{1}+i\sin\left(\tfrac{1}{2}\beta\right)(\vec{b}\cdot\vec{\sigma})\right). \tag{B.25}$$

Now multiply out the right side of (B.25), using (B.4). To get the angle γ take the trace of both sides (or identify the coefficients of $\mathbf{1}$) to find

$$\cos\left(\tfrac{1}{2}\gamma\right) = \cos\left(\tfrac{1}{2}\alpha\right)\cos\left(\tfrac{1}{2}\beta\right) - (\vec{a}\cdot\vec{b})\sin\left(\tfrac{1}{2}\alpha\right)\sin\left(\tfrac{1}{2}\beta\right). \tag{B.26}$$

To get the axis $\vec{c}$, identify the vectors of coefficients of the Pauli matrices:

$$\sin\left(\tfrac{1}{2}\gamma\right)\vec{c} = \sin\left(\tfrac{1}{2}\beta\right)\cos\left(\tfrac{1}{2}\alpha\right)\vec{b} + \sin\left(\tfrac{1}{2}\alpha\right)\cos\left(\tfrac{1}{2}\beta\right)\vec{a} - \sin\left(\tfrac{1}{2}\alpha\right)\sin\left(\tfrac{1}{2}\beta\right)(\vec{a}\times\vec{b}). \tag{B.27}$$

Note that (B.26) and (B.27) are trivially correct when $\vec{a}$ and $\vec{b}$ are parallel. A little geometrical thought reveals that they are also correct when α and β are both 180°. To try to see geometrically why they are correct more generally is to acquire a deep appreciation for the remarkable power of the representation of three-dimensional rotations in terms of two-dimensional unitary transformations. Other examples of the power of the representation are illustrated in the derivations of circuit identities in Section 2.6, in the characterization of the general 1-Qbit state in Appendix C, and in the construction of the Hardy state in Appendix D.

Appendix C

Structure of the general 1-Qbit state

The 1-Qbit computational-basis states $|0\rangle$ and $|1\rangle$ can be characterized to within an overall phase by the fact that they are eigenstates of the number operator $\mathbf{n}$ with eigenvalues 0 and 1 or, equivalently, that they are eigenstates of $1 - 2\mathbf{n} = \mathbf{Z} = \vec{z} \cdot \vec{\sigma}$ with eigenvalues 1 and -1.

Let $|\phi\rangle$ be any 1-Qbit state, and let $|\psi\rangle$ be the orthogonal state (unique to within an overall phase), satisfying $\langle\psi|\phi\rangle = 0$. Since $|0\rangle$ and $|1\rangle$ are linearly independent there is a unique linear transformation taking them into $|\phi\rangle$ and $|\psi\rangle$. But since $|\phi\rangle$ and $|\psi\rangle$ are an orthonormal pair (as are $|0\rangle$ and $|1\rangle$) this linear transformation preserves the inner product of arbitrary pairs of states, so it is a unitary transformation $\mathbf{u}$.

Since

$$|\phi\rangle = \mathbf{u}|0\rangle, \qquad |\psi\rangle = \mathbf{u}|1\rangle, \tag{C.1}$$

the operator $\mathbf{n}' = \mathbf{unu}^\dagger$ acts as a Qbit number operator on $|\phi\rangle$ and $|\psi\rangle$:

$$\mathbf{n}'|\phi\rangle = 0, \qquad \mathbf{n}'|\psi\rangle = |\psi\rangle. \tag{C.2}$$

Since, as shown in Appendix B, any 1-Qbit unitary transformation $\mathbf{u}$ is associated with a rotation $\mathbf{R}(\vec{m}, \theta)$, we have

$$\mathbf{n}' = \mathbf{unu}^\dagger = \tfrac{1}{2}\big(1 - \mathbf{u}(\vec{z} \cdot \vec{\sigma})\mathbf{u}^\dagger\big) = \tfrac{1}{2}(1 - \vec{z}\,' \cdot \vec{\sigma}), \tag{C.3}$$

where $\vec{z}\,' = R(\vec{m}, \theta)\vec{z}$.

Thus $\mathbf{n}'$, which functions as a number operator for the states $|\phi\rangle = \mathbf{u}(\vec{m}, \theta)|0\rangle$ and $|\psi\rangle = \mathbf{u}(\vec{m}, \theta)|1\rangle$, is constructed out of the component of the vector of operators $\boldsymbol{\sigma}$ along the direction $\vec{z}\,' = \mathbf{R}(\vec{m}, \theta)\vec{z}$ in exactly the same way that $\mathbf{n}$, the number operator for the computational basis states $|0\rangle$ and $|1\rangle$, is constructed out of the component along $\vec{z}$. This suggests that there might be nothing special about the choice of $|0\rangle$ and $|1\rangle$ to form the computational-basis states for each Qbit – that any pair of orthogonal states, $|0'\rangle = \mathbf{u}|0\rangle$ and $|1'\rangle = \mathbf{u}|1\rangle$, could serve equally well. Furthermore, it is at least a consistent possibility that to make an apparatus to measure the Qbits in this new basis we need do nothing more than apply the rotation $\mathbf{R}$ associated with $\mathbf{u}$ to the apparatus that served to measure them in the original basis.

This physical possibility is realized by some, but by no means all, of the physical systems that have been proposed as possible embodiments of Qbits. It is realized for certain atomic magnets – also called *spins* – which have the property that when the magnetization of such a spin is measured along any given direction, after the measurement the magnet is either maximally aligned along that direction or maximally aligned opposite to that direction. These two possible outcomes for a particular direction – conventionally taken to be $\vec{z}$ – are associated with the values 0 and 1 for the Qbit. After such a measurement the spin is left in the state $|0\rangle$ or $|1\rangle$. Any other state $|\phi\rangle$ and its orthogonal partner $|\psi\rangle$ specify an alternative direction, along which the magnetization might have been measured, associated with an alternative scheme for reading out values for the Qbits.

For this example the continuum of possible states available to a Qbit, compared with the pair of states available to a Cbit, reflects the continuum of ways in which one can read a Qbit (measuring its magnetization along any direction) as opposed to the single option available for reading a Cbit (finding out what value it actually has). For Qbits that are not spins, the richness lies in the possibility of applying an arbitrary unitary transformation to each Qbit, before measuring it in the computational basis. What makes spins special is that applying the unitary transformation to the Qbits (which is not always that easy to arrange) can be replaced by straightforwardly applying the corresponding rotation to every 1-Qbit measurement gate.

Appendix D

Spooky action at a distance

As a further exercise in applying the quantum-computational formalism to Qbits, and as a subject of interest in itself, though not directly related to quantum computation, I describe here a thought-provoking state of affairs illustrated with an example discovered by Lucien Hardy. (Similar thoughts are provoked by an example discovered by Daniel Greenberger, Michael Horne, and Anton Zeilinger, described in Section 6.6.)

Suppose that Alice and Bob each has one member of a pair of Qbits, which have been prepared in the 2-Qbit state

$$|\Phi\rangle = \frac{1}{\sqrt{12}}\big(3|00\rangle + |01\rangle + |10\rangle - |11\rangle\big). \tag{D.1}$$

A specification of how to prepare two Qbits in such a Hardy state, somewhat more transparent than the general procedure described in Section 1.11, is given after the extraordinary properties of the Hardy state are described. One easily verifies that the state $|\Phi\rangle$ can also be written as

$$|\Phi\rangle = \frac{1}{\sqrt{3}}\big(2|00\rangle - \mathbf{H}_a\mathbf{H}_b|11\rangle\big), \tag{D.2}$$

where we take $\mathbf{H}_a$ to act on the left (Alice's) Qbit and $\mathbf{H}_b$ to act on the right (Bob's) Qbit. Note the following four elementary properties of a pair of Qbits in the state $|\Phi\rangle$.

(i) If Alice and Bob each measures their own Qbit, then (D.1) shows that there is a nonzero probability ($\frac{1}{12}$) that both get the result 1.

(ii) If Alice and Bob each applies a Hadamard to their own Qbit then, since $\mathbf{H}^2 = 1$, the state (D.2) of the Qbits becomes

$$\mathbf{H}_a\mathbf{H}_b|\Phi\rangle = \frac{1}{\sqrt{3}}\big(2\mathbf{H}_a\mathbf{H}_b|00\rangle - |11\rangle\big) = \frac{1}{\sqrt{3}}\big(|00\rangle + |01\rangle + |10\rangle\big), \tag{D.3}$$

so if they measure their Qbits after each has applied a Hadamard, then the probability that both get the value 1 is zero.

(iii) If only Alice applies a Hadamard to her Qbit, then the state (D.2) of the two Qbits becomes

$$\mathbf{H}_a|\Phi\rangle = \frac{1}{\sqrt{3}}\big(2\mathbf{H}_a|00\rangle - \mathbf{H}_b|11\rangle\big). \tag{D.4}$$

Since $\mathbf{H}_a|00\rangle$ is a linear combination of $|00\rangle$ and $|10\rangle$, and since $\mathbf{H}_b|11\rangle$ is a linear combination of $|10\rangle$ and $|11\rangle$, the state $|01\rangle$ does not appear

Table D.1. Four ways to measure two Qbits in the Hardy state (D.1)

Gates		Result		Possible?
Alice	Bob	Alice	Bob	
1	**1**	1	1	Yes
1	**H**	1	0	No
H	**1**	0	1	No
H	**H**	1	1	No

in the expansion of $\mathbf{H}_a|\Phi\rangle$ in computational-basis states. So when the Qbits are subsequently measured the probability is zero that Alice will get the value 0 and Bob the value 1.

(iv) If only Bob applies a Hadamard to his Qbit, then by the same reasoning (except for the interchange of Alice and Bob) when the Qbits are subsequently measured the probability is zero that Alice will get the value 1 and Bob the value 0.

Taken together, these four cases seem to have some very strange implications. The cases are summarized in the four rows of Table D.1 above. On the left is indicated whether (**H**) or not (**1**) Alice or Bob sends their Qbit through a Hadamard gate before sending it through a measurement gate. In the center is listed the measurement outcome of interest for each case. The column on the right specifies whether that outcome can or cannot occur for that particular case.

To see what is strange, suppose that Alice and Bob each independently decides, by tossing coins, whether or not to apply a Hadamard to their Qbit before sending it through a measurement gate. There is a nonzero probability ($\frac{1}{4} \times \frac{1}{12} = \frac{1}{48}$) that neither applies a Hadamard and both measurement gates show 1 (see the first row of Table D.1). In the one time in 48 that this happens, it is tempting to conclude that each Qbit was, even before the coins were tossed, *capable* of producing a 1 when directly subjected to a measurement gate because, after all, each Qbit *did* produce a 1 when directly subjected to a measurement gate.

But if Alice's Qbit did indeed have such a capability, then, in the absence of spooky interactions between Bob's Hadamard and Alice's Qbit, her Qbit surely would have retained that capability, even if Bob's coin had come up the other way and he had applied a Hadamard to his own Qbit before measuring it. But if Alice's Qbit was indeed capable of registering a 1 when measured directly, then Bob's Qbit must have been incapable of registering a 0 if measured after a Hadamard, since (see the second row of Table D.1) when Bob applies a Hadamard before his measurement and Alice does not, it is impossible for Bob's measurement to give 0 while Alice's gives 1.

By the same reasoning (interchanging Alice and Bob and referring to the third row of Table D.1) we conclude that Alice's Qbit must also have been incapable of registering a 0 when measured after a Hadamard.

So in each of the slightly more than 2% of the cases in which neither Alice nor Bob applies Hadamards and both their measurement gates register 1, we conclude that if the tosses of both coins had come out the other way and both had applied Hadamards before measuring, then neither Qbit could have registered 0 when measured: both would have had to register 1. But according to the fourth row of Table D.1 this can never happen.

Although this particular argument was discovered by Lucien Hardy only in the early 1990s, similar situations (where the paradox is not as directly evident) have been known since a famous paper by John Bell appeared in 1964. Over the years passions have run high on the significance of this. Some claim that it shows that the value Alice or Bob finds upon measuring her or his Qbit *does* depend on whether or not the other, who, with his or her Qbit, could be far away, does or does not apply a Hadamard to his or her own Qbit before measuring it. They call this "quantum nonlocality" or "spooky action at a distance"– a translation of Einstein's disparaging *spukhafte Fernwirkungen.*

My own take on it is rather different. With any given pair of Qbits, Alice and Bob each either does or does not apply a Hadamard prior to their measurement. Only one of the four possible cases is actually realized. The other three cases *do not happen.* In a deterministic world it can make sense to talk about what *would* have happened if things had been other than the way they actually were, since the hypothetical situation can entail unique subsequent behavior. But in the intrinsically nondeterminstic case of measuring Qbits, one cannot infer, from what Alice's Qbit actually did, that it has a "capability" to do what it actually did, which it retains even in a hypothetical situation that did not, in fact, take place. To characterize the possible behavior of Alice's Qbit in a fictitious world requires more than just the irrelevance of Bob's decision whether or not to apply a Hadamard. It also requires that whatever it is that actually *is* relevant to Alice's outcome remains the same in both worlds and plays the same role in bringing about that outcome. But the reading of a measurement gate has an irreducible randomness to it: *nothing* need play a role in bringing it about.[1]

The real lesson here is that if one has a single pair of Qbits and various choices of gates to apply to them before sending them through a measurement gate, then it makes no sense to infer, from the actual

1 Conscience requires me to report here the existence of a small deviant subculture of physicists, known as Bohmians, who maintain that there is a deterministic substructure, unfortunately inaccessible to us, that underlies quantum phenomena. Needless to say, all Bohmians believe in real instantaneous action at a distance.

outcome of the measurement for the actual choice of gates, additional constraints, going beyond those implied by the initial state of the Qbits, on the hypothetical outcomes of measurements in the fictional case in which one made a different choice of gates. It is nonsense to insist that Alice's Qbit has to retain the "capability" to do what it actually did, if we imagine turning back the clock and doing it over again. Assigning a "capability" to Alice's Qbit prior to the measurement is rather like assigning it a state. But the pre-measurement state (D.1) is an entangled state, so Alice's Qbit has no state of its own.

One can, however, let Alice and Bob repeatedly play this game with many different pairs of Qbits, always preparing the Qbits in the same initial 2-Qbit state (D.1). It is then entirely sensible to ask whether the *statistics* of the values Bob finds upon measuring his Qbit depend on whether Alice applied a Hadamard transform to her Qbit. For Alice and Bob can accumulate a mass of data, and directly compare the statistics Bob got when Alice applied the Hadamard with those he got when she did not. If Bob got a different statistical distribution of readings depending on whether Alice did or did not apply a Hadamard to her faraway Qbit before she measured it, this would permit *non*spooky action at a distance which could actually be used to send messages. So it is important to note that Bob's *statistics* do not, in fact, depend on whether or not Alice applies a Hadamard.

We can show this under quite general conditions. Suppose that n Qbits are divided into two subsets, each of which may be independently manipulated (i.e. subjected to unitary transformations) prior to a measurement. Let the n_a Qbits on the left constitute one such group and the $n_b = n - n_a$ on the right, the other. Think of the first group as under the control of Alice and the second as belonging to Bob. If the n Qbits are always prepared in the state $|\Psi\rangle$, then if Alice and Bob separately measure their Qbits, the Born rule tells us that the joint probability $p(x, y)$ of Alice getting x and Bob y is

$$p_\Psi(x, y) = \langle\Psi|\mathbf{P}^a_x\mathbf{P}^b_y|\Psi\rangle, \tag{D.5}$$

where the projection operator $\mathbf{P}^a$ acts only on Alice's Qbits (i.e. it acts as the identity on Bob's) and $\mathbf{P}^b$ only on Bob's.

Suppose, now, that Alice acts on her Qbits with the unitary transformation $\mathbf{U}_a$ before making her measurement and Bob acts on his with $\mathbf{U}_b$. Then the state $|\Psi\rangle$ is changed into

$$|\Phi\rangle = \mathbf{U}_a\mathbf{U}_b|\Psi\rangle. \tag{D.6}$$

Now the probability of their measurements giving x and y, conditioned on their choices of unitary transformation, is

$$\begin{aligned} p_\Psi(x, y|\mathbf{U}_a, \mathbf{U}_b) &= \big\langle\Phi|\mathbf{P}^a_x\mathbf{P}^b_y|\Phi\big\rangle = \big\langle\Psi|\mathbf{U}_b^\dagger\mathbf{U}_a^\dagger\big(\mathbf{P}^a_x\mathbf{P}^b_y\big)\mathbf{U}_a\mathbf{U}_b|\Psi\big\rangle \\ &= \big\langle\Psi|\big(\mathbf{U}_a^\dagger\mathbf{P}^a_x\mathbf{U}_a\big)\big(\mathbf{U}_b^\dagger\mathbf{P}^b_y\mathbf{U}_b\big)|\Psi\big\rangle \end{aligned} \tag{D.7}$$

(where we have used the fact that all operators that act only on Alice's Qbits commute with all operators that act only on Bob's).

It follows from the fact that

$$\sum_x \mathbf{U}_a^\dagger \mathbf{P}_x^a \mathbf{U}_a = \mathbf{U}_a^\dagger \Big(\sum_x \mathbf{P}_x^a\Big)\mathbf{U}_a = \mathbf{U}_a^\dagger \mathbf{1} \mathbf{U}_a = \mathbf{1} \tag{D.8}$$

that Bob's marginal statistics do not depend on what Alice chose to do to her own Qbits:

$$p_\Psi(y|\mathbf{U}_a\mathbf{U}_b) = \sum_x p_\Psi(x, y|\mathbf{U}_a, \mathbf{U}_b) = \langle\Psi|\big(\mathbf{U}_b^\dagger \mathbf{P}_y^b \mathbf{U}_b\big)|\Psi\rangle = p_\Psi(y|\mathbf{U}_b), \tag{D.9}$$

which does not depend on the particular unitary transformation $\mathbf{U}_a$ chosen by Alice. Therefore the statistics of the measurement outcomes for any group of Qbits are not altered by anything done to other Qbits (provided, of course, that the other Qbits do not subsequently interact with those in the original group, for example by the application of appropriate 2-Qbit gates).

Like any 2-Qbit state, the state (D.1) leading to this remarkable set of data can be constructed with a single cNOT gate and three 1-Qbit unitary gates. Here is a construction that is somewhat more direct than the general construction given in Section 1.11. It exploits the connection between 1-Qbit unitary transformations and three-dimensional rotations developed in Appendix B.

It follows from (D.3) that

$$\begin{aligned}|\Phi\rangle &= \mathbf{H}_a\mathbf{H}_b \tfrac{1}{\sqrt{3}}\big(|00\rangle + |01\rangle + |10\rangle\big) \\ &= \mathbf{H}_a\mathbf{H}_b\Big(\sqrt{\tfrac{2}{3}}\mathbf{H}_b|00\rangle + \sqrt{\tfrac{1}{3}}|10\rangle\Big) \\ &= \mathbf{H}_a\Big(\sqrt{\tfrac{2}{3}}|00\rangle + \sqrt{\tfrac{1}{3}}\mathbf{H}_b|10\rangle\Big) \\ &= \mathbf{H}_a\mathbf{C}_{ab}^H\Big[\Big(\sqrt{\tfrac{2}{3}}|0\rangle + \sqrt{\tfrac{1}{3}}|1\rangle\Big)|0\rangle\Big] = \mathbf{H}_a\mathbf{C}_{ab}^H\mathbf{w}_a|00\rangle,\end{aligned} \tag{D.10}$$

where $\mathbf{w}$ is any 1-Qbit unitary transformation that takes $|0\rangle$ into $\sqrt{\frac{2}{3}}|0\rangle + \sqrt{\frac{1}{3}}|1\rangle$, and $\mathbf{C}^H$ is a 2-Qbit controlled-Hadamard gate:

$$\mathbf{C}_{10}^H|xy\rangle = \mathbf{H}_0^x|xy\rangle. \tag{D.11}$$

To construct a controlled-Hadamard $\mathbf{C}^H$ from a controlled-NOT $\mathbf{C}$, note that the NOT operation $\mathbf{X}$ is $\mathrm{x}\cdot\sigma$ while the Hadamard transformation is $\mathbf{H} = (1/\sqrt{2})(\mathbf{X}+\mathbf{Z}) = (1/\sqrt{2})(\mathrm{x}+\mathrm{z})\cdot\sigma$. It follows from the discussion of 1-Qbit unitaries in Appendix B that

$$\mathbf{H} = \mathbf{u}\mathbf{X}\mathbf{u}^\dagger, \tag{D.12}$$

where $\mathbf{u}$ is the 1-Qbit unitary associated with any rotation that takes x

into $(1/\sqrt{2})(\mathbf{x}+\mathbf{z})$. Since we also have $\mathbf{1} = \mathbf{u}\mathbf{u}^\dagger$, it follows that

$$\mathsf{C}^H_{10} = \mathsf{u}_0 \mathsf{C} \mathsf{u}_0^\dagger. \tag{D.13}$$

So (D.10) reduces to the compact form

$$|\Phi\rangle = \mathsf{H}_a \mathsf{u}_b \mathsf{C}_{ab} \mathsf{w}_a \mathsf{u}_b^\dagger |00\rangle.$$

If you want an explicit form for w, its matrix in the computational basis could be

$$\begin{pmatrix} \sqrt{\frac{2}{3}} & -\sqrt{\frac{1}{3}} \\ \sqrt{\frac{1}{3}} & \sqrt{\frac{2}{3}} \end{pmatrix}. \tag{D.14}$$

To get an explicit form for u, note that a rotation through $\pi/4$ about the y-axis takes $\mathbf{x}$ into $(1/\sqrt{2})(\mathbf{x}+\mathbf{z})$. The associated unitary transformation is

$$\mathsf{u} = \exp\big(i(\pi/8)\boldsymbol{\sigma}_y\big) = \cos(\pi/8)\mathbf{1} + i\sin(\pi/8)\boldsymbol{\sigma}_y. \tag{D.15}$$

Since the matrix for $\boldsymbol{\sigma}_y$ in the computational basis is

$$\begin{pmatrix} 0 & -i \\ i & 0 \end{pmatrix},$$

the matrix for u is

$$\begin{pmatrix} \cos(\pi/8) & \sin(\pi/8) \\ -\sin(\pi/8) & \cos(\pi/8) \end{pmatrix}. \tag{D.16}$$

Since the matrices for X and H are

$$\begin{pmatrix} 0 & 1 \\ 1 & 0 \end{pmatrix} \quad \text{and} \quad \frac{1}{\sqrt{2}}\begin{pmatrix} 1 & 1 \\ 1 & -1 \end{pmatrix}$$

you can easily confirm that these three matrices do indeed satisfy (D.12). Verifying this should give you an appreciation for the power of the method described in Appendix B.

Appendix E

Consistency of the generalized Born rule

A general state of $m + n$ Qbits can be written as

$$|\Psi\rangle_{m+n} = \sum_{x,y} \alpha_{xy} |x\rangle_m |y\rangle_n. \tag{E.1}$$

The most general form of the Born rule asserts that if just the m Qbits associated with the states $|x\rangle_m$ in (E.1) are measured, then with probability

$$p(x) = \sum_y |\alpha_{xy}|^2 \tag{E.2}$$

the result will be x, and after the measurement the state of all $m + n$ Qbits will be the product state

$$|x\rangle_m |\Phi_x\rangle_n, \tag{E.3}$$

where the (correctly normalized) state of the n unmeasured Qbits is given by

$$|\Phi_x\rangle_n = \frac{1}{\sqrt{p(x)}} \sum_y \alpha_{xy} |y\rangle_n. \tag{E.4}$$

This strongest form of the Born rule satisfies the reasonable consistency requirement that measuring r Qbits and then immediately measuring s more, before any other gates have had a chance to act, is equivalent to measuring all the $r + s$ Qbits together. An important consequence is that an n-Qbit measurement gate can be constructed by applying n 1-Qbit measurement gates to the n individual Qbits, as illustrated in Figure 1.8.

To establish this consistency condition, write the state of $r + s + u$ Qbits as

$$|\Psi\rangle_n = \sum_{x,y,z} \alpha_{xyz} |x\rangle_r |y\rangle_s |z\rangle_u. \tag{E.5}$$

If the $r + s$ Qbits are all measured together then a direct application of the rule tells us that the result will be xy with probability

$$p(xy) = \sum_z |\alpha_{xyz}|^2, \tag{E.6}$$

and that the post-measurement state of the Qbits will be

$$|x\rangle_r|y\rangle_s|\Phi_{xy}\rangle_u = |x\rangle_r|y\rangle_s \frac{1}{\sqrt{p(xy)}} \sum_z \alpha_{xyz}|z\rangle_u. \tag{E.7}$$

On the other hand if just the first r Qbits are measured then the rule tells us that the result will be x with probability

$$p(x) = \sum_{y,z} |\alpha_{xyz}|^2, \tag{E.8}$$

and that the post-measurement state will be

$$|x\rangle_r|\Phi_x\rangle_{s+u} = |x\rangle_r \frac{1}{\sqrt{p(x)}} \sum_{y,z} \alpha_{xyz}|y\rangle_s|z\rangle_u. \tag{E.9}$$

Given that the result of the first measurement is x, so that the post-measurement state is (E.9), a further application of the rule tells us that if the next s Qbits are measured, the result will be y with probability

$$p(y|x) = \sum_z \left|\alpha_{xyz}/\sqrt{p(x)}\right|^2, \tag{E.10}$$

and that the post-measurement state after the second measurement will be

$$|x\rangle_r|y\rangle_s|\Phi_{xy}\rangle_u, \tag{E.11}$$

where

$$|\Phi_{xy}\rangle_u = \frac{1}{\sqrt{p(y|x)}} \frac{1}{\sqrt{p(x)}} \sum_z \alpha_{xyz}|z\rangle_u. \tag{E.12}$$

Since the joint probability of getting x and then getting y is related to the conditional probability $p(y|x)$ by

$$p(xy) = p(x)p(y|x), \tag{E.13}$$

this final state and probability are exactly the same as the probability (E.6) and final state (E.7) associated with a direct measurement of all $r+s$ Qbits.

Appendix F

Other aspects of Deutsch's problem

Suppose that one attempted to solve Deutsch's problem, not by the trick that does the job in Chapter 2, but by doing the standard thing: starting with input and output registers in the state $|0\rangle|0\rangle$, applying a Hadamard to the input register, and then using the one application of $\mathbf{U}_f$ to associate with the two Qbits the state

$$|\psi\rangle = \frac{1}{\sqrt{2}}|0\rangle|f(0)\rangle + \frac{1}{\sqrt{2}}|1\rangle|f(1)\rangle. \tag{F.1}$$

A direct measurement of both Qbits reveals the value of f at either 0 or 1 (randomly), but gives no information whatever about the question under investigation, whether or not $f(0) = f(1)$.

Is there anything further one can do to two Qbits in the state (F.1) to learn whether or not $f(0) = f(1)$ (without any further application of $\mathbf{U}_f$)? The answer is yes, there is. But it works only half the time. Here is one such procedure.

For each of the four possibilities for the unknown function f, the corresponding forms for the state (F.1) are

$$f(0) = 0, \quad f(1) = 0: \qquad |\psi\rangle_{00} = \frac{1}{\sqrt{2}}\big(|0\rangle + |1\rangle\big)|0\rangle, \tag{F.2}$$

$$f(0) = 1, \quad f(1) = 1: \qquad |\psi\rangle_{11} = \frac{1}{\sqrt{2}}\big(|0\rangle + |1\rangle\big)|1\rangle, \tag{F.3}$$

$$f(0) = 0, \quad f(1) = 1: \qquad |\psi\rangle_{01} = \frac{1}{\sqrt{2}}\big(|0\rangle|0\rangle + |1\rangle|1\rangle\big), \tag{F.4}$$

$$f(0) = 1, \quad f(1) = 0: \qquad |\psi\rangle_{10} = \frac{1}{\sqrt{2}}\big(|0\rangle|1\rangle + |1\rangle|0\rangle\big). \tag{F.5}$$

We know that $|\psi\rangle$ has one of these four forms, and wish to distinguish between two cases:

Case 1: $|\psi\rangle = |\psi\rangle_{00}$ or $|\psi\rangle_{11}$; **Case 2:** $|\psi\rangle = |\psi\rangle_{01}$ or $|\psi\rangle_{10}$.

By applying Hadamards to both Qbits we change the four possible states to

$$\big(\mathbf{H}\otimes\mathbf{H}\big)|\psi\rangle_{00} = \frac{1}{\sqrt{2}}\big(|0\rangle|0\rangle + |0\rangle|1\rangle\big), \tag{F.6}$$

$$\big(\mathbf{H}\otimes\mathbf{H}\big)|\psi\rangle_{11} = \frac{1}{\sqrt{2}}\big(|0\rangle|0\rangle - |0\rangle|1\rangle\big), \tag{F.7}$$

$$\big(\mathbf{H}\otimes\mathbf{H}\big)|\psi\rangle_{01} = \frac{1}{\sqrt{2}}\big(|0\rangle|0\rangle + |1\rangle|1\rangle\big), \tag{F.8}$$

$$\big(\mathbf{H}\otimes\mathbf{H}\big)|\psi\rangle_{10} = \frac{1}{\sqrt{2}}\big(|0\rangle|0\rangle - |1\rangle|1\rangle\big). \tag{F.9}$$

Now measure both Qbits. If we have one of the Case-1 states, (F.6) or (F.7), we get 00 half the time and 01 half the time; and if we have one of the Case-2 states, (F.8) or (F.9), we get 00 half the time and 11 half the time. So regardless of what the state is, half the time we get 00 and learn nothing whatever, and half the time we get 01 or 11 and learn which case we are dealing with.

This way of dealing with Deutsch's problem – with a 50% chance of success – was noticed before the discovery of the 100%-effective method described in Chapter 2. One might wonder whether some more clever choice of operations on the state (F.1) could enable one always to make the discrimination. It is easy to show that this is impossible.

We wish to apply some general 2-Qbit unitary transformation $\mathbf{U}$ to $|\psi\rangle$ with the result that every possible outcome of a subsequent measurement must rule out one or the other of the two cases. For this to be so it must be that those computational-basis states that appear in the expansions of the states $\mathbf{U}|\psi\rangle_{00}$ and $\mathbf{U}|\psi\rangle_{11}$ cannot appear in the computational-basis expansions of the states $\mathbf{U}|\psi\rangle_{01}$ and $\mathbf{U}|\psi\rangle_{10}$, and vice versa, for otherwise there would be a nonzero probability of a measurement outcome that did not enable us to discriminate between the two cases. Consequently $\mathbf{U}|\psi\rangle_{00}$ and $\mathbf{U}|\psi\rangle_{11}$ must each be orthogonal to each of $\mathbf{U}|\psi\rangle_{01}$ and $\mathbf{U}|\psi\rangle_{10}$. But this is impossible, because unitary transformations preserve inner products, while (F.2)–(F.5) show that the inner product of any Case-1 state $|\psi\rangle_{ij}$ with any Case-2 state is $\frac{1}{2}$.

One can, in fact, show under very general circumstances that, starting with two Qbits in the state (F.1), one cannot do better than applying Hadamards to both before measuring: there *must* be at least a 50% chance that the measurement outcomes will not enable one to discriminate between Case 1 and Case 2. The proof that 50% is the best one can do provides an instructive illustration of many features of the quantum-mechanical formalism.

Suppose that we bring in n additional (*ancillary*) Qbits to help us out. These might be used to process the input and output registers further through some elaborate quantum subroutine, producing an arbitrary unitary transformation $\mathbf{W}$ that acts on all $n+2$ Qbits before a final measurement of the $n+2$ Qbits is made. (This, of course, reduces to the simpler case of no ancillary Qbits, if $\mathbf{W}$ acts as the identity except on the original two Qbits, hereafter called the pair.)

Let the ancillary Qbits start off in some state $|\chi\rangle_n$, which we can take to be $|0\rangle_n$. (Any other n-Qbit state is related to $|0\rangle_n$ by a unitary transformation in the ancillary subspace, which can be absorbed into $\mathbf{W}$.) Let the pair be in one of the four states $|\psi\rangle$ given in (F.2)–(F.5). After $\mathbf{W}$ acts the probability of a measurement giving x $(0 \leq x \leq 3)$ for the pair and y $(0 \leq y < 2^n)$ for the ancillary Qbits is

$$p_{|\psi\rangle}(x, y) = |\langle x, y|\mathbf{W}|\psi, 0\rangle|^2, \tag{F.10}$$

where it is convenient to write a $(2+n)$-Qbit state of the form $|\psi\rangle_2 \otimes |\chi\rangle_n$ as $|\psi, \chi\rangle$.

Note next that for arbitrary pair states $|\phi\rangle$

$$p_{|\phi\rangle}(x, y) = 0 \text{ if and only if } \langle x, y|\mathbf{W}|\phi, 0\rangle = 0, \tag{F.11}$$

so if $p_{|\phi\rangle}(x, y)$ vanishes for several different states $|\phi\rangle$, linearity requires it also to vanish for any state in the subspace they span. Therefore any measurement outcome that enables us to discriminate between Case 1 and Case 2 must have zero probability either for both of the states (F.2) and (F.3) and therefore for any state in the subspace they span, or for any state in the subspace spanned by the states (F.4) and (F.5). Now (F.2)–(F.5) reveal that the state

$$|\alpha\rangle = \tfrac{1}{2}\big(|00\rangle + |01\rangle + |10\rangle + |11\rangle\big) \tag{F.12}$$

belongs to *both* of these subspaces. So if there are any measurement outcomes x, y with

$$p_{|\alpha\rangle}(x, y) \neq 0, \tag{F.13}$$

then such outcomes are uninformative. Therefore the probability of a measurement outcome that fails to discriminate between Case 1 and Case 2 is at least

$$p_{\min} = \sum_{x,y}{}' \, p_{|\psi\rangle}(x, y), \tag{F.14}$$

where the prime indicates that the sum is restricted to those measurement outcomes x, y that satisfy (F.13).

Now it is easy to verify that every one of the four possible forms (F.2)–(F.5) for $|\psi\rangle$ is of the form

$$|\psi\rangle = \tfrac{1}{\sqrt{2}}\big(|\alpha\rangle + |\beta\rangle\big), \tag{F.15}$$

where $|\alpha\rangle$ is given in (F.12) and $|\beta\rangle$ is orthogonal to $|\alpha\rangle$. Since $|\psi\rangle$ has the form (F.15), we have from (F.14) and (F.10) that

$$p_{\min} = \tfrac{1}{2}\sum_{x,y}{}' \Big(p_{|\alpha\rangle}(x, y) + 2\,\mathrm{Re}[\langle\beta, 0|\mathbf{W}^\dagger|x, y\rangle\langle x, y|\mathbf{W}|\alpha, 0\rangle] + p_{|\beta\rangle}(x, y)\Big). \tag{F.16}$$

Although the sum in (F.16) is restricted to those x, y satisfying (F.13), we can extend it in each of the first two terms to all x, y since this adds either zero probabilities (first term) or (because of (F.11)) zero amplitudes (second term). The first term then gives

$$\sum_{\text{all } x,y} p_{|\alpha\rangle}(x, y) = 1, \tag{F.17}$$

while the second gives

$$2\,\mathrm{Re}\sum_{\text{all } x,y} \langle \beta, 0|\mathbf{W}^{\dagger}|x, y\rangle\langle x, y|\mathbf{W}|\alpha, 0\rangle = 2\,\mathrm{Re}\langle \beta, 0|\mathbf{W}^{\dagger}\mathbf{W}|\alpha, 0\rangle$$

$$= 2\,\mathrm{Re}\langle \beta, 0|\mathbf{1}|\alpha, 0\rangle = 0, \tag{F.18}$$

since $|\alpha\rangle$ and $|\beta\rangle$ are orthogonal. Hence

$$p_{\mathrm{min}} = \tfrac{1}{2}\left(1 + \sum_{x,y}{}' p_{|\beta\rangle}(x, y)\right) \geq \tfrac{1}{2}. \tag{F.19}$$

One must fail at least half the time.

Appendix G

The probability of success in Simon's problem

Section 2.5 gives a rough argument that the number of runs necessary to determine the n-bit number a in Simon's problem is of order n. Further analysis is needed to get a more accurate estimate of how many runs give a high probability of learning a.

If we invoke $\mathbf{U}_f$ m times, we learn m independently selected random numbers y, whose bits y_i satisfy

$$a \cdot y = \sum_{i=0}^{n-1} y_i a_i = 0 \pmod 2. \tag{G.1}$$

If we have $n - 1$ relations (G.1) for $n - 1$ linearly independent sets of y_i, then this gives us enough equations to determine a unique nonzero a. "Linearly independent" in this context means linear independence over the integers modulo 2; i.e. no subsets of the ys should satisfy $y \oplus y' \oplus y'' \oplus \cdots = 0 \pmod 2$. We have to invoke the subroutine enough times to give us a high probability of coming up with $n - 1$ linearly independent values of y.

Regardless of the size of n, for not terribly large x the probability becomes extremely close to 1 that a set of $n + x$ random vectors from an $(n - 1)$-dimensional subspace of the space of n-dimensional vectors, with components restricted to the modulo 2 integers 0 and 1, contains a linearly independent subset. This is obvious for ordinary vectors with continuous components, since the probability that a randomly selected vector in an $(n - 1)$-dimensional space lies in a specified subspace of lower dimensionality is zero – it is certain to have a nonzero component outside of the lower-dimensional subspace. The argument is trickier here because components are restricted to only two values: 1 or 0.

Introduce a basis in the full $(n - 1)$-dimensional subspace of all vectors y with $a \cdot y = 0$, so that a random vector in the subspace can be expressed as a linear combination of the basis vectors with coefficients that are randomly and independently 1 or 0. Arrange the resulting $(n + x)$ random vectors of ones and zeros into a matrix of $n + x$ rows and $n - 1$ columns. Since the row rank (the number of linearly independent rows) of a matrix is the same as the column rank, even when arithmetic is confined to the integers modulo 2, the probability that some subset of $n - 1$ of the $n + x$ $(n - 1)$-dimensional row vectors is linearly independent is the same as the probability that all $n - 1$ of the

$(n + x)$-dimensional column vectors are linearly independent. But it is easy to find a lower bound for this last probability.

Pick a column vector at random. The probability that it is nonzero is $1 - (1/2^{n+x})$. If so, take it as the first member of a basis in which we expand the remaining column vectors. The probability that a second, randomly selected column vector is independent of the first is $1 - (1/2^{n+x-1})$, since it will be independent unless every one of its (random) components along the remaining $n + x - 1$ vectors is zero. Continuing in this way, we conclude that the probability q of all $n - 1$ column vectors being linearly independent is

$$q = \left(1 - \frac{1}{2^{n+x}}\right)\left(1 - \frac{1}{2^{n+x-1}}\right)\cdots\left(1 - \frac{1}{2^{x+2}}\right). \tag{G.2}$$

(If you're suspicious of this argument, reassure yourself by checking that it gives the right q when $n = 3$, $a = 111$, and $x = 0$, by explicitly enumerating which of the 64 different sets of three ys, all satisfying $a \cdot y = 0$, contain two linearly independent vectors.)

Finally, to get a convenient lower bound on the size of q, note that if we have a set of non-negative numbers $a, b, c, \ldots$ whose sum is less than 1, then the product $(1 - a)(1 - b)(1 - c)\ldots$ exceeds $1 - (a + b + c + \cdots)$. (This is easily proved by induction on the number of numbers in the set.) The probability q is therefore greater than

$$1 - \frac{1}{2^{x+2}} - \frac{1}{2^{x+3}} - \cdots - \frac{1}{2^{x+n}}, \tag{G.3}$$

and this, in turn, is greater than

$$1 - \frac{1}{2^{x+1}}. \tag{G.4}$$

So if we want to determine a with less than one chance in a million of failure, it is enough to run the subroutine $n + 20$ times.

Appendix H

One way to make a cNOT gate

This more technical appendix is addressed to physicists curious about how one might, at least in principle, construct a cNOT gate, exploiting physically plausible interactions between two Qbits. Readers with no background in quantum physics will find some parts rather obscure. It is relevant only to readers curious about the possibilities for quantum-computational hardware, and plays no role in subsequent developments.

The controlled-NOT gate $\mathbf{C}_{10}$ with control Qbit 1 and target Qbit 0 can be written as

$$\mathbf{C}_{10} = \mathbf{H}_0 \mathbf{C}^Z \mathbf{H}_0, \tag{H.1}$$

where the controlled-Z operation is given by

$$\mathbf{C}^Z = \tfrac{1}{2}\big(\mathbf{1} + \mathbf{Z}_1 + \mathbf{Z}_0 - \mathbf{Z}_1\mathbf{Z}_0\big). \tag{H.2}$$

Because of its symmetry under interchange of the two Qbits, we may write $\mathbf{C}^Z$ without the subscripts distinguishing control and target. To within 1-Qbit Hadamard transformations, the problem of constructing a controlled-NOT gate is the same as that of constructing a controlled-Z gate.

Since $(\mathbf{C}^Z)^2 = \mathbf{1}$, $\mathbf{C}^Z$ satisfies the identity

$$\exp(i\mathbf{C}^Z\theta) = \cos\theta + i\mathbf{C}^Z \sin\theta. \tag{H.3}$$

We can therefore rewrite (H.2) as

$$\begin{aligned}\mathbf{C}^Z &= -i\exp\big(i\tfrac{\pi}{2}\mathbf{C}^Z\big) = -i\exp\big[i\big(\tfrac{\pi}{4}\big)(\mathbf{1} + \mathbf{Z}_1 + \mathbf{Z}_0 - \mathbf{Z}_1\mathbf{Z}_0)\big] \\ &= e^{-i(\pi/4)}\exp\big[i\big(\tfrac{\pi}{4}\big)(\mathbf{Z}_1 + \mathbf{Z}_0 - \mathbf{Z}_1\mathbf{Z}_0)\big].\end{aligned} \tag{H.4}$$

The point of writing $\mathbf{C}^Z$ in this clumsy way is that the unitary transformations one can construct physically are those of the form

$$\mathbf{U} = \exp(i\mathcal{H}t), \tag{H.5}$$

where $\hbar\mathcal{H}$ is the Hamiltonian that describes the external fields acting on the Qbits and the interactions between Qbits. So to within an overall constant phase factor we can realize a $\mathbf{C}^Z$ gate by letting the two Qbits interact through a Hamiltonian proportional to $\mathbf{Z}_1 + \mathbf{Z}_0 - \mathbf{Z}_1\mathbf{Z}_0$ for a precisely specified interval of time. If each Qbit is a spin-$\frac{1}{2}$, then (since $\mathbf{Z} = \sigma_z$) this Hamiltonian describes two such spins with a highly anisotropic interaction that couples only their z-components (*Ising interaction*) subject to a uniform magnetic field with a magnitude appropriately proportional to the strength of their coupling. This is perhaps the simplest example of how to make a cNOT gate.

Ising interactions, however, are rather hard to arrange. A much more natural interaction between two spins is the *exchange interaction*

$$\vec{\sigma}^{(1)} \cdot \vec{\sigma}^{(0)} = \sigma_x^{(1)}\sigma_x^{(0)} + \sigma_y^{(1)}\sigma_y^{(0)} + \sigma_z^{(1)}\sigma_z^{(0)}, \tag{H.6}$$

which is invariant under spatial rotations, as described in Appendix B.

One can also build a $\mathbf{C}^Z$ gate out of two spins interacting through (H.6), if one applies to each spin magnetic fields that are along the same direction but have different magnitudes and signs.[1]

What we must show is that to within an overall constant phase factor it is possible to express $\mathbf{C}^Z$ in the form

$$\mathbf{C}^Z = \exp(i\mathcal{H}t), \tag{H.7}$$

with a Hamiltonian $\mathcal{H}$ of the form

$$\mathcal{H} = \mathcal{J}\,\vec{\sigma}^{(1)} \cdot \vec{\sigma}^{(0)} + B_1\sigma_z^{(1)} + B_0\sigma_z^{(0)}, \tag{H.8}$$

for appropriate choices of $\mathcal{J}$ (known as the exchange coupling), of B_1 and B_0 (proportional to the magnetic fields acting on the two spins – hereafter we ignore the proportionality constant and refer to them simply as the "magnetic fields"), and of the time t during which the spins interact with each other and with the magnetic fields.

To see that the parameters in (H.8) can indeed be chosen so that $\mathcal{H}$ gives rise to $\mathbf{C}^Z$ through (H.7), recall first[2] that the operator $\frac{1}{2}(1 + \vec{\sigma}^{(1)} \cdot \vec{\sigma}^{(0)})$ acts as the swap operator on any 2-Qbit computational-basis state:

$$\tfrac{1}{2}\big(1 + \vec{\sigma}^{(1)} \cdot \vec{\sigma}^{(0)}\big)|xy\rangle = |yx\rangle. \tag{H.9}$$

It follows from (H.9) that the three states (called *triplet states*)

$$|11\rangle, \quad |00\rangle, \quad \tfrac{1}{\sqrt{2}}\big(|01\rangle + |10\rangle\big) \tag{H.10}$$

are eigenstates of $\vec{\sigma}^{(1)} \cdot \vec{\sigma}^{(0)}$ with eigenvalue 1, while the state

$$\tfrac{1}{\sqrt{2}}\big(|01\rangle - |10\rangle\big) \tag{H.11}$$

---

1 What follows was inspired by Guido Burkard, Daniel Loss, David P. DiVincenzo, and John A. Smolin, "Physical optimization of quantum error correction circuits," *Physical Review* B **60**, 11 404–11 416 (1999), `http://arxiv.org/abs/cond-mat/9905230`.

2 This was established in Equation (1.53). It is why the interaction is called the exchange interaction.

(called the *singlet state*) is an eigenstate of $\vec{\sigma}^{(1)} \cdot \vec{\sigma}^{(0)}$ with eigenvalue -3.[3]

The four states (H.10) and (H.11) are also eigenstates of $\frac{1}{2}(\sigma_z^{(1)} + \sigma_z^{(0)})$, the three triplet states (in the order in which they appear in (H.10)) having eigenvalues -1, 1, and 0, and the singlet state having eigenvalue 0.

Note also that the first two triplet states in (H.10) are eigenstates of $\frac{1}{2}(\sigma_z^{(1)} - \sigma_z^{(0)})$ with eigenvalue 0, while $\frac{1}{2}(\sigma_z^{(1)} - \sigma_z^{(0)})$ takes the third of the triplet states into the singlet state, and vice versa.

So the eigenstates of the Hamiltonian

$$\begin{aligned}\mathcal{H} &= \mathcal{J}\,\vec{\sigma}^{(1)} \cdot \vec{\sigma}^{(0)} + B_1\sigma_z^{(1)} + B_0\sigma_z^{(0)} \\ &= \mathcal{J}\,\vec{\sigma}^{(1)} \cdot \vec{\sigma}^{(0)} + B_+\tfrac{1}{2}(\sigma_z^{(1)} + \sigma_z^{(0)}) + B_-\tfrac{1}{2}(\sigma_z^{(1)} - \sigma_z^{(0)}),\end{aligned} \tag{H.12}$$

where

$$B_\pm = B_1 \pm B_0, \tag{H.13}$$

can be taken to be the first two of the triplet states (H.10) and two appropriately chosen orthogonal linear combinations of the third triplet state and the singlet state (H.11). The eigenvalues of $\mathcal{H}$ associated with the first and second triplet states are $\mathcal{J} - B_+$ and $\mathcal{J} + B_+$; those associated with the last two states are the eigenvalues of the matrix

$$\begin{pmatrix} \mathcal{J} & B_- \\ B_- & -3\mathcal{J} \end{pmatrix}$$

of $\mathcal{H}$ in the space spanned by the last two; i.e. $-\mathcal{J} \pm \sqrt{4\mathcal{J}^2 + B_-^2}$.

Now the four states (H.10) and (H.11) are also eigenstates of $\mathbf{C}^Z$, the first of the three triplet states having eigenvalue -1 and the other three having eigenvalue 1. Consequently these eigenstates of $\mathcal{H}$ are also eigenstates of $\mathbf{C}^Z$ with respective eigenvalues -1, 1, 1, and 1. We will therefore produce $\mathbf{C}^Z$ (to within a constant phase factor) if we can choose the exchange coupling $\mathcal{J}$, the magnetic fields B_1 and B_0, and the time t during which $\mathcal{H}$ acts to satisfy

$$-e^{it(\mathcal{J}-B_+)} = e^{it(\mathcal{J}+B_+)} = e^{it(-\mathcal{J}+\sqrt{4\mathcal{J}^2+B_-^2})} = e^{it(-\mathcal{J}-\sqrt{4\mathcal{J}^2+B_-^2})}. \tag{H.14}$$

The last equality is equivalent to

$$e^{2it\sqrt{4\mathcal{J}^2+B_-^2}} = 1, \quad \text{or} \quad e^{it\sqrt{4\mathcal{J}^2+B_-^2}} = \pm 1; \tag{H.15}$$

the first is equivalent to

$$e^{2itB_+} = -1, \quad \text{or} \quad e^{itB_+} = \pm i; \tag{H.16}$$

3 If $|0\rangle$ is the state $|\uparrow\rangle$ of spin-up along z, and $|1\rangle$ is $|\downarrow\rangle$, then the singlet state is the state of zero total angular momentum and the three triplet states are the states of angular momentum 1 with z-components $-\hbar$, 0, and $\hbar$.

and the second is equivalent to

$$e^{-2it\mathcal{J}} = e^{itB_+}e^{-it\sqrt{4\mathcal{J}^2+B_-^2}}. \tag{H.17}$$

The identities (H.15) and (H.16) require the right side of (H.17) to be $\pm i$. For the (positive) time t for which the gate acts to be as small as possible we should choose $-i$, which gives

$$\mathcal{J}t = \pi/4. \tag{H.18}$$

With this value of t we can satisfy (H.15) (with the minus sign) and (H.16) (with the plus sign) by taking $\sqrt{4\mathcal{J}^2 + B_-^2} = 4\mathcal{J}$ and $B_+ = 2\mathcal{J}$. So we can produce the gate $\mathbf{C}^Z$ (to within an overall constant phase factor) by taking the magnetic fields in the Hamiltonian (H.12) and the time for which it acts to be related to the exchange coupling by

$$B_+ = 2\mathcal{J}, \qquad B_- = 2\sqrt{3}\mathcal{J}, \qquad t = \tfrac{1}{4}\pi/\mathcal{J}, \tag{H.19}$$

or, in terms of the fields on each spin,

$$B_1 = (1+\sqrt{3})\mathcal{J}, \qquad B_0 = (1-\sqrt{3})\mathcal{J}, \qquad t = \tfrac{1}{4}\pi/\mathcal{J}. \tag{H.20}$$

Note the curious fact that although, as (H.2) makes explicit, the gate $\mathbf{C}^Z$ acts symmetrically on the two spins, the realization of $\mathbf{C}^Z$ by the unitary transformation $e^{i\mathcal{H}t}$ requires the fields acting on the spins to break that symmetry. Of course the symmetry survives in the fact that the alternative choice of fields $B_1 = (1-\sqrt{3})\mathcal{J}$, $B_0 = (1+\sqrt{3})\mathcal{J}$ works just as well.

Appendix I

A little elementary group theory

A set of positive integers less than N constitutes a *group* under multiplication modulo N if the set (a) contains 1, (b) contains the modulo-N inverse of any of its members, and (c) contains the the modulo-N products of all pairs of its members. A subset of a group meeting conditions (a)–(c) is called a *subgroup*. The number of members of a group is called the *order* of the group. An important result of the elementary theory of finite groups (Lagrange's theorem) is that the order of any of its subgroups is a divisor of the order of the group itself. This is established in the next three paragraphs.

If S is any subset of a group G (not necessarily a subgroup) and a is any member of G (which might or might not be in S), define aS (called a *coset* of S) to be the set of all members of G of the form $g = as$, where s is any member of S. (Throughout this appendix equality will be taken to mean equality modulo N.) Distinct members of S give rise to distinct members of aS, for if s and s' are in S and $as = as'$, then multiplying both sides by the inverse of a gives $s = s'$. So any coset aS has the same number of members as S itself.

If the subset S is a *subgroup* of G and s is a member of S, then every member of the coset sS must be in S. Since sS has as many distinct members as S has, $sS = S$. If two cosets aS and bS of a subgroup S have a common member then there are members s and s' of S that satisfy $as = bs'$, so $(as)S = (bs')S$. But $(as)S = a(sS) = aS$, and similarly $(bs')S = bS$. Therefore $aS = bS$: two cosets of a *subgroup* are either identical or have no members in common.

If S is a subgroup and a is a member of G, then since 1 is in S, a is in the coset aS. Since every member of G is thus in some coset, and since the cosets of a subgroup are either identical or disjoint, it follows that the distinct cosets of a subgroup S partition the whole group G into disjoint subsets, each of which has the same number of members as S does. Consequently the total number of members of G must be an integral multiple of the number of members of any of its subgroups S: *the order of any subgroup S is a divisor of the order of the whole group G.*

Of particular interest is the subgroup given by all the distinct powers of any particular member a of G. Since G is a finite set, the set of distinct powers of a is also finite, and therefore for some n and m with $n > m$ we must have $a^n = a^m$, or $a^{(n-m)} = 1$. The *order* of a is defined to be the smallest nonzero k with $a^k = 1$. The subset $a, a^2, \ldots, a^k$ of G is a subgroup of G, since it contains $1 = a^k$ and the inverses and products of all its members. It is called the subgroup *generated* by a, and its order is the order k of a. Since the order of any subgroup of G divides the order of G, we conclude that the order of any member of G divides the order of G.

Appendix J

Some simple number theory

J.1 The Euclidean algorithm

We wish to find the greatest common divisor of two numbers f and c, with $f > c$. The Euclidean algorithm is the iterative procedure that replaces f and c by $f' = c$ and $c' = f - [f/c]c$, where $[x]$ is the largest integer less than or equal to x. Evidently any factors common to f and c are also common to f' and c' and vice versa. Furthermore, f' and c' decrease with each iteration and each iteration keeps $f' > c'$, until the procedure reaches $c' = 0$. Let f_0 and c_0 be the values of f and c at the last stage before $c' = 0$. They have the same common factors as the original f and c, and f_0 is divisible by c_0, since the next stage is $c_0' = 0$. Therefore c_0 is the greatest common divisor of f and c.

J.2 Finding inverses modulo an integer

We can use the Euclidean algorithm to find the inverse of an integer c modulo an integer $f > c$, when f and c have no common factors. In this case iterating the Euclidean algorithm eventually leads to $c_0 = 1$. This stage must have arisen from a pair f_1 and c_1 satisfying $1 = f_1 - mc_1$ for some integer m. But f_1 and c_1 are given by explicit integral linear combinations of the pair at the preceding stage, f_2 and c_2, which in turn are explicit integral linear combinations of f_3 and c_3, etc. So one can work backwards through the iterations to construct integers j and k with $1 = jf + kc$. Since k cannot be a multiple of f, we can express k as $lf + d$ with $1 \le d < f$ and with l an integer (negative, if k is negative); d is then the inverse of c modulo f.

J.3 The probability of common factors

The probability of two random numbers having no common factors is greater than $\frac{1}{2}$, for the probability is $\frac{3}{4}$ that they are not both divisible by 2, $\frac{8}{9}$ that they are not both divisible by 3, $\frac{24}{25}$ that they are not both divisible by 5, etc. The probability that they share no prime factors at all is

$$\begin{aligned}\prod_{\text{primes}} (1 - 1/p^2) &= 1\Big/ \prod_{\text{primes}} (1 + 1/p^2 + 1/p^4 + \ldots)\\ &= 1/\left(1 + 1/2^2 + 1/3^2 + 1/4^2 + 1/5^2 + 1/6^2 + \cdots\right)\\ &= 6/\pi^2 = 0.6079\ldots \qquad \text{(J.1)}\end{aligned}$$

If the numbers are confined to a finite range this argument gives only an estimate of the probability, but it is quite a good estimate if the range is large.

Appendix K

Period finding and continued fractions

We illustrate here the mathematics of the final (post-quantum-computational) stage of Shor's period-finding procedure. The final measurement produces (with high probability) an integer y that is within $\frac{1}{2}$ of an integral multiple of $2^n/r$, where n is the number of Qbits in the input register, satisfying $2^n > N^2 > r^2$. Deducing the period r of the function f from such an integer y makes use of the theorem that if x is an estimate for the fraction j/r that differs from it by less than $1/2r^2$, then j/r will appear as one of the partial sums in the continued-fraction expansion of x.[1] In the case of Shor's period-finding algorithm $x = y/2^n$. If j and r happen to have no factors in common, r is given by the denominator of the partial sum with the largest denominator less than N. Otherwise the continued-fraction expansion of x gives r_0: r divided by whatever factor it has in common with the random integer j. If several small multiples of r_0 fail to be a period of f, one repeats the whole procedure, getting a different submultiple r_1 of r. There is a good chance that r will be the least common multiple of r_0 and r_1, or a not terribly large multiple of it. If not, one repeats the whole procedure a few more times until one succeeds in finding a period of f. We illustrate this with two examples.

Example 1. (Successful the first time.) Suppose we know that the period r is less than $2^7 = 128$ and that $y = 11\,490$ is within $\frac{1}{2}$ of an integral multiple of $2^{14}/r$. What is r?

Example 2. (Two attempts required.) Suppose we know that the integer r is less than 2^7 and that 11 343 and 13 653 are both within $\frac{1}{2}$ of integral multiples of $2^{14}/r$. What is r?

In either example the fraction j/r for some (random) integer j will necessarily be one of the partial sums (defined below) of the continued-fraction expansion of $y/2^{14}$, where y is one of the cited five-digit integers. The partial sum with the largest denominator less than 128 is the one we are looking for. Once we have found the answer we can easily check that it is correct.

1 Theorem 184, page 153, G. H. Hardy and E. M. Wright, *An Introduction to the Theory of Numbers*, 4th edition, Oxford University Press (1965).

The continued-fraction expansion of a real number x between 0 and 1 is

$$x = \cfrac{1}{a_0 + \cfrac{1}{a_1 + \cfrac{1}{a_2 + \cdots}}} \tag{K.1}$$

with positive integers $a_0, a_1, a_2, \ldots$ Evidently a_0 is the integral part of $1/x$. Let x_1 be the fractional part of $1/x$. Then it follows from (K.1) that

$$x_1 = \cfrac{1}{a_1 + \cfrac{1}{a_2 + \cfrac{1}{a_3 + \cdots}}} \tag{K.2}$$

so a_1 is the integral part of $1/x_1$. Letting x_2 be the fractional part of $1/x_1$, one can continue this iterative procedure to extract a_2 as the integral part of $1/x_2$, and so on.

By the partial sums of the continued fraction (K.1), one means

$$\frac{1}{a_0}, \quad \cfrac{1}{a_0 + \cfrac{1}{a_1}}, \quad \cfrac{1}{a_0 + \cfrac{1}{a_1 + \cfrac{1}{a_2}}}, \quad \text{etc.} \tag{K.3}$$

One can deal with both examples using an (unprogrammed) pocket calculator. One starts with $1/x = 2^{14}/y$ in the display and subtracts the integral part a_0, noting it down. One then inverts what remains, to get $1/x_1$, and repeats the process until one has accumulated a long enough list of a_j.

Analysis of example 1. We know that $r < 128$ and that $x = 11\,490/2^{14}$ is within $\frac{1}{2}2^{-14}$ of j/r for integers j and r. Playing with a calculator tells us that

$$11\,490/2^{14} = 0.701\,293\,945\,3\ldots = \cfrac{1}{1 + \cfrac{1}{2 + \cfrac{1}{2 + \cfrac{1}{1 + \cfrac{1}{7 + \cfrac{1}{35 + \cdots}}}}}} \tag{K.4}$$

If we drop what comes after the 35 and start forming partial sums we quickly get to a denominator bigger than 128. If we also drop $\frac{1}{35}$, we find that

$$11\,490/2^{14} \approx \cfrac{1}{1+\cfrac{1}{2+\cfrac{1}{2+\cfrac{1}{1+\cfrac{1}{7}}}}} \tag{K.5}$$

which works out[2] to $\frac{54}{77}$. Since 77 is the only multiple of 77 less than 128, $r = 77$. And indeed,

$$2^{14} \times \tfrac{54}{77} = 11\,490.079\ldots,$$

which is within $\frac{1}{2}$ of 11 490.

Analysis of example 2. We know that the integer r is less than 128 and that $x = 11\,343/2^{14}$ and $x' = 13\,653/2^{14}$ are both within $\frac{1}{2}2^{-14}$ of integral multiples of $1/r$. The calculator tells us that

$$11\,343/2^{14} = \cfrac{1}{1+\cfrac{1}{2+\cfrac{1}{3+\cfrac{1}{1+\cfrac{1}{419+\cdots}}}}} \tag{K.6}$$

Since 419 is bigger than 128 we can drop the $\frac{1}{419}$ to get

$$\cfrac{1}{1+\cfrac{1}{2\frac{1}{4}}} \tag{K.7}$$

which gives $\frac{9}{13}$, and indeed

$$2^{14} \times \tfrac{9}{13} = 11\,342.769\ldots, \tag{K.8}$$

which is within $\frac{1}{2}$ of 11 343. The number r is thus a multiple of 13 less than 128, of which there are nine. Had we the function f at hand (which we do in the case of interest) we could try all nine to determine the period, but to illustrate what one can do when there are too many possibilities to try them all, we take advantage of the second piece

2 A more systematic way to get this is to use the famous but not transparently obvious recursion relation for the numerators p and denominators q of the partial sums: $p_n = a_n p_{n-1} + p_{n-2}$, and $q_n = a_n q_{n-1} + q_{n-2}$, with $q_0 = a_0$, $q_1 = 1 + a_0 a_1$ and $p_0 = 1$, $p_1 = a_1$. One easily applies these to the sequence $a_0, a_1, a_2, \ldots = 1, 2, 2, 1, 1, 7, 35, \ldots$, stopping when one gets to a denominator larger than 100.

of information, which could have been produced by running Shor's algorithm a second time.

We also have

$$13\,653/2^{14} = \cfrac{1}{1+\cfrac{1}{4+\cfrac{1}{1+\cfrac{1}{1364+\cdots}}}} \tag{K.9}$$

Since 1364 is bigger than 128 we can drop the $\frac{1}{1364}$ to get

$$\cfrac{1}{1+\cfrac{1}{5}} \tag{K.10}$$

which gives $\frac{5}{6}$, and indeed

$$2^{14} \times \tfrac{5}{6} = 13\,653.333\ldots, \tag{K.11}$$

which is within $\frac{1}{2}$ of 13 653. So r is also a multiple of 6 less than 100. Since 6 and 13 have no common factors the least multiple of both is $6 \times 13 = 78$. Since there is no multiple of 78 less than 100 other than 78 itself, $r = 78$.

Appendix L

Better estimates of success in period finding

In Section 3.7 it is shown that with a probability of at least 0.4, a single application of Shor's period-finding procedure produces an integer y that is within $\frac{1}{2}$ of an integral multiple of $2^n/r$, where r is the period sought. Since $2^n > N^2 > r^2$, $y/2^n$ is within $1/(2r^2)$ of j/r for some integer j, and therefore, by the theorem cited in Appendix K, j/r and hence a divisor of r (r divided by any factors it may have in common with j) can be found from the continued-fraction expansion of $y/2^n$.

What is crucial for learning a divisor of r is that the estimate for j/r emerging from Shor's procedure be within $1/2r^2$ of a multiple of $1/r$. Now when N is the product of two odd primes p and q, as it is in the case of RSA encryption, then the required period r is not only less than N, but also less than $\frac{1}{2}N$. This is because $\frac{1}{2}(q-1)$ is an integer, so it follows from Fermat's little theorem,

$$b^{p-1} \equiv 1 \pmod{p}, \tag{L.1}$$

that

$$b^{(p-1)(q-1)/2} \equiv 1 \pmod{p}. \tag{L.2}$$

For the same reason it follows from

$$b^{q-1} \equiv 1 \pmod{q} \tag{L.3}$$

that

$$b^{(p-1)(q-1)/2} \equiv 1 \pmod{q}. \tag{L.4}$$

But since p and q are prime, the fact that $b^{(p-1)(q-1)/2} - 1$ is divisible by both p and q means that it must be divisible by the product pq, and therefore

$$b^{(p-1)(q-1)/2} \equiv 1 \pmod{pq}. \tag{L.5}$$

So if

$$b^r \equiv 1 \pmod{pq} \tag{L.6}$$

and r exceeded $\frac{1}{2}N$, then we would also have

$$b^{r-(p-1)(q-1)/2} = 0 \pmod{pq}, \tag{L.7}$$

and since $r - \frac{1}{2}(p-1)(q-1) > r - \frac{1}{2}N > 0$, (L.5) would give a positive power of b smaller than r that was congruent to 1 modulo pq, so r could not be the period (which is the *least* such power).

It follows that even if y is not the closest integer to an integral multiple of $2^n/r$, if it is within 2 of such an integral multiple, then

$$|y/2^n - j/r| < 2/N^2 < 1/2r^2. \tag{L.8}$$

So for each j/r the algorithm will succeed in providing a divisor of r not only if the measured y is the closest integer to $2^n j/r$, but also if it is the second, third, or fourth closest. Gerjuoy has estimated that this increases the probability of a successful run to about 0.9.[1]

Bourdon and Williams have refined this to 0.95 for large N and r.[2] They also show that if one modifies the hardware, adding a few more Qbits to the input register so that $n > 2n_0 + q$, then for rather small q the probability of finding a divisor of r from the output of a single run of the quantum computation can be made quite close to 1.

1 Edward Gerjuoy, "Shor's factoring algorithm and modern cryptography. An illustration of the capabilities inherent in quantum computers," *American Journal of Physics* 73, 521–540 (2005), `http://arxiv.org/abs/quant-ph/0411184`.

2 P. S. Bourdon and H. T. Williams, "Sharp probability estimates for Shor's order-finding algorithm," `http://arxiv.org/abs/quant-ph/0607148`.

Appendix M

Factoring and period finding

We establish here the only hard part of the connection between factoring and period finding: that the probability is at least $\frac{1}{2}$ that if a is a random member of G_{pq} for prime p and q, then the order r of a in G_{pq} satisfies both

$$r \text{ even} \tag{M.1}$$

and

$$a^{r/2} \not\equiv -1 \pmod{pq}. \tag{M.2}$$

(In Section 3.10 it is shown that given such an a and its order r, the problem of factoring $N = pq$ is easily solved.)

Note first that the order r of a in G_{pq} is the least common multiple of the orders r_p and r_q of a in G_p and in G_q. That r must be *some* multiple of both r_p and r_q is immediate, since $a^r \equiv 1 \pmod{pq}$ implies that $a^r \equiv 1 \pmod{p}$ and $a^r \equiv 1 \pmod{q}$. Furthermore, *any* common multiple r' of r_p and r_q satisfies $a^{r'} \equiv 1 \pmod{pq}$, because if $a^{r'} = 1 + mp$ and $a^{r'} = 1 + nq$, then $mp = nq$. But since the primes p and q have no common factors this requires $m = kq$ and $n = kp$, and hence $a^{r'} = 1 + kpq \equiv 1 \pmod{pq}$. Since r is the least integer with $a^r \equiv 1 \pmod{pq}$, r must be the least common multiple of r_p and r_q.

Consequently condition (M.1) can fail only if r_p and r_q are both odd. Condition (M.2) can fail only if r_p and r_q are both odd multiples of the *same* power of 2. For if r_p contains a higher power of 2 than r_q, then since r is a common multiple of r_p and r_q, it will remain a multiple of r_q if a single factor of 2 is removed from it, and therefore $a^{r/2} \equiv 1 \pmod{q}$. But this is inconsistent with a failure of condition (M.2), which would imply that $a^{r/2} \equiv -1 \pmod{q}$.

So a necessary condition for failure to factor $N = pq$ is that r_p and r_q are either both odd, or both odd multiples of the same power of 2. The first condition is absorbed into the second if we agree that the powers of 2 include $2^0 = 1$. Our effort to factor N can fail only if we have picked a random a for which r_p and r_q are both odd multiples of the same power of 2.

To calculate an upper bound for the probability of failure p_{f}, note first that the modulo-p and modulo-q orders, r_p and r_q, of a are the same as the mod-p and mod-q orders of the numbers a_p and a_q in G_p and G_q, where

$$a \equiv a_p \pmod{p}, \qquad a \equiv a_q \pmod{q}. \tag{M.3}$$

Furthermore, every number a in G_{pq} is associated through (M.3) with a unique pair from G_p and G_q. For if $a_p = b_p$ and $a_q = b_q$ then $a - b$ is a multiple of both p and q, and therefore, since p and q are distinct primes, $a - b$ is a multiple of pq itself, so $a \equiv b \pmod{pq}$.

Since the $(p-1)(q-1)$ different members of G_{pq} are thus in one-to-one correspondence with the number of distinct pairs, one from the $p-1$ members of G_p and one from the $q-1$ members of G_q, the modulo-p and modulo-q orders r_p and r_q of a random integer a in G_{pq} will have exactly the same statistical distribution as the orders r_p and r_q of randomly and independently selected integers in G_p and G_q. So to show that the probability of failure is at most $\frac{1}{2}$, we must show that the probability is at most $\frac{1}{2}$ that the orders r_p and r_q of such a randomly and independently selected pair are both odd multiples of the same power of 2.

We do this by showing that for any prime p, no more than half of the numbers in G_p can have orders r_p that are odd multiples of *any* given power of 2. (Given this, if $P_p(j)$ and $P_q(j)$ are the probabilities that random elements of G_p and G_q have orders that are odd multiples of 2^j, then the probability of failure p_f is less than $\sum_{j\geq 0} P_p(j)P_q(j) \leq \frac{1}{2}\sum_{j\geq 0} P_q(j) = \frac{1}{2}$.) This follows from the fact that if the order $p-1$ of G_p is an odd multiple of 2^k for some $k \geq 0$, then exactly half the elements of G_p have orders that are odd multiples of 2^k. This in turn follows from the theorem that if p is a prime, then G_p has at least one *primitive* element b of order $p-1$, whose successive powers therefore generate the entire group. Given this theorem – which is proved at the end of this appendix – we complete the argument by showing that the orders of the odd powers of any such primitive b are odd multiples of 2^k, but the orders of the even powers are not.

If r_0 is the order of b^j with j odd, then

$$1 \equiv (b^j)^{r_0} \equiv b^{jr_0} \pmod{p}, \tag{M.4}$$

so jr_0 must be a multiple of $p-1$, the order of b. Since j is odd r_0 must contain at least as many powers of 2 as does $p-1$. But since the order r_0 of any element must divide the order $p-1$ of the group, r_0 cannot contain more powers of 2 than $p-1$ does. So r_0 is an odd multiple of 2^k. On the other hand if j is even, then b^j satisfies

$$(b^j)^{(p-1)/2} = \left(b^{p-1}\right)^{j/2} \equiv 1 \pmod{p}, \tag{M.5}$$

so the order r_0 of b^j divides $(p-1)/2$. Therefore $p-1$ contains at least one more power of 2 than does r_0.

This concludes the proof that the probability is at least $\frac{1}{2}$ that a random choice of a in G_{pq} will satisfy both of the conditions (M.1) and (M.2) that lead, with the aid of an efficient period-finding routine, to an easy factorization of $N = pq$, as described in Section 3.10.

What remains is to prove that when p is prime, G_p contains at least one number of order $p-1$. The relevant property of the multiplicative group of integers $\{1, 2, 3, \ldots, p-1\}$ modulo a *prime* is that together with 0 these integers also constitute a group under *addition*. This provides all the structure necessary to ensure that a polynomial of degree d has at most d roots.[1] We can exploit this fact as follows.

Write the order $s = p-1$ of G_p in terms of its prime factors q_i:

$$s = p - 1 = q_1^{n_1} \cdots q_m^{n_m}. \tag{M.6}$$

For each q_i, the equation $x^{s/q_i} - 1 = 0$ has at most s/q_i solutions, and since $s/q_i < s$, the number of elements in G_p, there must be elements a_i in G_p satisfying

$$a_i^{s/q_i} \not\equiv 1 \pmod{p}. \tag{M.7}$$

Given such an a_i, define

$$b_i = a_i^{s/(q_i^{n_i})}. \tag{M.8}$$

We next show that the order of b_i is $q_i^{n_i}$. This is because

$$b_i^{(q_i^{n_i})} \equiv a_i^s \equiv 1 \pmod{p}, \tag{M.9}$$

so the order of b_i must divide $q_i^{n_i}$ and therefore be a power of q_i, since q_i is prime. But if that order were any power of q_i less than n_i, then we would have $a_i^{s/q_i^k} \equiv 1 \pmod{p}$ with $k \geq 1$, which contradicts (M.7).

Because each b_i has order $q_i^{n_i}$, the product $b_1 b_2 \cdots b_m$ has order $q_1^{n_1} q_2^{n_2} \cdots q_m^{n_m} = p - 1$. This follows from the fact that if two numbers in G_p have orders that are coprime, then the order of their product is

1 This is easily proved by induction on the degree of the equation, using the fact that every nonzero integer modulo p has a multiplicative inverse modulo p. It is obviously true for degree 1. Suppose that it is true for degree $m-1$ and a polynomial $P(x)$ of degree m satisfies $P(a) = 0$. Then $P(x) = 0$ implies $P(x) - P(a) = 0$. Since $P(x) - P(a)$ has the form $\sum_j c_j(x^j - a^j)$, the factor $x - a$ can be extracted from each term, leading to the form $(x-a)Q(x)$, where $Q(x)$ is a polynomial of degree $m-1$. So if $x \neq a$ then $P(x) = 0$ requires $Q(x) = 0$, and this has at most $m-1$ distinct solutions by virtue of the inductive assumption.

the product of their orders.[2] Therefore since $q_1^{n_1}$ and $q_2^{n_2}$ are coprime, $b_1 b_2$ has order $q_1^{n_1} q_2^{n_2}$. But since $q_1^{n_1} q_2^{n_2}$ and $q_3^{n_3}$ are coprime, it follows that $b_1 b_2 b_3$ has order $q_1^{n_1} q_2^{n_2} q_3^{n_3}$. Continuing in this way, we conclude that $b_1 b_2 \cdots b_m$ has order $q_1^{n_1} q_2^{n_2} \cdots q_m^{n_m} = s = p - 1$.

2 Let u, v, and w be the orders of c, d, and cd. Since $c^u \equiv 1 \pmod{p}$ and $(cd)^w \equiv 1 \pmod{p}$, it follows that $d^{wu} \equiv 1 \pmod{p}$. So the order v of d divides wu, and since v and u have no common factors, v divides w. In the same way one concludes that u divides w. Therefore, since v and u are coprime, w must be a multiple of uv. Furthermore, $(cd)^{uv} \equiv c^{uv} d^{vu} \equiv 1 \pmod{p}$, so uv must be a multiple of w. Therefore $w = uv$.

Appendix N

Shor's 9-Qbit error-correcting code

Shor demonstrated that quantum error correction was possible using the two orthogonal 9-Qbit codeword states

$$|\bar{0}\rangle = 2^{-3/2}\big(|000\rangle + |111\rangle\big)\big(|000\rangle + |111\rangle\big)\big(|000\rangle + |111\rangle\big),$$
$$|\bar{1}\rangle = 2^{-3/2}\big(|000\rangle - |111\rangle\big)\big(|000\rangle - |111\rangle\big)\big(|000\rangle - |111\rangle\big). \qquad \text{(N.1)}$$

These can be viewed as an extension of the simple 3-Qbit codewords we examined in Section 5.2, making it possible to deal with 1-Qbit phase errors, as well as bit-flip errors. An encoding circuit for the 9-Qbit code – with an obvious resemblance to Figure 5.1 for the 3-Qbit code – is shown in Figure N.1.

The form (5.18) of a general 1-Qbit corruption simplifies slightly when the state $|\Psi\rangle$ is a superposition of the codeword states (N.1), for it follows from (N.1) that

$$\begin{aligned}\mathbf{Z}_0|\Psi\rangle &= \mathbf{Z}_1|\Psi\rangle = \mathbf{Z}_2|\Psi\rangle,\\ \mathbf{Z}_3|\Psi\rangle &= \mathbf{Z}_4|\Psi\rangle = \mathbf{Z}_5|\Psi\rangle,\\ \mathbf{Z}_6|\Psi\rangle &= \mathbf{Z}_7|\Psi\rangle = \mathbf{Z}_8|\Psi\rangle.\end{aligned} \qquad \text{(N.2)}$$

As a result, the general form of a 1-Qbit corruption of $|\Psi\rangle$ contains only 22 independent terms (rather than $28 = (3 \times 9) + 1$):

$$|e\rangle|\Psi\rangle \to \left(|d\rangle + |c\rangle\mathbf{Z}_0 + |c'\rangle\mathbf{Z}_3 + |c''\rangle\mathbf{Z}_6 + \sum_{i=1}^{9}\big(|a_i\rangle\mathbf{X}_i + |b_i\rangle\mathbf{Y}_i\big)\right)|\Psi\rangle. \qquad \text{(N.3)}$$

We diagnose the error syndrome with eight commuting Hermitian operators that square to unity:

$$\begin{gathered}\mathbf{Z}_0\mathbf{Z}_1,\quad \mathbf{Z}_1\mathbf{Z}_2,\quad \mathbf{Z}_3\mathbf{Z}_4,\quad \mathbf{Z}_4\mathbf{Z}_5,\quad \mathbf{Z}_6\mathbf{Z}_7,\quad \mathbf{Z}_7\mathbf{Z}_8,\\ \mathbf{X}_0\mathbf{X}_1\mathbf{X}_2\mathbf{X}_3\mathbf{X}_4\mathbf{X}_5,\quad \mathbf{X}_3\mathbf{X}_4\mathbf{X}_5\mathbf{X}_6\mathbf{X}_7\mathbf{X}_8.\end{gathered} \qquad \text{(N.4)}$$

All six Z-operators trivially commute with each other as do the two X-operators, and any of the six Z-operators commutes with any of the two X-operators because in every case the number of anticommutations between a Z_i and an X_j is either zero or two.

One easily confirms from (N.1) that $|\bar{0}\rangle$, $|\bar{1}\rangle$, and hence any superposition $|\Psi\rangle$ of the two, are invariant under all eight operators in (N.4). Each one of the 22 corrupted terms in (N.3) is also an eigenstate of

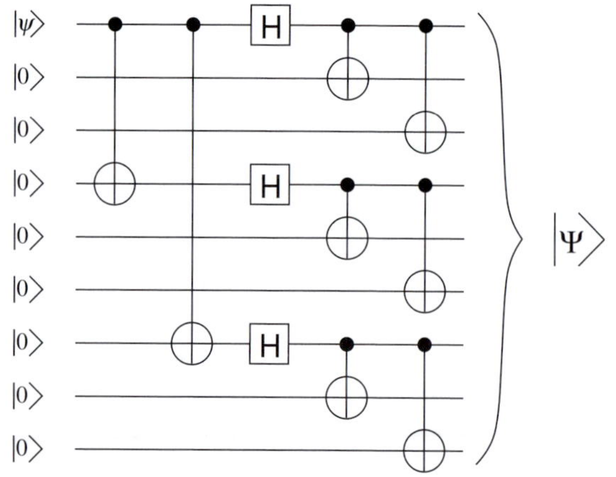

Fig N.1 A circuit that transforms the 1-Qbit state $|\psi\rangle = \alpha|0\rangle + \beta|1\rangle$ into its 9-Qbit encoded form $|\Psi\rangle = \alpha|\bar{0}\rangle + \beta|\bar{1}\rangle$, where $|\bar{0}\rangle$ and $|\bar{1}\rangle$ are given in (N.1.) Note the relation to the simpler 3-Qbit encoding circuit in Figure 5.1.

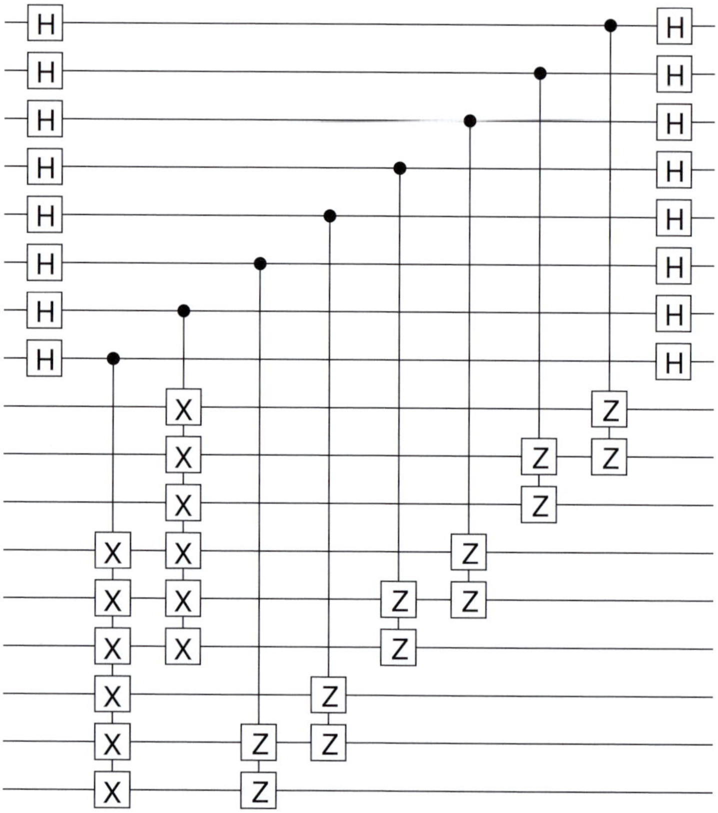

Fig N.2 A circuit to measure the "error syndrome" for Shor's 9-Qbit code. The nine Qbits are the nine lower wires. The circuit is of the type illustrated in Figure 5.7, but with eight ancillary Qbits (the eight upper wires) associated with the measurement of the eight commuting operators in (N.4), $\mathbf{Z}_0\mathbf{Z}_1$, $\mathbf{Z}_1\mathbf{Z}_2$, $\mathbf{Z}_3\mathbf{Z}_4$, $\mathbf{Z}_4\mathbf{Z}_5$, $\mathbf{Z}_6\mathbf{Z}_7$, $\mathbf{Z}_7\mathbf{Z}_8$, $\mathbf{X}_0\mathbf{X}_1\mathbf{X}_2\mathbf{X}_3\mathbf{X}_4\mathbf{X}_5$, and $\mathbf{X}_3\mathbf{X}_4\mathbf{X}_5\mathbf{X}_6\mathbf{X}_7\mathbf{X}_8$. Measurement of the eight ancillas projects the state of the nine lower Qbits into the appropriate simultaneous eigenstate of those eight operators.

the eight operators in (N.4) with eigenvalues 1 or -1, because each of the eight operators either commutes (resulting in the eigenvalue 1) or anticommutes (resulting in the eigenvalue -1) with each of the $\mathbf{X}_i$, $\mathbf{Y}_i$, and $\mathbf{Z}_i$. And each of the 22 terms in (N.3) gives rise to a distinct pattern of negative eigenvalues for the eight operators.

(a) The three errors $\mathbf{Z}_0$, $\mathbf{Z}_3$, and $\mathbf{Z}_6$ are distinguished from the $\mathbf{X}_i$ and $\mathbf{Y}_i$ by the fact that they commute with every one of the six Z-operators in (N.4). These three $\mathbf{Z}_i$ can be distinguished from each other because $\mathbf{Z}_0$ anticommutes with one of the two X-operators, $\mathbf{Z}_6$ anticommutes with the other, and $\mathbf{Z}_3$ anticommutes with both.

(b) All nine errors $\mathbf{X}_i$ are distinguished both from the $\mathbf{Z}_i$ and from the $\mathbf{Y}_i$ by the fact that they commute with both X-operators. They can be distinguished from each other because $\mathbf{X}_0$, $\mathbf{X}_2$, $\mathbf{X}_3$, $\mathbf{X}_5$, $\mathbf{X}_6$, and $\mathbf{X}_8$ each anticommutes with a single one of the six Z-operators in (N.4) (respectively $\mathbf{Z}_0\mathbf{Z}_1$, $\mathbf{Z}_1\mathbf{Z}_2$, $\mathbf{Z}_3\mathbf{Z}_4$, $\mathbf{Z}_4\mathbf{Z}_5$, $\mathbf{Z}_6\mathbf{Z}_7$, and $\mathbf{Z}_7\mathbf{Z}_8$) while $\mathbf{X}_1$, $\mathbf{X}_4$, and $\mathbf{X}_7$ each anticommutes with two distinct Z-operators (respectively $\mathbf{Z}_0\mathbf{Z}_1$ and $\mathbf{Z}_1\mathbf{Z}_2$, $\mathbf{Z}_3\mathbf{Z}_4$ and $\mathbf{Z}_4\mathbf{Z}_5$, and $\mathbf{Z}_6\mathbf{Z}_7$ and $\mathbf{Z}_7\mathbf{Z}_8$).

(c) Finally, the nine errors $\mathbf{Y}_i$ have the same pattern of commutations with the Z-operators in (N.4) as the corresponding $\mathbf{X}_i$ operators, permitting them to be distinguished from each other in the same way. They can be distinguished from the $\mathbf{X}_i$ operators by their failure to commute with at least one of the two X-operators in (N.4).

So, as with the other codes we have examined, the simultaneous measurement of the eight commuting operators in (N.4) projects the corrupted state onto a single one of the terms in (N.3), and the set of eigenvalues reveals which term it is. One then applies the appropriate inverse unitary transformation to restore the uncorrupted state. A circuit that diagnoses the 9-Qbit error syndrome is shown in Figure N.2.

Appendix O

Circuit-diagrammatic treatment of the 7-Qbit code

As a further exercise in the use of circuit diagrams, we rederive the properties of the 7-Qbit error-correcting code, using the method developed in Chapter 5 to establish that the circuit in Figure 5.11 gives the 5-Qbit codewords.

We start with the observation that the seven mutually commuting operators $\mathbf{M}_i$, $\mathbf{N}_i$ $(i = 0, 1, 2)$ in (5.42), and $\overline{\mathbf{Z}}$ in (5.49), each with eigenvalues ± 1, have a set of 2^7 nondegenerate eigenvectors that form an orthonormal basis for the entire seven-dimensional codeword space. In particular the two codeword states $|\overline{0}\rangle$ and $|\overline{1}\rangle$ are the unique eigenstates of all the $\mathbf{M}_i$ and $\mathbf{N}_i$ with eigenvalues 1, and of $\overline{\mathbf{Z}}$ with eigenvalues 1 and -1, respectively.

It follows from this that if a circuit produces a state $|\Psi\rangle$ that is invariant under all the $\mathbf{M}_i$ and $\mathbf{N}_i$ then $|\Psi\rangle$ must be a superposition of the codeword states $|\overline{0}\rangle$ and $|\overline{1}\rangle$, and if $|\Psi\rangle$ is additionally an eigenstate of $\overline{\mathbf{Z}}$ then, to within factors $e^{i\varphi}$ of modulus 1, $|\Psi\rangle$ must be $|\overline{0}\rangle$ or $|\overline{1}\rangle$ depending on whether the eigenvalue is 1 or -1.

Figure O.1 shows that the state $|\Psi\rangle$ produced by the circuit in Figure 5.10 is indeed invariant under $\mathbf{M}_0 = \mathbf{X}_0\mathbf{X}_4\mathbf{X}_5\mathbf{X}_6$. This figure demonstrates that when $\mathbf{M}_0$ is brought to the left through all the gates in the circuit it acts directly as $\mathbf{Z}_0$ on the input state on the left, which is invariant under $\mathbf{Z}_0$. The caption explains why essentially the same argument applies to the other $\mathbf{M}_i$: when brought all the way to the left, $\mathbf{M}_1$ reduces to $\mathbf{Z}_1$ acting on the input state, and $\mathbf{M}_2$ reduces to $\mathbf{Z}_2$. Figure O.2 similarly establishes the invariance of $|\Psi\rangle$ under the three $\mathbf{N}_i$.

Figure O.3 establishes that the effect of $\overline{\mathbf{Z}} = \mathbf{Z}_0\mathbf{Z}_1\mathbf{Z}_2\mathbf{Z}_3\mathbf{Z}_4\mathbf{Z}_5\mathbf{Z}_6$ acting on the right is the same as that of $\mathbf{Z}_3\mathbf{Z}_4\mathbf{Z}_5\mathbf{Z}_6$ acting on the left. But since $\mathbf{Z}_6$, $\mathbf{Z}_5$, and $\mathbf{Z}_4$ all act on the 1-Qbit states $|0\rangle$ this leaves only $\mathbf{Z}_3$ which converts $|\psi\rangle$ to $\mathbf{Z}|\psi\rangle$, which multiplies by $(-1)^x$ when $|\psi\rangle = |x\rangle$. This shows that, as required, $\overline{\mathbf{Z}}|\Psi\rangle = (-1)^x|\Psi\rangle$ when $|\psi\rangle = |x\rangle$.

Figure O.4 establishes that the effect of $\overline{\mathbf{X}} = \mathbf{X}_0\mathbf{X}_1\mathbf{X}_2\mathbf{X}_3\mathbf{X}_4\mathbf{X}_5\mathbf{X}_6$ acting on the right is the same as that of $\mathbf{Z}_0\mathbf{Z}_1\mathbf{Z}_2\mathbf{X}_3$ acting on the left. But since $\mathbf{Z}_0$, $\mathbf{Z}_1$, and $\mathbf{Z}_2$ all act on the 1-Qbit states $|0\rangle$ this leaves only $\mathbf{X}_3$ which interchanges $|1\rangle$ and $|0\rangle$ when $|\psi\rangle = |x\rangle$. This shows that

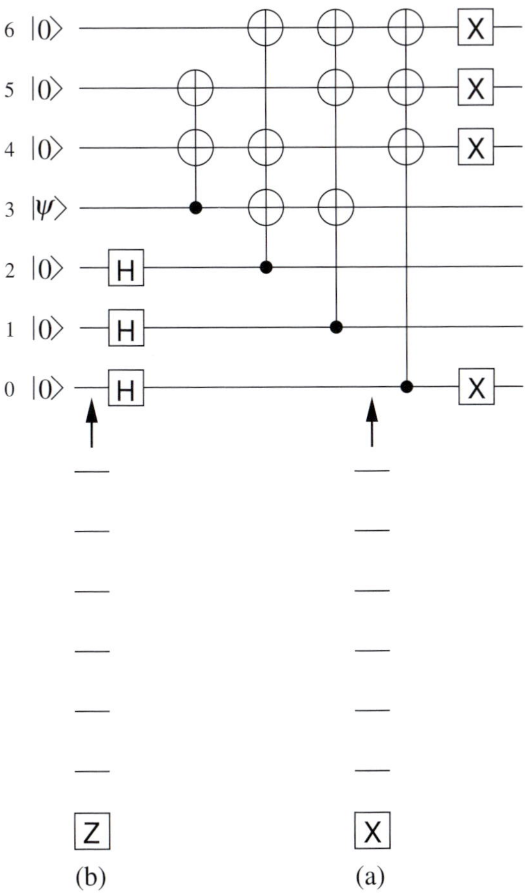

Fig O.1 Demonstration that the state $|\Psi\rangle$ constructed by the circuit in Figure 5.10 is invariant under $\mathbf{M}_0 = \mathbf{X}_0\mathbf{X}_4\mathbf{X}_5\mathbf{X}_6$. We exploit the fact that bringing an $\mathbf{X}$, acting on the control Qbit of a cNOT, from one side of the cNOT to the other introduces an additional $\mathbf{X}$ acting on the target Qbit (and the fact that an $\mathbf{X}$ acting on the target Qbit commutes with the cNOT). Bringing the $\mathbf{X}$ acting on Qbit 0 to the left of the three cNOT gates, represented by the controlled triple-NOT on the right, introduces $\mathbf{X}$ operators on all three target Qbits, which combine with the three $\mathbf{X}$ already acting on those Qbits to produce unit operators. So all four $\mathbf{X}$ gates on the right reduce to $\mathbf{X}_0$, as indicated in inset (a). That $\mathbf{X}_0$ can be moved further to the left through $\mathbf{H}_0$, if it is changed into $\mathbf{Z}_0$, as shown in inset (b). So $\mathbf{M}_0$ acting on the extreme right is equivalent to $\mathbf{Z}_0$ acting on the extreme left. Since $\mathbf{Z}_0$ leaves the 1-Qbit state $|0\rangle$ invariant, $|\Psi\rangle$ is invariant under $\mathbf{M}_0$. A similar argument applies to $\mathbf{M}_1 = \mathbf{X}_1\mathbf{X}_3\mathbf{X}_5\mathbf{X}_6$: the $\mathbf{X}_i$ all commute with the first controlled triple-NOT on the right, and then produce a single $\mathbf{X}_1$ when moved through the middle controlled triple-NOT, resulting in $\mathbf{Z}_1$ when moved the rest of the way to the left. Similarly, $\mathbf{M}_2 = \mathbf{X}_2\mathbf{X}_3\mathbf{X}_4\mathbf{X}_6$ produces $\mathbf{Z}_2$ when moved all the way to the left.

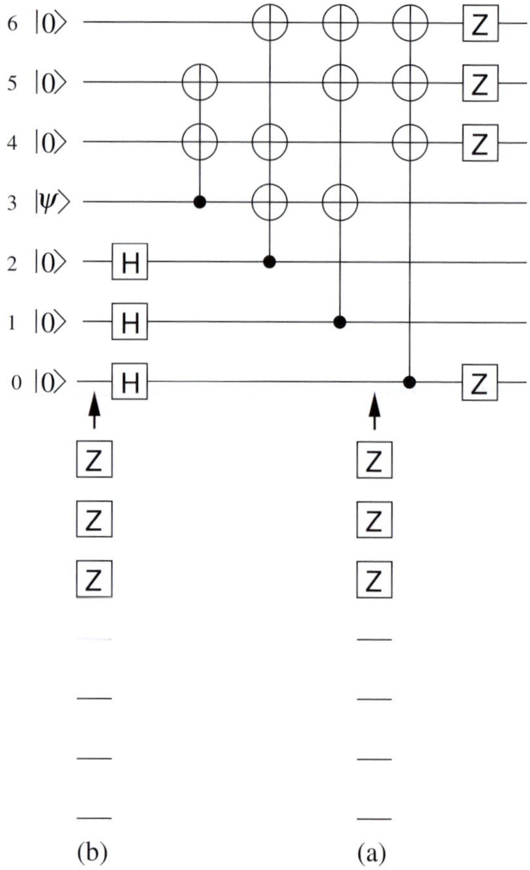

Fig O.2 Demonstration that the state $|\Psi\rangle$ constructed by the circuit in Figure 5.10 is invariant under $\mathbf{N}_0 = \mathbf{Z}_0\mathbf{Z}_4\mathbf{Z}_5\mathbf{Z}_6$. We exploit the fact that bringing a $\mathbf{Z}$, acting on the target Qbit of a cNOT, from one side of the cNOT to the other introduces an additional $\mathbf{Z}$ acting on the control Qbit (and the fact that a $\mathbf{Z}$ acting on the control Qbit commutes with the cNOT). So bringing $\mathbf{Z}_4$, $\mathbf{Z}_5$, and $\mathbf{Z}_6$ to the left of all three cNOT gates represented by the controlled triple-NOT on the right introduces three $\mathbf{Z}$ operators on the control Qbit 0, which combine with the $\mathbf{Z}_0$ already acting to produce the unit operator, reducing the collection of four $\mathbf{Z}$ gates on the left to the three $\mathbf{Z}$ acting on Qbits 4, 5, and 6, as indicated in (a). Those $\mathbf{Z}$ can be moved all the way to the left, always producing a pair of $\mathbf{Z}$ gates on the control Qbits of the multiple cNOT gates they move through, until they act directly on the input state as $\mathbf{Z}_4\mathbf{Z}_5\mathbf{Z}_6$, which leaves it invariant. A similar argument shows that $\mathbf{N}_1 = \mathbf{Z}_1\mathbf{Z}_3\mathbf{Z}_5\mathbf{Z}_6$ acting on the extreme right is the same as $\mathbf{Z}_5\mathbf{Z}_6$ acting on the extreme left, and that $\mathbf{N}_2 = \mathbf{Z}_2\mathbf{Z}_3\mathbf{Z}_4\mathbf{Z}_6$ on the right is the same as $\mathbf{Z}_4\mathbf{Z}_6$ on the left.

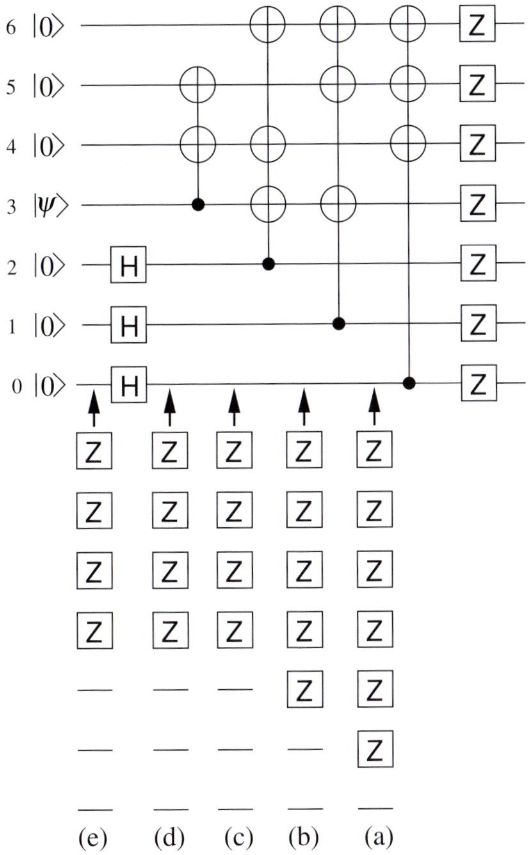

Fig O.3 Demonstration that $\overline{\mathbf{Z}} = \mathbf{Z}_0\mathbf{Z}_1\mathbf{Z}_2\mathbf{Z}_3\mathbf{Z}_4\mathbf{Z}_5\mathbf{Z}_6$ acting on the right of the circuit in Figure 5.10 is the same as $\mathbf{Z}_3\mathbf{Z}_4\mathbf{Z}_5\mathbf{Z}_6$ acting on the left. Since $\mathbf{Z}_4$, $\mathbf{Z}_5$, and $\mathbf{Z}_6$ all act as the identity on the 1-Qbit states $|0\rangle$ this leaves only $\mathbf{Z}_3$ which converts $|\psi\rangle$ to $\mathbf{Z}|\psi\rangle$. This results in a factor of $(-1)^x$ when $|\psi\rangle = |x\rangle$, showing that $\overline{\mathbf{Z}}|\Psi\rangle = (-1)^x|\Psi\rangle$ when $|\psi\rangle = |\overline{x}\rangle$.

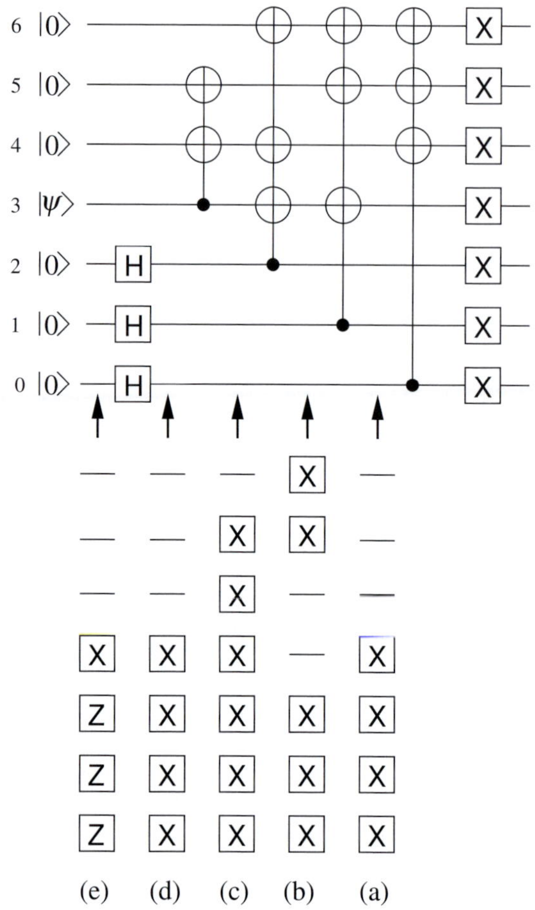

Fig O.4 Demonstration that $\overline{\mathbf{X}} = \mathbf{X}_0\mathbf{X}_1\mathbf{X}_2\mathbf{X}_3\mathbf{X}_4\mathbf{X}_5\mathbf{X}_6$ acting to the right of the circuit in Figure 5.10 is the same as $\mathbf{X}_3\mathbf{Z}_2\mathbf{Z}_1\mathbf{Z}_0$ acting to the left. Since $\mathbf{Z}_2$, $\mathbf{Z}_1$, and $\mathbf{Z}_0$ all act as the identity on the 1-Qbit states $|0\rangle$ this leaves only $\mathbf{X}_3$ which converts $|\psi\rangle$ to $\mathbf{X}|\psi\rangle$. When $|\psi\rangle = |x\rangle$ this interchanges $|0\rangle$ and $|1\rangle$, and therefore $\overline{\mathbf{X}}$ interchanges the corresponding states produced by the circuit.

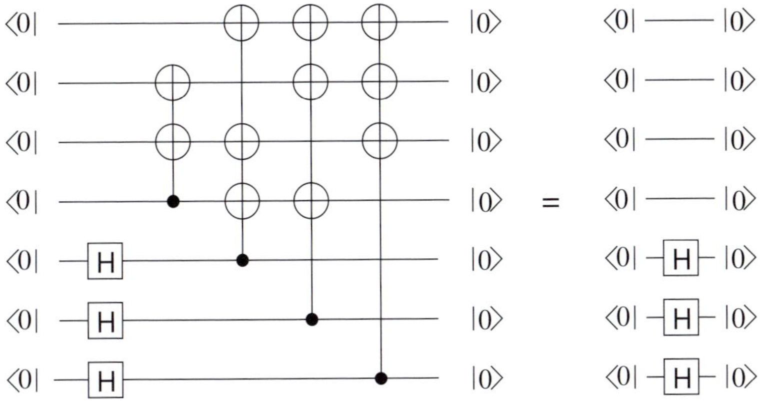

Fig O.5 Demonstration that the state produced by the circuit in Figure 5.10 when $|\psi\rangle = |0\rangle$ has an inner product with the state $|0\rangle_7$ that is $1/2^{3/2}$, thereby establishing that the state is precisely $|\overline{1}\rangle$ without any additional phase factor. We sandwich the circuit of Figure 5.10 between $|0\rangle_7$ and ${}_7\langle 0|$, following the procedure developed in Figure 5.19. Since all the cNOT gates have $|0\rangle$ for their control bits, they all act as the identity. The diagram simplifies to the form on the right, consisting of four inner products $\langle 0|0\rangle = 1$ and three matrix elements $\langle 0|\mathbf{H}|0\rangle = 1/\sqrt{2}$. So the inner product is indeed $1/2^{3/2}$.

$\overline{\mathbf{X}}$ interchanges the corresponding states produced by the circuit. It also establishes that if $|\Psi\rangle$ differs by a phase factor $e^{i\varphi}$ from $|\overline{0}\rangle$ when $|\psi\rangle = |0\rangle$, then it will differ by the *same* phase factor from $|\overline{1}\rangle$ when $|\psi\rangle = |1\rangle$.

It remains to show that when $|\psi\rangle = |0\rangle$ in Figure 5.10, the resulting state is given by $|\overline{0}\rangle$ without any nontrivial phase factor $e^{i\varphi}$. Since $|0\rangle_7$ appears in the expansion of $|\overline{0}\rangle$ with the amplitude $1/2^{3/2}$, we must show that when the input to the circuit in Figure 5.10 is $|0\rangle_7$ the inner product of the output with $|0\rangle_7$ is $1/2^{3/2}$, without any accompanying nontrivial $e^{i\varphi}$. This is established in a circuit-theoretic manner in Figure O.5, as explained in the caption.

Appendix P

On bit commitment

Alice prepares n Qbits in a computational basis state $|x\rangle$, applies a certain n-Qbit unitary transformation $\mathbf{U}$ to the Qbits, and then gives them to Bob. If Bob knows that all 2^n values of x are equally likely, what can he learn from the Qbits about Alice's choice of $\mathbf{U}$?

The answer is that he can learn nothing whatever about $\mathbf{U}$. The most general thing he can do to acquire information is to adjoin m ancillary Qbits to the n Alice gave him (m could be zero), subject them all to a quantum computation that brings about an $(n+m)$-Qbit unitary transformation $\mathbf{W}$, and then measure all $n+m$ Qbits. The state prior to the measurement will be

$$|\Psi_x\rangle = \mathbf{W}\Big(\big(\mathbf{U}|x\rangle\big) \otimes |\Phi\rangle\Big), \tag{P.1}$$

where $|\Phi\rangle$ is the initial state of the m ancillas and all 2^n values of x from 0 to $2^n - 1$ are equally likely. The probability of Bob getting z when he measures all $n+m$ Qbits is

$$p(z) = (1/2^n)\sum_x \langle z|\Psi_x\rangle\langle\Psi_x|z\rangle = (1/2^n)\langle z|\sum_x\Big(|\Psi_x\rangle\langle\Psi_x|\Big)|z\rangle. \tag{P.2}$$

We have

$$|\Psi_x\rangle\langle\Psi_x| = \mathbf{W}\Big(\big(\mathbf{U}|x\rangle\langle x|\mathbf{U}^\dagger\big) \otimes \big(|\Phi\rangle\langle\Phi|\big)\Big)\mathbf{W}^\dagger, \tag{P.3}$$

and since

$$\sum_x |x\rangle\langle x| = \mathbf{1}, \tag{P.4}$$

we then have

$$\sum_x |\Psi_x\rangle\langle\Psi_x| = \mathbf{W}\Big(\big(\mathbf{U}\mathbf{U}^\dagger\big)\otimes\big(|\Phi\rangle\langle\Phi|\big)\Big)\mathbf{W}^\dagger = \mathbf{W}\Big(\mathbf{1}\otimes\big(|\Phi\rangle\langle\Phi|\big)\Big)\mathbf{W}^\dagger. \tag{P.5}$$

We see from (P.2) and (P.5) that $\mathbf{U}$ has dropped out of the probability $p(z)$, so the outcome of Bob's final measurement provides no information whatever about Alice's unitary transformation.

In the application to bit commitment in Section 6.3, Alice's unitary transformation $\mathbf{U}$ is either the n-Qbit identity or the tensor product of n 1-Qbit Hadamards, $\mathbf{H}^{\otimes n}$, and the random n-Qbit state $|x\rangle$ arises from the tensor product of n 1-Qbit states, each of which is randomly $|0\rangle$ or $|1\rangle$.

One might wonder whether Bob could do better by measuring some subset of all the Qbits at an intermediate stage of the computation, and then applying further unitary transformations to the unmeasured Qbits conditional upon the outcome of that measurement. But this, by an inversion of the Griffiths–Niu argument in Section 3.6, would be equivalent to first applying an appropriate multi-Qbit controlled unitary gate, and only then measuring the control Qbits. That gate can be absorbed in $\mathbf{W}$ and the subsequent measurement of its control Qbits deferred to the end of the computation. So this possibility is covered by the case already considered.

Index